Systems Biology: Introduction to Metabolic Control Analysis

Herbert M. Sauro
University of Washington
Seattle, WA

Ambrosius Publishing

Mosaic image modified from Daniel Steger's Tikz image (http://www.texample.net/tikz/examples/mosaic-from-pompeii/)

Front-Cover: Metabolic pathway image from the Interactive Pathways Explorer v3, iPath 3 https://pathways.embl.de/ and KEGG Pathway Atlas (with kind permission from iPath and KEGG. Norway).

The image depicts part of the glycolytic and Krebs cycle. Ref: Ivica Letunic, Takuji Yamada, Minoru Kanehisa, and Peer Bork (2008), iPath: interactive exploration of biochemical pathways and networks. Trends in Biochemical Sciences 33(3), p101-103

University Disclaimer: Any views, opinions, data, documentation and other information presented in this book are solely those of the author and do not represent those of the University of Washington.

Contents

Preface

This book is an introduction to control in biochemical pathways, specially Metabolic Control Analysis (MCA). The book should be suitable for undergraduates in their early (Junior, USA, second year UK) to mid years at college. The book can also serve as a reference guide for researchers and teachers. The free software used in the book can be found at tellurium.analogmachine.org.

This is a book I've wanted to write for many years. I worked on MCA as part of my PhD with David Fell and still like MCA's ability to help understand how perturbations propagate through a network and rationalize the many non-intuitive aspects of metabolic dynamics. The book covers the basics of MCA, a framework developed originally by Heinrich, Rapoport, Kacser, Burns and Savageau. I've tried to take an operational style wherever possible rather than a purely algebraic approach to proofs. I've found the operational style much more illuminating and biologically insightful. There are however a number of topics missing from this edition. These include control in complex branched systems, a detailed look at the effect of sequestration, including metabolic channeling, and hierarchical control analysis that includes genetic regulation as part of the analysis. More advanced mathematical topics that involve extensive use of matrix algebra is also omitted. These may be included, together with the recent frequency domain extensions, in a later edition.

As with my earlier text books I have decided to publish this book myself via a service called Createspace that is part of Amazon. Over the years I've considered publishing text books via bonafide publishers but have found the contracts they offer to be far too restrictive. Two restrictions in particular stand out, the loss of copyright on the text as well as any figures and the inability to rapidly update the text when either errors are found or new material needs to be added.

There are many people and organizations who I should thank but foremost must be my infinitely patient wife, Holly, who has put up with the many hours I have spent working alone in our basement and who contributed significantly to editing this book. I am also most grateful to the National Science Foundation and the National Institutes of Health who paid my summer salary so that I could allocate the time to write, edit and research. I would also like to thank the many undergraduates, graduates and colleagues who have directly or indirectly contributed to this work. In particular I want to thank my two teachers, David Fell and the late Henrik Kacser who I had the privilege to work with as a graduate student and postdoctoral fellow respectively. I had many hours of fruitful and stimulating conversations with Luis Azerenza, Frank Bruggeman, Jim Burns, Athel Cornish-Bowden, Reinhart Heinrich, Jannie Hofmeyr, Boris Kholedenko, Edda Klipp, Pedro Mendes, Rankin Small, Stefan Schuster, Michael Savageau, Jacky Snoep, Bas Teusink, and last but not least Hans Westerhoff. In early 2000s I had the good fortune to be introduced to engineering control theory by John Doyle, Brian Ingalls, and Mustafa Khammash; all three had a significant influence on my understanding of control theory. They had the rare knack of being able to very clearly explaining complex ideas.

I wish to sincerely thank the authors of the TEX system, MikTeX (2.9), TikZ (3.0.1a), PGFPlots (1.15) and WinEdt (9.0) for making available such amazing tools to technical authors.

Finally, I should thank Michael Corral (http://www.mecmath.net/) and Mike Hucka (sbml.org) whose LaTeX work inspired some of the styles I used in the text.

I am sure there will be typographical errors in the text, although I've tried my best to eliminate as many as possible. If you find something amiss, please forward details to hsauro@uw.edu.

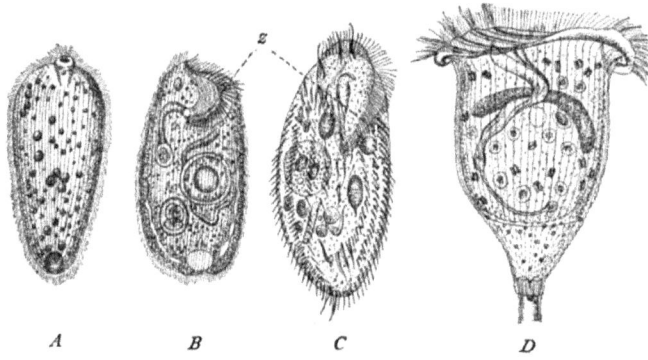

Figure 1 Some of my favourite animals. Types of Ciliates: From The Protozoa, Gary Nathan Calkins, Macmillan, 1910.

August 2018
Seattle, WA

HERBERT M. SAURO

Expected Prerequisites

To keep the printing costs down there is little in the way of introductory content. As a result, certain topics and concepts are assumed to be familiar by the reader. These include:

Kinetics

- Basic Chemical Kinetics: Difference between rates of change, reaction rates. Mass-action kinetics, the equilibrium constant, mass-action ratio and the disequilibrium constant.

- Basic Enzyme Kinetics: Irreversible and reversible Michaelis-Menten kinetics for single substrate, single product reactions. Product inhibition, competitive and uncompetitive inhibition. Cooperativity and allostery (eg MWC model), and the Hill equation.

Biology

- Knowledge of metabolic pathways, the role of enzymes in metabolic pathways, knowledge of allosteric control. Knowledge of protein signaling networks for more advanced topics.

- Stoichiometric networks, stoichiometric matrix, $\mathbf{N}$. Mass-balance equations to formulate metabolic models, the systems equation $\frac{d\mathbf{s}}{dt} = \mathbf{N}\boldsymbol{v}(\mathbf{s}(\boldsymbol{p}), \boldsymbol{p})$.

- Concept of the steady state, concentrations and fluxes, boundary and floating species, effect of parameter and species perturbations on the steady state.

Mathematics

- Elementary differential calculus, understanding partial derivatives and differential equations, logarithms.

- For more advanced topics some knowledge of matrices and vectors, matrix addition, multiplication and inversion.

- For some advanced topics, knowledge of eigenvalues, eigenvectors and de Movire's theorem

Recommended reading:

Sauro, HM (2014) Systems Biology: Introduction to Pathway Modeling. Ambrosius Publishing, Seattle. ISBN-10: 0982477376

Notation

Following other authors, an upper case letter such as S indicates the name of a molecular species while the lower case italic variant, s, indicates the concentration of species S. This is to avoid the use of square brackets, as in [S] which can result in unnecessary clutter in some equations.

Bold text will tend to be used either for emphasis or when introducing new terminology.

1

Traditional Concepts in Metabolic Regulation

1.1 Introduction

This book is about the control of biochemical systems with a focus on metabolic and signaling pathways. The ability to control reaction rates and concentrations in a changing environment is one of the characteristics of living systems. Cells must monitor prevailing conditions and make appropriate decisions. Cells make sure, for example, that adequate phosphate and redox potentials are available at all times. They also have to ensure that major transitions from one state to another (for example cell division) avoid any disruption to subsystems that are essential to cell viability. These activities presumably require a great deal of coordination and control. Indeed, years of research has uncovered a myriad number of feedback and feedfoward control loops together with many less obvious means of control.

It is worth examining some of the history of how we came to understand control in biological cells. The first thing to note is that understanding control in any complex system is challenging. It was difficult in the past and it is difficult now. Man's propensity to grasp the many factors involved in a complex system is limited. As a result, reasoning about complex systems cannot be done by intuition alone, it requires expertise and the application of approaches from mathematics, engineering and computer science.

1.2 Early Quantitative Efforts

During the early part of the 20th century, it became apparent that chemical processes in biological cells were the result of sequences of separate chemical transformations. The first sequence of steps discovered, later to be called a 'pathway', was yeast glycolysis. Subsequently, many other pathways were discovered such as the Calvin and Krebs cycle, including pathways involved in amino acid biosynthesis and degradation. As early as the 1930s, a number of individuals began taking a theoretical interest in the dynamic properties of such pathways. Much of the early work focused on the question of limiting factors. This may have originated from a statement by Blackman [5] in 1905 who stated as an axiom: "when a process is conditioned as to its rapidity by a number of separate factors, the rate of the process is limited by the pace of the slowest factor". This implied that the understanding of a complex system could be accomplished by identifying the limiting factor; and so the idea of the rate-limiting step, the pacemaker, the bottleneck, and master reaction was born.

The Pacemaker

Although the idea of a pacemaker reaction in a pathway was extremely attractive, there were opponents to the idea even as early as the 1930s. Burton [14] was probably one of the first to point out that: "In the steady state of reaction chains, the principle of the master reaction has no application". Hearon [40] made a more general mathematical analysis and developed strict rules for the prediction of mastery in a linear sequence of enzyme-catalysed reaction. Webb [117] gave a severe criticism of the concept of the pacemaker and of its blind application to solving problems of regulation in metabolism. Waley [115] made a simple but clear analysis of linear reaction chains that showed that rate-limitingness was a shared commodity. Later, authors from the biochemical community, such as Higgins [45] but particularly Heinrich and Rapoport [43], supported the same conclusion with more advanced analysis. In parallel with this work, other communities were coming to the same conclusion. Most notably Sewell Wright, a geneticist, wrote a treatise on 'Physiological and Evolutionary Theories of Dominance' [119], where he discussed the limiting factors in relation to hypothesized networks controlled by 'genes'. This work was taken up by Kacser and Burns [54] in Edinburgh and was developed into a major theory of control in pathways. Heinrich and Rapoport [43] simultaneously accomplished the same feat but from a more biochemical perspective. Finally, Savageau [91] in the United States, an engineer by training, developed the same approach and reached similar conclusions.

1.3   Prevailing Ideas

While there was considerable theoretical and some experimental work that suggested that the concept of the pacemaker enzyme was erroneous, the biochemical community, for what ever reason, ignored these results. Instead, they developed their own framework for understanding the operating principles of cellular networks. This framework was derived largely through an intuitive approach using faulty analogies, and based neither on experimental evidence or mathematical reasoning. This ultimately led to a number of unfortunate misunderstandings in how cellular networks operate, misconceptions that still prevail today.

One of the chief concepts in the traditional control framework is the pacemaker or rate-limiting step. The rate-limiting step is thought to be located near the start of a pathway and because it is rate-limiting, the pathway is controlled entirely by this one key step. In addition, it is proposed that rate-limiting steps are likely to be the site for allosteric regulation. There are a number of criteria that are used to identify the possible rate-limiting step, though there is no real definitive test. These criteria include:

- The rate-limiting step is the slowest step in the pathway.

- The rate-limiting step has the lowest substrate-affinity (highest K_m), meaning that the reaction velocity is the lowest when saturating substrate concentrations are present for all enzymes.

- The rate-limiting step will be the regulated step.

- The rate-limiting step is an irreversible reaction.

- The rate-limiting step is usually the first step in the pathway.

- The rate-limiting step is far from equilibrium.

No single criterion could positively identify a rate limiting step, but the cross-over theorem is one that was considered important. The technique worked as follows. A metabolic pathway is perturbed by adding an inhibitor that changes the activity of one of the enzyme catalyzed steps, and the metabolite concentrations before and after the inhibited step are measured. If the inhibited step is rate limiting, then the metabolites upstream would increase and those downstream decrease. The technique was originally developed by Britton Chance [16] in the 1950s as a means to study the electron transport chain in mitochondria. The advantage here was that many of the intermediates had characteristic absorption spectra so could be easily measured. The method was used to identify the sites where electron transfer was being coupled to ATP production. Although applicable to the electron transport chain (Fell, 1996), its subsequent use to identify sites of regulation in metabolic pathways has been considered on theoretical grounds to be untrustworthy (Heinrich et al, 1974).

Regulatory Enzymes and Feedback Regulation

A key concept in traditional metabolic control is that the reaction step where feedback regulation acts is the rate-limiting step. Rate-limiting enzymes could therefore be identified by locating the regulated steps. For example, a classic rate-limiting step in glycolysis is phosphofructokinase. Understanding how pathways are controlled however is much more subtle than this. Repeated measurements [41, 94, 22, 13, 106, 84, 71, 111, 64, 70] have shown for example that phosphofructokinase is in fact not rate-limiting even though it is heavily regulated. Since the development of recombinant technology in the late 70s, the ability to control enzyme levels has become relatively easy. There are many experiments reported where over expression of a regulated step resulted in no change in the pathway flux even though such steps were considered rate-limiting. Even in the face of considerable experimental evidence, the idea that regulated steps are rate limiting continues to persist. As we will see in subsequent chapters, such an idea is inconsistent with logic and experimental evidence even though most undergraduate textbooks (except for Lehninger) and Wikipedia (a major source of (miss)information for students) continue to support the concept.

1.4 A Modern Understanding of Metabolism

Although the metabolic parts list is almost complete as witnessed by the development of genomic scale metabolic reconstructions [74], our understanding of how metabolism operates is still conceptually primitive and incomplete. The last four decades has seen some progress but still many researchers use the classical conceptual framework of metabolic control. The reminder of the book will focus on modern ideas of metabolic control.

Further Reading

1. Fell D A (1996) Understanding the Control of Metabolism. Ashgate Publishing. ISBN-10: 185578047X

2. Fell D A (1992) Metabolic control analysis: a survey of its theoretical and experimental development. Biochemical J, 286, 313-330

3. Morandini P (2009) Rethinking metabolic control. Plant Science, 176(4), 441-451

2

Elasticities

2.1   Introduction

Enzymes catalyze virtually all the chemical transformations of metabolism. They coordinate all the primary activities of a cell, ranging from energy transformations and storage, through to maintenance of cellular structure and integrity. They directly manage the expression and maintenance of host DNA, including replication. Enzymes clearly serve an essential and fundamental role in the activity of a cell, and for this reason we can regard them as fundamental units of life. If we are to understand how cellular systems work, an appreciation of the properties of these fundamental units is obviously essential. This chapter will focus on the properties of the isolated enzymes, and later we will consider intact pathways.

Since enzymes are the functional units of metabolism, it is important to understand how an enzyme responds to changes in its environment. An important part of Metabolic Control Analysis (MCA) is a consideration of this question. In MCA the measure that describes this response is called the **elasticity coefficient**. Elasticity coefficients are so important to MCA that the remainder of the chapter is devoted to their discussion.

Elasticities describe how sensitive a reaction rate is to changes in reactant, product and effector concentrations. They represent the degree to which changes are transmitted from the immediate environment to the reaction rate. From a systems perspective they are critical components in understanding how a disturbance, such as the introduction of an inhibitor applied at one or more points in a cellular pathway, propagates to the rest of the system. It is the magnitude and signs of the elasticities that determine how far and at what strength the disturbance travels. Elasticities are therefore central in helping us understand how networks function. In this chapter we will focus on describing the properties of elasticities, and how they can be computed and used to describe changes at a reaction step.

To study the properties of an individual enzyme, the usual experimental procedure is to purify the enzyme and study it *in vitro*. Once purified and isolated, the environment of the enzyme can be controlled and in principle, the concentrations of all the participating molecules manipulated at will. Individual substrates, effectors etc., can be selectively changed and any change in reaction rate recorded. In this manner, the response of the rate to changes in factors that might affect the reaction rate can be studied.

Consider an experiment where we investigate the response of the reaction rate to changes in substrate concentration. In the experiment the concentration of substrate can be changed and the change in reaction rate observed. A plot of the reaction rate versus the substrate concentration would form a continuous curve (Figure 2.1).

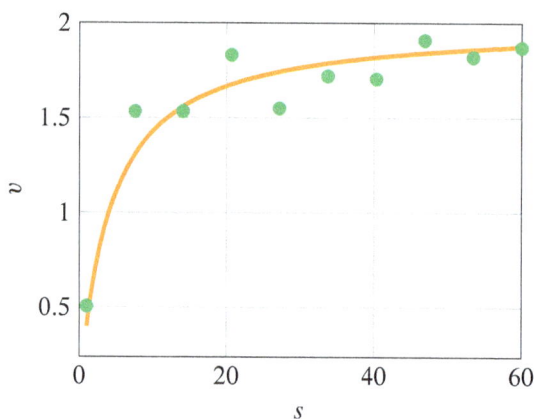

Figure 2.1 Experiment showing reaction rate, v, for an enzyme catalyzed reaction as a function of substrate concentration, s. Markers indicate hypothetic data points. Continuous curve plotted using $2s/(4 + s)$

To determine the influence an effector such as a substrate has on a reaction rate, we can carry out the following experiment. Let us denote the concentration of substrate by the symbol s, and the rate of reaction by v. The experiment proceeds in two steps. The first step involves measuring the rate of reaction, v, at some substrate concentration of interest, say s. In the second part of the experiment, the concentration of substrate is increased by an amount given by δs,[1] and the experiment repeated at the new concentration of $s + \delta s$. The increase in s is likely to cause a change in the rate of reaction from v to v_{new}. The difference between the two rates, $v_{\text{new}} - v$ is the change in rate as a result of the change, δs. We denote this change in rate by δv. Depending on the particular enzyme, the effectors, the substrates and products, the change we observe in the rate may be small or large. In order to judge the relative effectiveness of any particular effector, we can form the ratio:

$$\frac{\delta v}{\delta s}$$

This will give us the change in v per unit change in s. By measuring this ratio for each factor that might affect the rate, we can gauge which ones have more or less of an effect.

There are however, two problems with this ratio. The first is that its value depends on the size of the change we make to s. This is particularly true if the response of v to changes in s is non-linear (as most enzyme rate responses are). The second problem is that the ratio depends on the units we

---

[1]The δ means 'a small change'

choose to measure the rate and concentration. A possible solution to the later problem would be for all experimenters to employ a standard set of units, but this would be almost impossible to achieve in practice. A much easier way around this problem is to eliminate the units altogether by scaling the ratio with the rate and concentration. We can eliminate the concentration units by dividing the change, δs, by the concentration of s, i.e. $\delta s/s$. Likewise, we can eliminate the reaction units by dividing by v. Therefore, rather than measure $\delta v/\delta s$, it is more sensible to measure:

$$\frac{\delta v}{\delta s}\frac{s}{v}$$

This still leaves us with the first problem, which is that the value of the ratio varies with the amount of change we make to s. We could all decide on a standard change to make in s, say doubling s, and measuring the change in v. This would be difficult to achieve in practice however, and ultimately has limited value. A better way is as follows.

Assume that the substrate concentration has been set to a value s. At this concentration, the enzyme will show a reaction rate of v. If we make a change δs to s, then this will cause a change in rate δv and we can compute the ratio, $\delta v/\delta s$. We could make the change smaller and remeasure the change in v and compute the ratio again. If we were to continue making δs smaller and smaller, the ratio given by $\delta v/\delta s$ will slowly approach a limiting value. This value is the slope of the curve v vs. s, at the point s. Those familiar with the calculus will recognize that in reducing δs to a smaller and smaller increment, the ratio, $\delta v/\delta s$, has reached a limiting value called the derivative:

$$\frac{\delta v}{\delta s} \quad \longrightarrow \quad \frac{dv}{ds} \tag{2.1}$$
$$\text{as } \delta s \to 0$$

The ratio, $\delta v/\delta s$ tends to the derivative dv/ds, as δs tends to zero. The differential has a precise meaning, it is the slope of the curve at the point s and, significantly for us, it has a *unique* value at this point.

As before, we can scale dv/ds to eliminate the measuring units to obtain:

$$\frac{dv}{ds}\frac{s}{v}$$

This expression represents the scaled slope of the response curve at s, and is called the **elasticity coefficient** of the rate of reaction, v, with respect to the concentration of metabolite S. It can be used to measure how responsive a reaction rate is to changes in the concentration of any effector, in this case it happens to be the concentration of substrate, S. We could also have changed the concentration of the product, P, or the concentration of an effector. In either case we would be able to measure an elasticity. This means there will be as many elasticity coefficients for a particular enzyme as there are effectors that might influence the reaction rate. Thus, not only will an enzyme be characterized by a substrate elasticity, but also by a product elasticity and any effector elasticities. In addition, other factors which might affect the reaction rate, such as pH, ionic strength and so on, will also have associated elasticity coefficients. Any particular enzyme will thus be fully characterized at a particular operating point when all its elasticities have been measured or computed.

Elasticities must be measured under *in vivo* conditions

In practice, if an enzyme is purified with the intention of measuring its elasticities, then the concentrations of the substrates and products, the pH, ionic strength and so on should be faithfully recreated in order to mimic the *in vivo* condition. If this is not done, the measured values for the elasticities will not reflect the elasticities *in vivo* and their usefulness is lost. As will be revealed in the next chapter, the elasticities are the building blocks with which we can begin to understand the properties of intact pathways.

2.2 Elasticity Coefficients

The **elasticity coefficient** is defined according to the following expression:

$$\varepsilon_{s_i}^v = \left(\frac{\partial v}{\partial s_i} \frac{s_i}{v} \right)_{s_j, s_k, \dots} = \frac{\partial \ln v}{\partial \ln s_i} \approx v\% / s_i\% \tag{2.2}$$

The symbol for an elasticity is the Greek epsilon, ε.[2]

> **Unitless:** Due to the scaling, the elasticity is a dimensionless quantity.

The elasticity measures how responsive a reaction rate is to changes in the concentration of an effector, in this case the concentration of effector s_i. Any effector can be changed to see how it affects the reaction rate. We could have changed the concentration of the product, effector or anything else that might affect the reaction rate. The larger the elasticity, the greater the effect the effector has on the reaction rate.

Sometimes the **unscaled elasticity** is useful (See Chapter 9) and we designate this using the symbol $\mathcal{E}$.

> **Unscaled Elasticity:**
>
> $$\mathcal{E}_{s_i}^v = \left(\frac{\partial v}{\partial s_i} \right)_{s_j, s_k, \dots} \tag{2.3}$$

When writing the elasticity symbol, ε or $\mathcal{E}$, a subscript is often used to indicate the modulating factor (s_i), and a superscript to indicate the effect that is being measured (v)

The subscripts, $s_j, s_k, \dots$ in the definition (2.2) indicate that any species or factor that could also influence the reaction rate **must be held constant** at their current value when species s_i is changed. This is also implied in the use of the **partial derivative** symbol, ∂, rather than the derivative symbol, d. In normal usage, these subscripts are often left out as the partial derivative symbol is usually sufficient to indicate what is meant.

Since the elasticity is defined in terms of a derivative we can derive elasticities by differentiating the rate law and scaling. The elasticity is closely related to the **kinetic order**, sometimes called the reaction order. For simple mass-action chemical kinetics, the kinetic order is the power to which a species is raised in the kinetic rate law. Reactions with zero-order, first-order, and second-order are commonly found in chemistry, and in each case the kinetic order is zero, one and two, respectively.

---

[2]Those familiar with quantitative economics will have come across a similar concept.

For a reaction such as:

$$2H_2 + O_2 \rightarrow 2H_2O$$

assuming that the irreversible mass-action rate law is given by:

$$v = k \, h_2^2 \cdot o_2$$

the kinetic order with respect to hydrogen is two and oxygen one. In this case the kinetic order also corresponds to the stoichiometric amount of each molecule although this may not always be true.

Example 2.1 shows the elasticities for zero, first, second, and n^{th} order reactions. Given that the elasticity is defined in terms of a derivative, it is possible, if the rate law is known, to compute an elasticity by differentiation (See Example 2.1). From the example we see that the elasticity reduces to the kinetic order for simple mass-action kinetics.

In biochemical systems theory, elasticities are also called the **apparent kinetic order**.

> **Kinetic Order:** The elasticity for a reactant in an elementary reaction is equal to the kinetic order of the reactant.

Example 2.1

Determine the elasticities with respect to species s, for the following mass-action rate laws by differentiating and scaling each rate law:

1. $v = k$

   Elasticity: $\varepsilon_s^v = \dfrac{\partial v}{\partial s} \dfrac{s}{v} = 0$

2. $v = ks$

   Elasticity: $\varepsilon_s^v = \dfrac{\partial v}{\partial s} \dfrac{s}{v} = \dfrac{s \, k}{k s} = 1$

3. $v = ks^2$

   Elasticity: $\varepsilon_s^v = \dfrac{\partial v}{\partial s} \dfrac{s}{v} = \dfrac{2kss}{ks^2} = 2$

4. $v = ks^n$

   Elasticity: $\varepsilon_s^v = \dfrac{\partial v}{\partial s} \dfrac{s}{v} = \dfrac{nks^{n-1}s}{ks^n} = n$

Operational Interpretation

The definition of the elasticity (2.2) also gives us a useful operational interpretation.

> **Operational Definition:** The elasticity is the fractional change in reaction rate in response to a fractional change in a given reactant or product, while keeping all other reactants, products, and other effectors constant.

Since the elasticity is expressed in terms of fractional changes, it is also possible to get an approximate value for the elasticity by considering **percentage changes.** For example, if we increase the substrate concentration of a particular reaction by 2% and the reaction rate increases by 1.5%, then the elasticity is given approximately by $1.5/2 = 0.75$. The elasticity is however only strictly defined (See equation (2.2)) for infinitesimal changes and not finite percentage changes. However, so long as the changes are small, the finite approximation is a good estimate for the true elasticity.

For species that cause reaction rates to increase, the elasticity is **positive**, while for species that cause the reaction rate to decrease, the elasticity is **negative**. Therefore, reactants generally have positive elasticities and products generally have negative elasticities (Figure 2.2).

> Effectors that **increase** a reaction rate will have a **positive** elasticity.
>
> Effectors that **decrease** a reaction rate will have a **negative** elasticity.

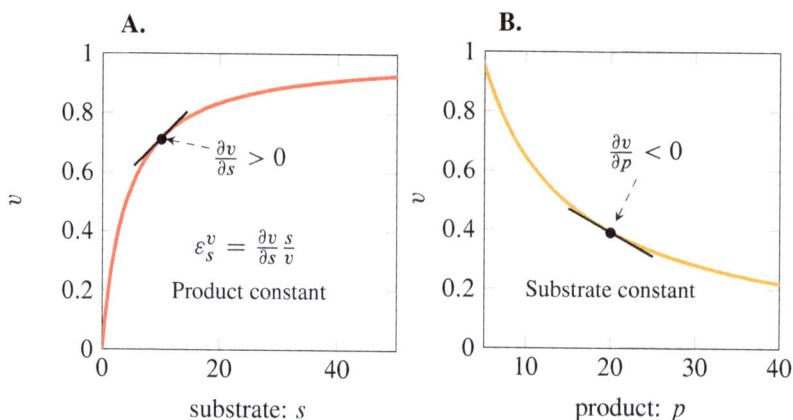

Figure 2.2 A. Reaction rate versus substrate. Increases in the substrate cause an increase in the rate. A positive slope yields a positive elasticity. **B.** Reaction rate versus product (assuming a positive rate from reactant to product). Increases in product result in a decrease in reaction rate; a negative slope yields a negative elasticity. Curves generated by assuming $v = s/(2+s)$ and $2/(1+(0.1+0.2p))$, respectively.

Example 2.2

How many elasticities are there for the following reversible mass-action reactions and what are their likely signs?

a) $A \rightarrow B$

There are two elasticities, ε_a^v which is likely to be positive and ε_b^v which will be negative.

b) $2A + B \rightarrow 3C$

There are three elasticities, ε_a^v which will be positive, ε_b^v which will also be positive, and ε_c^v which will be negative.

Numerical Estimation

We saw in example (2.1) how elasticities can be computed algebraically by differentiating the rate law and scaling. An elasticity can also be determined numerically by making a small change (say 5%) to the chosen reactant concentration and measuring the change in reaction rate. For example, assume that the reference reaction rate is v_o, and the reference reactant concentration, s_o. If we increase the reactant concentration by[3] Δs_o and observe the new reaction rate as v_1, then the elasticity can be estimated by using Newton's difference quotient:

$$\varepsilon_s^v \simeq \frac{v_1 - v_o}{\Delta s_o} \frac{s_o}{v_o} = \frac{v_1 - v_o}{v_o} \bigg/ \frac{s_1 - s_o}{s_o}$$

Newton's quotient method relies on making one perturbation to s_o, Δs_o. A much better estimate for the elasticity can be obtained by doing two separate perturbations in s_o. One perturbation to **increase** s_o, and another to **decrease** s_o. In each case the new reaction rate is recorded; this is called the three-point estimation method. For example, if v_1 is the reaction rate when we increase s_o, and v_2 is the reaction rate when we decrease s_o, then we can use the following three-point formula to estimate the elasticity:

$$\varepsilon_s^v \simeq \frac{1}{2} \frac{v_1 - v_2}{s_1 - s_o} \left(\frac{s_o}{v_o} \right)$$

These approximate numerical methods are particularly useful when computing elasticities using software.

Example 2.3

Estimate the elasticity using Newton's difference quotient and the three-point estimation method. Compare the results with the exact value derived algebraically:

Let $v = s/(0.5 + s)$. Find the elasticity when $s = 0.6$.

a) Algebraic Evaluation

Differentiation the rate laws yields: $0.5/(s + 0.5)^2$, scaling by s/v gives the elasticity as $0.5/(0.5 + s)$. At a value of 0.6 for s, the exact value for the elasticity is: 0.4546

b) Difference Quotient

Let us use a step size of 5%. Therefore $h = 0.05 \times 0.6 = 0.03$ from which $s_1 = 0.63$, and $s_o = 0.6$. From these values we can compute v_1 and v_o. $v_o = 0.6/(0.5 + 0.6) = 0.5454$, $v_1 = 0.63/(0.5 + 0.63) = 0.5575$. From these values the estimated elasticity is given by: $\varepsilon_s^v = ((0.5575 - 0.5454)/0.5454) / ((0.63 - 0.6)/0.6) = 0.443$

Compared to the exact value the error is 0.0116, or **2.55 % error**.

c) Three-Point Estimation

In addition to calculating v_1 in the last example, we must also compute v_2. To do this we subtract h from s_o to give $v_2 = 0.533$. The three-point estimation formula gives us: $\varepsilon_s^v = 0.5 \frac{0.5575 - 0.5327}{0.03} \frac{0.6}{0.5454} = 0.4549$

Compared to the exact value the error is only 0.0033, or **0.7 % error**, a significant improvement over the difference quotient method.

The degree of error in the difference quotient method will depend on the value of s, which in turn determines the degree of curvature (or nonlinearity) at the chosen point. The more curvature there is, the more inaccurate the estimate. The value in this example was chosen where the curvature is high, therefore the error was larger.

---

[3] Δs_o means a change to s_o

In example (2.1), the elasticities were constant values. However for more complex rate law expressions, this need not be the case (see example (2.4)), and the elasticity will change in response to changes in the reactant and product concentrations. Consequently, when measuring the elasticity numerically or experimentally, one has to choose a particular operating point, most commonly the *in vivo* state.

Example 2.4

Determine the elasticities for the following rate laws with respect to s, by differentiating and scaling:

1. $v = k(s + 1)$

   Elasticity: $\varepsilon_s^v = \dfrac{\partial v}{\partial s} \dfrac{s}{v} = k \dfrac{s}{k(s + 1)} = \dfrac{s}{(s + 1)}$

2. $v = k/(s + 1)$

   Elasticity: $\varepsilon_s^v = \dfrac{\partial v}{\partial s} \dfrac{s}{v} = -\dfrac{k}{(1 + s)^2} \dfrac{s}{k/(s + 1)} = -\dfrac{s}{s + 1}$

3. $v = s/(s + 1)$

   Elasticity: $\varepsilon_s^v = \dfrac{\partial v}{\partial s} \dfrac{s}{v} = \dfrac{1}{(s + 1)^2} \dfrac{s}{s/(s + 1)} = \dfrac{1}{s + 1}$

4. $v = ks(s + 1)$

   Elasticity: $\varepsilon_s^v = \dfrac{\partial v}{\partial s} \dfrac{s}{v} = k(1 + 2s) \dfrac{s}{ks(s + 1)} = 1 + \dfrac{s}{s + 1}$

The examples illustrate that for more complex rate laws, the elasticity becomes a function of the effector concentration.

Experimental Estimation

Experimentally, we can measure an elasticity using the following procedure. Consider a simple reaction such as A → B, where we wish to measure the elasticity of the reaction rate with respect to A. We must first select an operating point for A and B. This choice will depend on the system under study. For example, perhaps we are interested in the value of the substrate elasticity for an enzyme catalyzed reaction when the substrate and product concentration are at their K_m levels. Once the operating point is chosen, the reaction is started and the rate of reaction measured. It is important that during the measurement, only a small amount of substrate is consumed and product produced, otherwise the estimate for the elasticity will not be accurate. We now begin the experiment again but this time the substrate concentration is increased by a small amount, and the product concentration is reset to its value in the first experiment. The reaction is started and the new reaction rate measured. The fractional change in reaction rate and substrate is recorded and the ratio computed to give the substrate elasticity. In principle, the same kind of experiment could be performed on the product, this time keeping the substrate concentration constant.

Simple protocol for estimating the substrate elasticity

1. Set substrate and product concentrations to their operating points.

2. Record the reaction rate at the operating point.

3. Restore all concentrations to their original starting points.

4. Increase the concentration of substrate by a small amount.

5. Record the new reaction rate.

6. Compute the elasticity by dividing the fractional change in reaction rate by the fractional change in substrate concentration.

7. At all times, maintain other substrate, product and effector concentrations at the operating point.

The algebraic definition of the elasticity automatically suggests ways to estimate their values that include algebraic differentiation of the rate law (if the rate laws is available), and numerical computation of values by simulation or even by experiment. In the next section the use of log/log plots will be used to empirically determine an elasticity from an enzyme kinetics experiment.

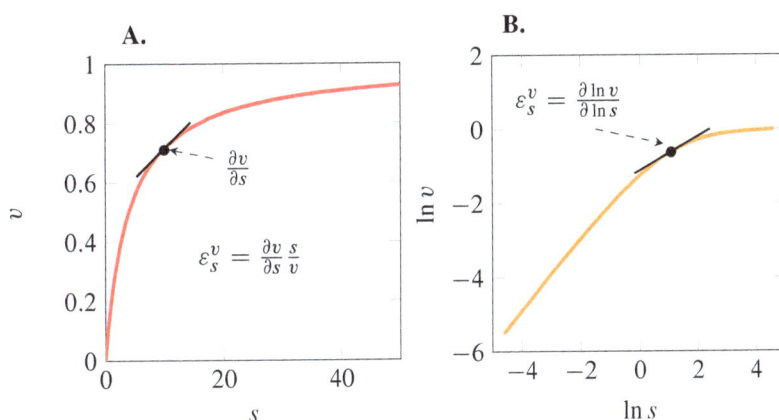

Figure 2.3 **A.** The slope of the reaction rate versus the reactant concentration scaled by both the reactant concentration and reaction rate yields the elasticity, ε_s^v. **B.** If the log of the reaction rate and log of the reactant concentration are plotted, the elasticity can be read directly from the slope of the curve. Curves are generated by assuming $v = s/(2 + s)$.

Log Form

The definition of the elasticity in equation (2.2) shows the elasticity expressed using a log notation:

$$\varepsilon_s^v = \frac{\partial \ln v}{\partial \ln s}$$

This notation is frequently used in the literature. In Figure 2.3, the left panel is a simple plot of rate versus substrate concentration and shows a typical response for an enzyme as it saturates at high substrate concentration. The right-hand panel shows the same plot but now the axes are logarithmic.

As we will see, the slope of the curve on the right-panel is a direct measure of the elasticity. The origin of this effect is worth explaining to those unfamiliar with the calculus.

If we examine the growth pattern of a micro-organism, we will often find that it follows a pattern of the kind, $y = a^x$. What this means is that the number of microorganisms increases by a fixed proportion per unit time. Often such data is plotted on a semi-logarithmic scale rather than the usual linear scale as it helps to emphasize the fact that the relative growth under these conditions is the same throughout the growth phase. To explain this statement, a numerical example will be useful. Of the two sequences of numbers:

$$\begin{array}{cccccc} 100, & 150, & 200, & 250, & 300, & \dots \\ 100, & 150, & 225, & 337.5, & 506.25, & \dots \end{array}$$

the first shows a regular increase of 50 units and the second a regular increase of 50 percent. from one number to the next. On a linear scale, the points representing the first sequence appear as equal distances from each other and those representing the second sequence at increasing distances. If instead, we take the logarithms to base 10 of these numbers as in the following sequence:

$$\begin{array}{cccccc} 2, & 2.176, & 2.301, & 2.398, & 2.477, & \dots \\ 2, & 2.176, & 2.352, & 2.528, & 2.704, & \dots \end{array}$$

then on the logarithmic scale, it is the second sequence that gives points at equal distances from each other while the first sequence shows points at decreasing distances along the axis. It would seem, therefore, that equal distances between points on a linear scale indicate equal *absolute* changes in the variable, and equal distances between points on a logarithmic scale indicate equal *proportional* changes in the variable. Before taking the logarithm, the second sequence increased by 50% each time. In log form however, it increased by a constant absolute amount of 1.176.

More formally we can describe this effect as follows. Consider a variable y to be some function $f(x)$, that is $y = f(x)$. If x increases from x to $(x + h)$, then the change in the value of y will be given by $f(x + h) - f(x)$. The **proportional** change however, is given by:

$$\frac{f(x + h) - f(x)}{f(x)}$$

The **rate of proportional change** at the point x is given by the above expression divided by the step change in the x value, namely h:

Rate of proportional change =

$$\lim_{h \to 0} \frac{f(x + h) - f(x)}{h f(x)} = \frac{1}{f(x)} \lim_{h \to 0} \frac{f(x + h) - f(x)}{h} = \frac{1}{y} \frac{dy}{dx}$$

From calculus we know that $d \ln y / dx = (1/y)\, dy/dx$, therefore the rate of proportional change equals:

$$\frac{d \ln y}{dx}$$

This is a measure of the rate of *proportional* change of the variable y, or function $f(x)$. Just as dy/dx measures the gradient of the curve, $y = f(x)$ plotted on a linear scale, $d \ln y / dx$ measures the slope of the curve when plotted on a semi-logarithmic scale, that is the rate of proportional change. For example, a value of 0.05 means that the curve increases at 5% per unit x.

We can apply the same argument to the case when we plot a function on both x and y logarithmic scales. In such a case, the following result is true:

$$\frac{d \ln y}{d \ln x} = \frac{x}{y} \frac{dy}{dx}$$

This shows the relationship between the log form and non-log form of the elasticity. If experimental data is derived from a rate experiment on an isolated enzyme, the data can be plotted in log space (Figure 2.4) and the elasticity read directly from the slope of the line.

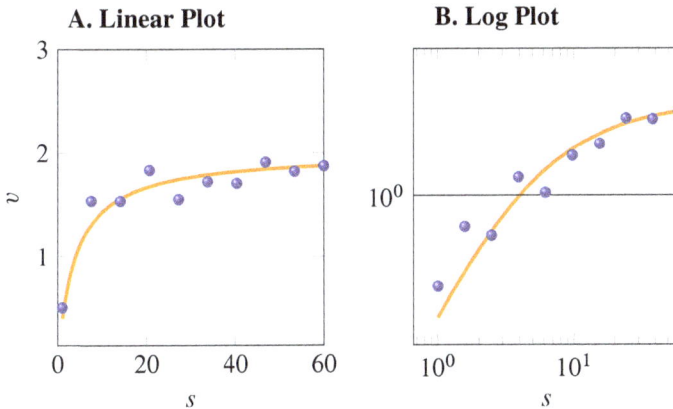

Figure 2.4 A: Plot of reaction rate versus substrate concentration. Measurements include errors. **B:** Same data but plotted in log space. The elasticity can be read directly from the slope of the curve. Curves are generated by assuming $v = 2s/(4+s)$.

As already mentioned, the log form of the elasticity expresses the ratio of relative changes. In approximate terms, we can say that for an x % change in the concentration of a molecular species, the elasticity will give the percentage change, v %, in the reaction rate. For this reason the elasticity is sometimes expressed as a ratio of percentage changes:

$$\varepsilon_{s_i}^v \approx \frac{\% \text{ change in } v}{\% \text{ change in } s_i} \tag{2.4}$$

For example, if the concentration of a substrate is increased from 1.5 mM to 1.95 mM, then the percentage increase in substrate concentration is 30 %. If at the same time, the reaction rate of the enzyme increased from 55 $\mu M g^{-1} \text{min}^{-1}$ to 12 $\mu M g^{-1} \text{min}^{-1}$, then the percentage increase in rate must be 24 %. Therefore the elasticity can be estimated approximately from the ratio 24/30, which is equal to 0.8; that is, the enzyme rate changes almost in proportion to a change in substrate. If the enzyme were acting *in vivo* and a disturbance upstream caused the concentration of substrate to rise, then this enzyme would respond by increasing its rate almost in proportion to the change in substrate concentration.

2.3 Mass-action Kinetics

Computing the elasticities for mass-action kinetics is straightforward. For a reaction such as $v = ks$, it was shown earlier (2.1) that $\varepsilon_s^v = 1$. For a generalized irreversible mass-action law such as:

$$v = k \prod_i s_i^{n_i}$$

the elasticity for species s_i is n_i. For simple mass-action kinetic reactions, the kinetic order and elasticity are therefore identical and independent of species concentration.

Consider the simple reversible mass-action reaction rate law:

$$v = k_1 s - k_2 p \tag{2.5}$$

The elasticities for the substrate and product can be determined as before by differentiating and scaling:

$$\varepsilon_s^v = \frac{k_1 s}{k_1 s - k_2 p} = \frac{v_f}{v} \tag{2.6}$$

$$\varepsilon_p^v = -\frac{k_2 p}{k_1 s - k_2 p} = -\frac{v_r}{v} \tag{2.7}$$

In the above equations v_f is the forward rate ($k_1 s$), v_r is the reverse rate ($k_2 p$), and v is the net rate. Note that ε_s^v is **positive** and ε_p^v **negative**. ε_p^v is negative because increases in product concentration will slow down the net forward rate.

Consider the reversible reaction $S \rightleftharpoons P$ with forward rate constant k_1 and reverse rate constant k_2. At equilibrium it must be true that $k_1 s - k_2 p = 0$, so that by rearrangement it can be shown that:

$$\frac{k_1}{k_2} = \frac{p}{s} = K_{eq}$$

If we divide top and bottom of equation (2.6) by k_1 and s, and equation (2.7) by k_2 and p, we can define the following terms: the ratio $k_1/k_2 = K_{eq}$ which is also the ratio of p/s **at equilibrium**; the ratio of the product to reactant, not necessarily at equilibrium, is defined using the ratio: $p/s = \Gamma$, also called the **mass-action ratio**, and finally $\Gamma/K_{eq} = \rho$ where ρ is called the **disequilibrium ratio**. These terms are summarized below:

$$K_{eq} = \frac{p}{s} \qquad \frac{p}{s} = \Gamma \qquad \frac{\Gamma}{K_{eq}} = \rho \tag{2.8}$$

Given these definitions (2.8), the elasticities can be expressed in the form shown in equations (2.9).

$$\varepsilon_s^v = \frac{1}{1 - \Gamma/K_{eq}} = \frac{1}{1 - \rho}$$

$$\varepsilon_p^v = -\frac{\Gamma/K_{eq}}{1 - \Gamma/K_{eq}} = -\frac{\rho}{1 - \rho} \tag{2.9}$$

ρ	$\varepsilon_s^v = 1/(1-\rho)$	$\varepsilon_p^v = -\rho/(1-\rho)$
0.98	50	-49
0.9	10	-9
0.5	2	-1
0.2	1.25	-0.25
0.1	1.111	-0.111
0.01	1.01	-0.01

Table 2.1 Selected values for the elasticities and the disequilibrium ratio, ρ: Note: $\|\varepsilon_s^v\| > \|\varepsilon_p^v\|$.

ρ is a useful quantity to use in these expressions because far from equilibrium $\rho \approx 0$, while close to equilibrium $\rho \approx 1$. As a result, the elasticity expressions can vary over a wide range of values. For example, far from equilibrium ($\rho \simeq 0$) ε_s^v will lie close to 1.0, while ε_p^v will be close to -0.0. When operating close to equilibrium however ($\rho \approx 1$), the same elasticities will tend to $+\infty$ and $-\infty$, respectively. This behavior is depicted in Figure 2.5.

The mass-action ratio, ρ can be written as:

$$\rho = \frac{\Gamma}{K_{eq}} = \Gamma \frac{k_2}{k_1} = \frac{pk_2}{sk_1}$$

Since the forward rate, $v_f = k_1 s$ and the reverse rate, $v_r = k_2 p$, it follows that:

$$\rho = \frac{v_r}{v_f}$$

Finally using results (2.6) and (2.7), we can state the result:

$$\frac{\varepsilon_p^v}{\varepsilon_s^v} = -\frac{v_r}{v_f} = -\rho$$

This connects the ratio of the elasticities to the disequilibrium ratio and the ratio of the forward and reverse rates.

It also follows from equations (2.9) that the sum of the elasticities for reversible mass-action kinetic rate laws is always one:

$$\varepsilon_s^v + \varepsilon_p^v = 1 \tag{2.10}$$

This means that if one of the elasticities is known, the other can be easily determined by subtraction.

Equation (2.10) is significant for another reason. Since ε_p^v is negative, the absolute magnitude of ε_s^v will **always** be larger than the absolute value for ε_p^v when dealing with mass-action kinetics. That is:

$$\|\varepsilon_s^v\| > \|\varepsilon_p^v\|$$

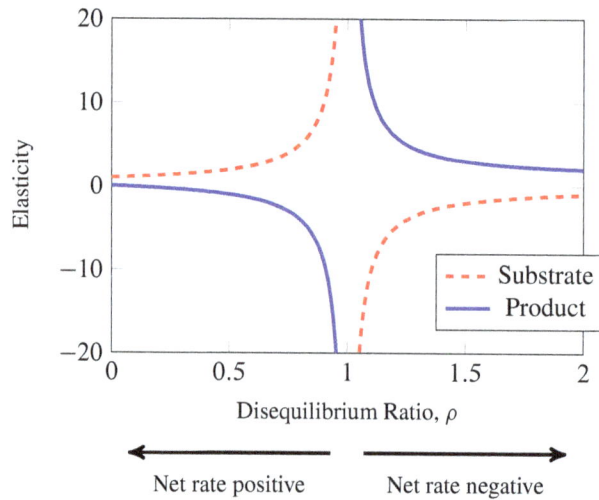

Figure 2.5 Elasticities as a function of the disequilibrium ratio, ρ.

For small elasticity values, the relative difference between the elasticities can be significant. This means that changes in substrate concentrations can have a much greater effect on the reaction velocity than changes in product concentrations.

> For simple mass-action kinetics, changes in substrate concentrations will have a much greater effect on the reaction velocity than changes in product concentrations.

The propagation of signals along a pathway is **determined** by the elasticity values. Given that substrate elasticities are larger than product elasticities, signal propagation tends to amplify when traveling downstream compared to signals traveling upstream which tend to be attenuated.

For the general reversible mass-action rate law:

$$v = k_1 \prod s_i^{n_i} - k_2 \prod p_i^{m_i} \tag{2.11}$$

The elasticities can be shown to equal:

$$\varepsilon_{s_i}^v = \frac{n_i}{1 - \rho}$$

$$\varepsilon_{p_i}^v = -\frac{m_i \rho}{1 - \rho} \tag{2.12}$$

2.4 Enzyme Catalyzed Reactions

The irreversible Briggs-Haldane equation[4] is given by:

$$v = \frac{V_m \, s}{K_m + s}$$

where V_m is the maximal rate and K_m the substrate concentration that yields half the maximal rate. It is straightforward to determine the algebraic elasticity with respect to the subtract concentration. The derivative $\partial v / \partial s$ is given by:

$$\frac{\partial v}{\partial s} = \frac{V_m \, K_m}{(K_m + s)^2}$$

Scaling by v and s yields the elasticity equation:

$$\varepsilon_s^v = \frac{K_m}{K_m + s}$$

The substrate elasticity shows a range of values (Figure 2.6) from zero at high substrate concentrations to one at low substrate concentrations. When the enzyme is near saturation it is naturally unresponsive to further changes in substrate concentration, hence the elasticity is near zero. The reaction behaves as a zero-order reaction at this point. When the elasticity is close to one at low s, the reaction behaves with first-order kinetics. In addition, the reaction order changes depending on the substrate concentration.

It is interesting to note that when $s = K_m$, the elasticity is equal to one half ($\varepsilon_{s=K_m}^v = 0.5$)

Enzyme Elasticity.   We can also compute the elasticity with respect to enzyme concentration since *in vivo* enzyme concentrations can change. Given that $V_m = e_t \, k_{cat}$, where e_t is the total enzyme concentration and k_{cat} the catalytic constant, the enzyme elasticity is derived as follows:

$$\frac{\partial v}{\partial e_t} = \frac{k_{cat} \, K_m}{K_m + s}$$

Scaling by e_t and v yields:

$$\frac{\partial v}{\partial e_t} \frac{e_t}{v} = \frac{k_{cat} \, s}{K_m + s} e_t \frac{K_m + s}{e_t \, k_{cat} \, s} = 1$$

Hence the **enzyme elasticity is one**:

$$\varepsilon_e^v = 1 \tag{2.13}$$

This is not surprising because the reaction rate is first-order with respect to the enzyme concentration.

For the reversible Briggs-Haldane equation, the pattern of elasticities is more complex. It is still however the case that $\varepsilon_e^v = 1$. Details can be found in the companion book [86] but Table 2.2 summaries the results. Two aspects are worth pointing out. Near equilibrium, as with simple mass-action kinetics, the substrate and product elasticities approach positive and negative infinity, respectively. Secondly, there are subtle competition effects between the substrate and product that effect the substrate and product elasticities.

---

[4]But also often called the Michaelis-Menten equation.

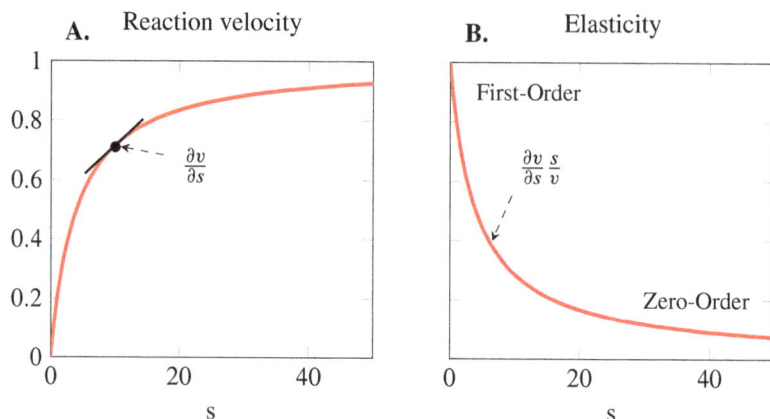

Figure 2.6 A. Left panel: the reaction velocity for an irreversible Michaelis-Menten rate law as a function of substrate concentration. The curve is also marked by the slope $\partial v/\partial s$. **B.** Right panel: the substrate elasticity is plotted as a function of substrate concentration. $K_m = 4$ and $V_m = 1$. Note the elasticity, ε_s^v, starts at one, then decreases to zero as s increases.

Table 2.2 Values of Elasticities Depending on Saturation and Equilibrium Conditions

For the reversible rate law: $V_m/K_s(s - p/K_{eq})/(1 + s/K_s + p/K_p)$

Equilibrium State	Degree of Saturation	Elasticities	
Near Equilibrium	All degrees of saturation	$\varepsilon_s^v \gg 1$; $\varepsilon_p^v \ll -1$; $\varepsilon_s^v + \varepsilon_p^v \approx 1$	
Far from Equilibrium	$s \ll K_s$ and $p \ll K_p$	$\varepsilon_s^v \approx 1$; $\varepsilon_p^v \approx 0$	
Far from Equilibrium	$s \gg K_s$ and $p \ll K_p$	$\varepsilon_s^v \approx 0$; $\varepsilon_p^v \approx 0$	
Far from Equilibrium	$p \gg K_p$ (Any substrate level)	$\varepsilon_s^v \approx -\varepsilon_p^v$	
	if $s/K_s \ll p/K_p$	$\varepsilon_s^v \approx 1$	$\varepsilon_p^v \approx -1$
	$s/K_s \approx p/K_p$	$\varepsilon_s^v \approx 0.5$	$\varepsilon_p^v \approx -0.5$
	$s/K_s > p/K_p$	$\varepsilon_s^v < 0.5$	$\varepsilon_p^v > -0.5$

2.5 Cooperativity

Given the Hill equation:

$$v = \frac{V_m\, s^n}{K_d + s^n} = \frac{V_m\, s^n}{K_H^n + s^n} \tag{2.14}$$

where K_H is the concentration of ligand that yields half the maximal rate. The elasticity coefficient, ε_s^v may be derived directly from the Hill equation (2.14). Differentiating and scaling the Hill equation yields the following elasticity both in terms of the dissociation constant, K_d and the half maximal activity constant, K_H:

$$\varepsilon_s^v = \frac{n\, K_d}{K_d + s^n} = \frac{n}{1 + \left(\dfrac{s}{K_H}\right)^n} \tag{2.15}$$

The elasticity of a reaction obeying the Hill equation has a value equal to n at low substrate concentrations ($s \ll K_d$). In contrast, irreversible Michaelian enzymes at low substrate concentrations have an elasticity value of one. Therefore an enzyme obeying the Hill equation shows a much higher elasticity to the substrate concentration compared to a Michaelian enzyme. Like a Michaelian enzyme, the value of the elasticity falls off rapidly as the substrate concentration increases, reaching zero as the enzyme becomes saturated. Figure 2.7 illustrates this response for $n = 4$ and $K_d = 1$. An interesting feature in Figure 2.7 is the delayed fall in the elasticity at low substrate concentrations.

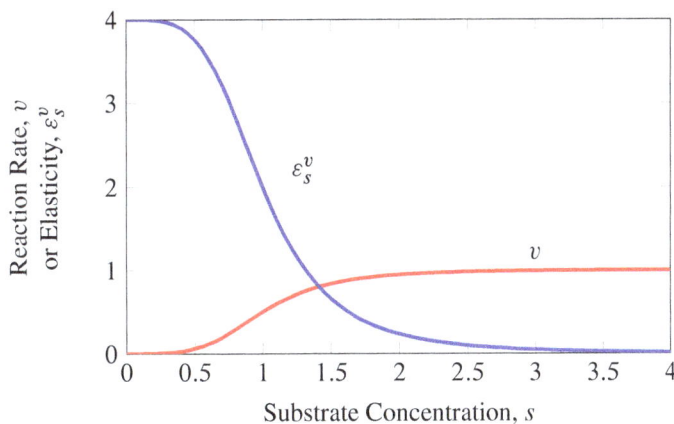

Figure 2.7 Plot showing the response of the rate and elasticity for the Hill model, with $n = 4$ and $K_d = 1$.

Similar equations can be derived for other cooperative models as well as allosteric control. The important message is that cooperative and allosteric control illicite high values for elasticity coefficients which can have a significant impact on pathway behavior. Some of these effects will be covered in detail in later chapters. Much more detail on the elasticities of a great variety of rate laws can be found in the companion book [86].

2.6 Local Equations

The elasticity coefficient is of central importance to metabolic control analysis (MCA). Just as the Michaelian constants are essential to describing the rate of an enzyme-catalysed reaction, the elasticities are equally essential to describing the behavior of whole pathways. Before we discuss this topic in the next chapter, some direct uses of the elasticities will be given here.

Recall that the elasticity coefficient is given by:

$$\varepsilon_s^v = \frac{\partial v}{\partial s}\frac{s}{v}$$

This definition can be rearranged and an approximate equation written in the form:

$$\frac{\delta v}{v} \approx \varepsilon_s^v \frac{\delta s}{s}$$

This relation is approximate because the changes considered are finite, and the definition of an elasticity applies strictly to infinitesimal changes. The equation describes how, given a fractional change in some effector S, the resulting fractional change in rate can be computed. For example, if the elasticity of an enzyme reaction towards an effector S is 0.8, then given a fractional change in s of 0.05 (a 5% change in s), the fractional change in rate is given by:

$$0.8 \times 0.05 = 0.04$$

In other words, a 5% change in s leads to a 4% change in reaction rate.

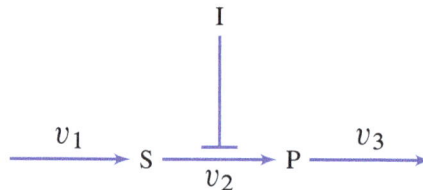

Figure 2.8 Species I inhibits reaction v_2 in addition to potential affects from S and P.

The diagram (Figure 2.8) shows a fragment from a larger pathway. The central reaction step has three effectors which could potentially change the rate v_2 – these are S, P and an inhibitor, I. Let us consider a disturbance[5] somewhere in the pathway but *not* originating at the reaction step under consideration. This disturbance will ultimately cause changes in each of the effectors by amounts δs, δp and δi. These changes will also be accompanied by a change to the reaction rate by an amount, δv. There are two immediate questions we can ask: (i) what is the relationship between the change in the effectors and the change in rate? and, (ii) what is the contribution that each change in effector makes to the final change in rate?

The answers to these questions are straightforward to obtain. Provided that the changes are small, then the fractional change in rate, $\delta v/v$, is defined by the sum of the individual contributions:

$$\frac{\delta v}{v} \approx \varepsilon_s^v \frac{\delta s}{s} + \varepsilon_p^v \frac{\delta p}{p} + \varepsilon_i^v \frac{\delta i}{i}$$

---

[5]This could be due to one of a number of causes, a change in enzyme expression, nutrient supply change, hormonal change, etc.

For example, let us assume the following values for the elasticities, $\varepsilon_s^v = 0.4$; $\varepsilon_p^v = -0.5$; $\varepsilon_i^v = -0.2$, and assuming the following changes in effectors, $\delta s/s = 0.05$; $\delta p/p = 0.03$; $\delta i/i = 0.01$, then the fractional change in rate through the step is given by:

$$\frac{\delta v}{v} \approx 0.4 \times 0.05 + (-0.5) \times 0.03 + (-0.2) \times 0.01 = 0.003$$

The rate has only changed by 0.3 %, much of the potential increase that could have been obtained by the change in S has been reduced by strong product inhibition. To answer the question – what contribution does each effector make to the final change in rate – is given simply by examining the individual changes. Thus out of the total absolute change in rate, the change brought about by S contributed 54 %, while the change in product and inhibitor contributed -41 % and -5 %, respectively. Clearly the change in inhibitor was not an important factor in this particular case.

The degree to which each effector has an affect on the reaction rate is an important consideration. Although in pathway diagrams we may see many feedback loops, we rarely see the quantitative contribution that the effector makes in relation to other potential effectors.

In general, for a reaction step embedded in a pathway and acted upon by m effectors, the change in rate due to changes in all effectors is given by the relation:

$$\frac{dv}{v} = \sum_{j=1}^{m} \varepsilon_{s_j}^v \frac{ds_j}{s_j}$$

where the symbol, $\sum$ means 'sum of'. In the equation the small but finite changes have been replaced by differentials so that the relation is exact. If the concentration of enzyme is also changed, then we may also add the enzyme elasticity to the sum, as in:

Local Equation:

$$\frac{dv}{v} = \sum_{j=1}^{m} \varepsilon_{s_j}^v \frac{ds_j}{s_j} + \varepsilon_e^v \frac{de}{e} \tag{2.16}$$

Relation (2.16) is probably one of the most important mathematical relations used in MCA, and we will come across its application in subsequent chapters. In mathematical terms, the relationship is a modification of the standard total derivative. This states that if y is some function f of m variables, x_i, that is:

$$y = f(x_1, x_2, \ldots, x_m),$$

Then the total derivative of y is given by:

$$dy = \frac{\partial y}{\partial x_1} dx_1 + \frac{\partial y}{\partial x_2} dx_2 + \ldots + \frac{\partial y}{\partial x_m} dx_m$$

Scaling both sides of this equation will lead a form similar to the local equation (2.16).

2.7 General Elasticity Rules

Just as there are rules for differential calculus, there are similar rules for computing elasticities. These rules can be used to simplify the derivation of elasticities for complex rate law expressions.

Table 2.3 shows some common elasticity rules, where a designates a constant, and x the variable. For example, the first rule says that the elasticity of a constant is zero.

1.	$\varepsilon(a) = 0$
2.	$\varepsilon(x) = 1$
3.	$\varepsilon(f(x) \pm g(x)) = \varepsilon(f(x)) \frac{f(x)}{f(x)+g(x)} \pm \varepsilon(g(x)) \frac{g(x)}{f(x)+g(x)}$
4.	$\varepsilon(x^a) = a$
5.	$\varepsilon(f(x)^a) = a\varepsilon(f(x))$
6.	$\varepsilon(f(x)\,g(x)) = \varepsilon(f(x)) + \varepsilon(g(x))$
7.	$\varepsilon(f(x)/g(x)) = \varepsilon(f(x)) - \varepsilon(g(x))$

Table 2.3 Transformation rules for determining the elasticity of a function, a = constant, x = variable.

We can illustrate the use of these rules with a simple example. Consider the reversible mass-action rate law (2.5):

$$v = k_1\,s - k_2\,p$$

To determine the elasticity we first apply rule 3 to give:

$$\varepsilon_s^v = \varepsilon_s(k_1\,s)\frac{k_1\,s}{k_1\,s - k_2\,p} - \varepsilon_s(k_2\,p)\frac{-k_2\,p}{k_1\,s - k_2\,p}$$

where $\varepsilon_s(f)$ means the elasticity of expression f with respect to variable s.

We transform the elasticity terms by applying additional rules. Apply rule 6 to the expression $\varepsilon_s(k_1\,s)$ to give:

$$\varepsilon_s(k_1\,s) = \varepsilon_s(k_1) + \varepsilon_s(s)$$

We can now apply rule 1 to the first term on the right, and rule 2 to the second term on the right to give:

$$\varepsilon_s(k_1\,s) = 0 + 1$$

Since we're evaluating the elasticity of s, p in this situation is a constant, therefore:

$$\varepsilon_s(k_2\,p) = \varepsilon_s(k_2) + \varepsilon_s(p) = 0 + 0$$

Combining these results yields:

$$\varepsilon_s^v = \frac{k_1\,s}{k_1\,s - k_2\,p}$$

which corresponds to the first equation in (2.6). Now consider a simple irreversible enzyme kinetic rate equation:

$$v = \frac{V_m\,s}{K_m + s}$$

where V_m is the maximal velocity and K_m the substrate concentration at half maximal velocity. The elasticity for this equation can be derived by first using the quotient rule (rule 7) which gives:

$$\varepsilon_s^v = \varepsilon(V_m\,s) - \varepsilon(K_m + s)$$

The rules can now be applied to each of the sub-elasticity terms. For example, we can apply rule 6 to the first term, $n\varepsilon(V_m\ s)$, and rule 3 to the second term, $\varepsilon(K_m + s)$, to yield:

$$\varepsilon_s^v = (\varepsilon(V_m) + \varepsilon(s)) - \left(\varepsilon(K_m)\frac{K_m}{K_m + s} + \varepsilon(s)\frac{s}{K_m + s}\right)$$

Applying rules 1 and 2 allows us to simplify ($\varepsilon(V_m) = 0; \varepsilon(Km) = 0; \varepsilon(s) = 1$) the equation to:

$$\varepsilon_s^v = 1 - \left(\frac{s}{K_m + s}\right)$$

or:

$$\varepsilon_s^v = \frac{K_m}{K_m + s}$$

Example 2.5

Determine the elasticity expression for the rate laws with respect to s, using log-log rules:

1. $v = k(s + 1)$

   Begin with the product rule 6 followed by rule 1:

   $$\varepsilon_s^v = \varepsilon(k) + \varepsilon(s + 1) = \varepsilon(s + 1)$$

   Next use the summation rule 3, 2, and 1:

   $$\varepsilon_s^v = \varepsilon(s + 1) = \varepsilon(s)\frac{s}{s + 1} + \varepsilon(1)\frac{1}{s + 1}$$

   $$= \frac{s}{s + 1} + 0 = \frac{s}{s + 1}$$

2. $v = k/(s + 1)$

   Begin with the quotient rule 7 followed by Rule 1, 3 and 2:

   $$\varepsilon_a^v = \varepsilon(k) - \varepsilon(s + 1) = 0 - \varepsilon(s)\frac{s}{s + 1}$$

   $$= -\frac{s}{s + 1}$$

3. $v = s(s + 1)$

   Begin with the product rule 6:

   $$\varepsilon_s^v = \varepsilon(s) + \varepsilon(s + 1)$$

   Next use Rule 2, 3 and 1:

   $$\varepsilon_s^v = 1 + \frac{s}{s + 1}$$

```
(* Define elasticity evaluation rules *)
el[x_, x_]    := 1
el[k_, x_]    := 0
el[Log[u_,x_] := el[Log[u],x] = el[u,x]/Log[u]
el[Sin[u_],x_] := el[Sin[u],x] = u el[u,x]Cos[u]/Sin[u]
el[Cos[u_],x_] := el[Sin[u],x] = -u el[u,x]Sin[u]/Cos[u]
el[u_*v_,x_]   := el[u*v,x] = el[u,x] + el[v,x]
el[u_/v_,x_]   := el[u/v,x] = el[u,x] - el[v,x]
el[u_+v_,x_]   := el[u+v,x] = el[u,x]u/(u+v) + el[v,x]v/(u+v)
el[u_-v_,x_]   := el[u-v,x] = el[u,x]u/(u-v) - el[v,x]v/(u-v)
el[u_^v_,x_]   := el[u^v,x]  = v (el[u,x] + el[v,x] Log[u])
```

Figure 2.9 Elasticity rules expressed as a Mathematica script.

To make matters even simpler, we can define the elasticity rules using an algebraic manipulation tool such as Mathematica (http://www.wolfram.com/) to automatically derive the elasticities [118]. To do this we must first enter the rules in Table 2.3 into Mathematica. The script shown in Figure 2.9 shows the same rules (with a few additional ones) in Mathematica format.

The notation f[x_,y_] := g() means define a function that takes two arguments, x_ and y_. The underscore character in the argument terms is essential. Note also the symbol ':' in the assignment operator.

Typing el[k1 s - k2 p, s] into Mathematica will result in the output:

k1 s/(-k2 p + k1 s)

2.8 Summary

The elasticity coefficient is a measure of how sensitive the rate of a reaction is to changes in its environment. The factors of interest are the concentrations of substrates, products, effectors and enzyme.

There will be as many elasticities as there are effectors of the reaction. The elasticity is strictly defined in terms of a partial derivative, which means that it measures the change in rate when one effector is changed. For example, the substrate elasticity is measured when all other effectors are held constant except for the substrate concentration. Algebraically this is achieved by partial differentiation, and experimentally by clamping the appropriate effector concentrations.

The elasticity coefficient can be written in various equivalent forms, each reflecting a different emphasis:

$$\varepsilon_{s_j}^v = \left(\frac{\partial v_i / v_i}{\partial s_j / s_j} \right)_{s_k, s_l, \dots} = \frac{s_j}{v_i} \left(\frac{\partial v_i}{\partial s_j} \right)_{s_k, s_l, \dots} = \left(\frac{\partial \ln v_i}{\partial \ln s_j} \right)_{s_k, s_l, \dots}$$

The first form is the ratio of fractional changes, the second form the scaled slope on a linear plot, and the third form the slope of a log/log plot.

The notation, $s_k, s_l \dots$ means that these effectors are held constant during the measurement of the partial derivative. An approximate form of the elasticity is given by:

$$\varepsilon_s^v \approx \frac{\% \text{ change in } v}{\% \text{ change in } s}$$

which can be used to estimate an elasticity when changes in reaction rate and effector are known. Elasticities have a number of important properties:

- The elasticity coefficient is **not** a constant but depends on the concentrations of all effectors that might affect the reaction rate. An elasticity is not like a K_m or K_i; the Michaelian constants are characteristic for a particular enzyme and effector, reflecting the enzymes' kinetic mechanism and interaction energy with the effector. Kinetic constants do not in general depend on the concentrations of the effectors; elasticities do.

- In general, an elasticity is a function of both the kinetic characteristics of an enzyme, and the concentration of all the various effectors that might interact with the enzyme.

- For the standard irreversible Michaelian mechanism, the elasticity of a substrate at saturating levels is zero, and when the substrate is below its K_m, the elasticity is unity. When the substrate concentration is equal to the K_m, the elasticity has a value of 0.5.

Given a change in concentration of a effector, it is possible to use the elasticity coefficient to predict (approximately) the change in rate, thus:

$$\frac{\delta v}{v} \approx \varepsilon_s^v \frac{\delta s}{s}$$

If more than one effector is changing at a time, then the approximate change in rate is given by the sum of the individual contributions:

$$\frac{\delta v}{v} \approx \sum_j \varepsilon_{s_j}^v \frac{\delta s_j}{s_j}$$

It is very important to appreciate that the elasticities used in the above equation must be measured at the prevailing state of the effectors. It makes no sense to use an elasticity that has previously been measured at a substrate concentration of 2mM, and then to use the same elasticity *value* at a substrate concentration of 20mM.

Further Reading

1. Fell D A (1996) Understanding the Control of Metabolism. Portland Press, ISBN: 185578047X

2. Heinrich R and Schuster S (1996) The Regulation Of Cellular Systems. Springer; 1st edition, ISBN: 0412032619

3. Sauro HM (2012) Enzyme Kinetics for Systems Biology. 2nd Edition, Ambrosius Publishing ISBN: 978-0982477335

Exercises

1. What is the relevance of elasticity coefficients in understanding network dynamics?

2. State the operational interpretation of an elasticity.

3. Why is the elasticity coefficient expressed in terms of a partial derivative? What does it mean in terms of an experimental operation?

4. An experiment indicates that a given effector x has an elasticity of -0.5 with respect to the rate of a reaction. State two key aspects that this elasticity describes.

5. What is the elasticity with respect to the species a given the rate law $v = ka^3$?

6. Work out algebraically the elasticity for the rate law, $v = k_1 s + k_2$. Describe its properties at high and low levels of s.

7. Derive the elasticity expression with respect to x for the following:
   a) $v = x^2 + 1$
   b) $v = x^2 + x$
   c) $v = x/(x^2 + 1)$

8. Describe one technique for numerically estimating an elasticity.

9. Given a change in $\delta v/v$ equal to 0.04, and if $\varepsilon_s^v = 0.1$, what is the change in $\delta s/s$? If the concentration of s was 2.5mM, what is the absolute change in s?

10. If the concentration of s is 3mM, and the corresponding elasticity of an enzyme with a rate law, $V_{max} s/(K_m + s)$ is 0.6, what is the K_m of the enzyme, assuming $V_m = 1$? What would be the elasticity at 8mM? Why does the elasticity change?

11. What does the term Γ/K_{eq} measure?

12. Describe what value the disequilibrium ratio tends to as a reaction nears equilibrium.

13. For a mass-action reversible reaction, describe what happens to the substrate and product elasticities as the reaction approaches equilibrium.

14. Derive the two equations in (2.9).

15. Describe the significance of equation (2.10).

16. Using the elasticity rules in Table 2.3, derive the elasticity for the following equation indicating all intermediate steps.
$$v = s^n/(K_m + s^n)$$

3

Introduction to Biochemical Control

3.1   What do we mean by Control?

For most people the word control means the ability to influence, command or to restrain a situation or process[1]. In this chapter the term control will be used in a similar sense to describe how much influence a given reaction step in a network has on the system. To make matters simpler, the system will be considered at steady state so that control will refer to how much influence a given reaction step has on the steady state. That is, how fluxes and concentrations are influenced. The measure of much control a reaction step has over the steady state will be called the **control coefficient**. Our initial definition of a control coefficient is as follows:

> The amount of control (i.e. influence) that a particular reaction step has on a flux or species concentration is called the **control coefficient**.

Most reaction steps in a cell are controlled by proteins. One question to ask is, how much influence does a given protein have on the system's steady state? Experimentally, such control can be measured by changing the concentration of an enzyme or changing its activity via an inhibitor and measuring the effect on the steady state flux and species concentrations. We can change the concentration of a protein in various ways such as using irreversible inhibitors, changing the promoter consensus sequence on the gene that codes for the protein, employing antisense RNA to reduce the expression level, or using dCas9 guided by a specific guide RNA.

In addition to investigating how individual reaction steps control the fluxes and concentrations in a network, we are also interested in how external factors influence the network. Examples of external factors include the level of nutrients, hormones, and of particular interest to human health, thera-

---

[1]In engineering, control theory refers to the body of knowledge concerned with the design and study of systems that can perform specific tasks or achieve a particular objective.

$$X_o \xrightarrow{v_1} S_1 \xrightarrow{v_2} S_2 \xrightarrow{v_3} S_3 \xrightarrow{v_4} S_4 \xrightarrow{v_5} X_5$$

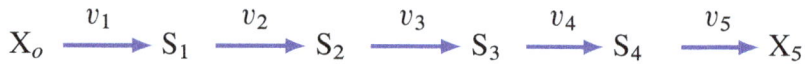

Figure 3.1 Five step linear pathway. X_o and X_5 are assumed to be fixed in the pathway model.

peutic drugs. In these situations, rather than using the word control, we will use the term response. Thus, a biological cell will have a response to a particular infusion of a drug. The degree of influence an external factor has on a biological system will be described using **response coefficients**.

> The degree of influence a particular external input has on a pathway is called the **response coefficient**.

We will consider response coefficients in a later chapter, the focus here will be on control coefficients.

A Reminder on Notation

It is worth describing again the notation that will be used in this and subsequent chapters.

1. Upper case letters such as S or X refer to the name of a molecular species.

2. Lower case letters such as s or x refer to the concentration of the molecules species.

3. The upper case letter E refers to the name of an enzyme.

4. The lower case letter e refers to the concentration of the enzyme, E.

5. The symbol Δ means a change, for example Δs means a change in the concentration of S.

6. The symbol δ means a small change, for example δe_1 means a small change to enzyme E_1.

7. The symbol J refers to the steady state flux through a pathway.

8. The symbol v_i refers to the rate of reaction through the ith reaction step.

3.2 Control Coefficients

Control coefficients are used to describe how much influence (i.e. control) a given reaction step has on the steady state flux or species concentration level. It is common to measure this influence by changing the concentration of the enzyme that catalyzes the reaction. To describe control coefficients in more detail, let us consider a thought experiment.

The following discussion will be centered on the simple linear pathway shown in Figure 3.1. Let us assume that the species pools, S_1 to S_4 in the pathway are empty (zero concentration), and that X_o and X_5 are *fixed* species forming the system boundary.

To make matters simpler, assume that the concentration of right-hand boundary pool, X_5 is set to zero. In order to have a net flux through the pathway, the external metabolite, X_o, must have a positive value, perhaps 1 mM. This is the situation at time zero. Let us allow the pathway to evolve in time. The first thing that happens is that the reaction catalyzed by the first enzyme begins to convert X_o into product S_1. Since we assume that X_o is fixed, the concentration of X_o is unaffected by this rate of consumption. However, the product S_1 is a floating species and as time goes on, its concentration will rise. As the concentration of S_1 increases, two things will happen. First, the second enzyme will begin to convert S_1 into S_2, and secondly S_1 will begin to inhibit its own production rate by the first enzyme on account of product inhibition. The first reaction will therefore begin to rise at a **slower** rate.

Since the second enzyme is now producing S_2, S_2 starts to increase. S_2 in turn will stimulate the third enzyme to begin making S_3, but it also begins to inhibit the second enzyme. And so on down the chain, all concentrations begin to rise and all enzymes show a positive rate. The concentrations of the floating species and the reaction rates cannot however go on rising forever. We have already seen that as the species concentrations rise, they begin to inhibit the enzymes that produce them. The net effect of these many interactions is that the concentrations slowly settle to a constant value such that the rates at which they are being made is exactly balanced by the rates at which they are being consumed. The rate of the first enzyme must balance the rate of the second enzyme, that is $v_1 = v_2$, but the second and third rates will also be in balance, so that $v_2 = v_3$. This must mean that the rate through the first enzyme must be the same as the rate through the third enzyme, $v_1 = v_3$. In fact all rates across each enzyme will equal each other, that is:

$$v_1 = v_2 = v_3 = v_4 = v_5$$

This state is the steady state, where the concentrations of all the metabolites settle to some value and no longer evolve in time, and the rate through each step is the **same**.

The fact that the rate across each enzyme is the same also means that there is a constant flow of material through the pathway, which we call the **flux**, symbolized by J. At steady state, there are no 'slow' rates or 'fast' rates, they are all the **same**.

In a linear pathway at steady state, the rates of reaction are equal to each other and non-zero. At the same time all floating species are unchanging.

The steady state rate through the pathway is called the flux, J.

Given the pathway at steady state, we can consider some additional thought experiments such as the effect of perturbations on the steady state. Let us change the concentration of one of the enzymes and see what happens to the steady state concentrations and the flux, J. Let us double the concentration of enzyme, E_2, that catalyzes the second step. The immediate effect is to increase the rate, v_2, through the step. This in turn results in more S_2 being produced and more S_1 consumed, S_2 will therefore rise and S_1 will fall. The rise in S_2 will cause the reaction rates through each step downstream to increase as S_3 and S_4 start to rise. Assuming that the first enzyme is product inhibited by S_1, the fall in S_1 will cause a **rise** in the rate through v_1. The net effect of all these changes is that the net flux through the pathway will **increase**, all species concentrations downstream of v_2 will **increase** and S_1 will **decrease**. Figure 3.2 illustrates a simulation that shows the change in flux through E_2 as the pathway approaches steady state, followed by the effect of a perturbation in E_2 at $t = 0.2$.

To see how effective the change in enzyme concentration is, we can take the ratio of the change in

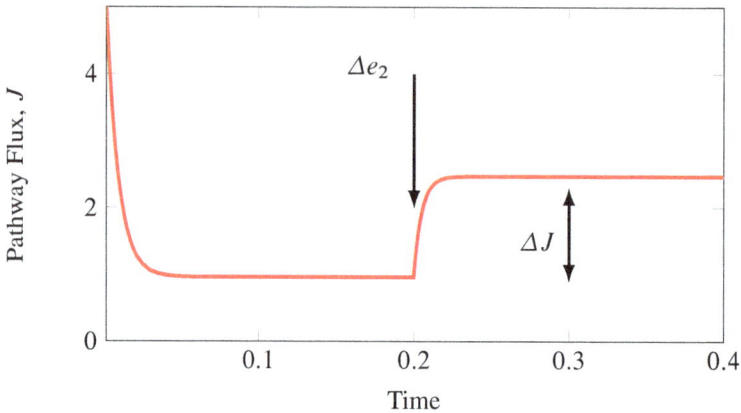

Figure 3.2 Effect of a perturbation in E_2 at $t = 0.2$ on the flux through the pathway, Figure 3.1. Note the initial transition to steady state between $t = 0$ and $t = 0.2$. At $t = 0.2$ a change Δe_2 is made to E_2 resulting in a change ΔJ in the steady state pathway flux. Parameters are given in the Tellurium Script: 3.2 at the end of the chapter.

flux or species concentration to the change in enzyme:

$$\frac{\Delta J}{\Delta e_2}, \quad \frac{\Delta s_1}{\Delta e_2}, \dots \frac{\Delta s_4}{\Delta e_2}$$

where Δ means 'a change in'. However, because enzyme kinetic rate laws are usually nonlinear, the degree of influence we measure will depend on the size of the ΔE. Therefore instead of making large changes to the enzyme concentration, we should make small changes, for example:

$$\frac{\delta J}{\delta e_2}, \quad \frac{\delta s_1}{\delta e_2}, \dots \frac{\delta s_4}{\delta e_2}$$

where δ means 'a small change'. We can be more precise mathematically if we make the changes infinitesimally small. Our measurement of influence then becomes:

$$\frac{d J}{d e_2}, \quad \frac{d s_1}{d e_2}, \dots \frac{d s_4}{d e_2}$$

Finally, if we want to make the measurement useful to experimentalists, we can remove the units by scaling the derivatives, such that:

$$\frac{d J}{d e_2} \frac{e_2}{J}, \quad \frac{d s_1}{d e_2} \frac{e_2}{s_1}, \dots \frac{d s_4}{d e_2} \frac{e_2}{s_4}$$

Obviously in an experiment we cannot make infinitesimal changes, but we can make changes sufficiently small (but still measurable) that we can approximate the derivatives. The scaled derivatives are called **control coefficients**, and we will define both the flux and concentration control coefficients as follows:

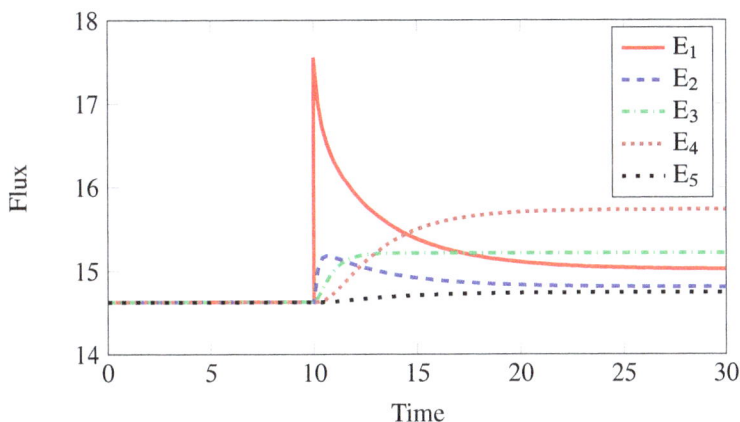

Figure 3.3 Effect of perturbing each enzyme by 20% in the linear pathway, Figure 3.1. Note that each enzyme affects the system differently both in the transient response and in the final steady state response. Parameters given in the Tellurium script: 3.3.

Definition of the Flux Control Coefficient:

$$C_{e_i}^J = \frac{dJ}{de_i}\frac{e_i}{J} = \frac{d\ln J}{d\ln e_i} \approx \frac{J\%}{e_i\%} \tag{3.1}$$

Definition of the Concentration Control Coefficient:

$$C_{e_i}^{s_j} = \frac{ds_j}{de_i}\frac{e_i}{s_j} = \frac{d\ln s_j}{d\ln e_i} \approx \frac{s_j\%}{e_i\%} \tag{3.2}$$

Example 3.1

A given enzyme catalyzed reaction in a metabolic pathway has a flux control coefficient equal to 0.2:

$$C_e^J = 0.2$$

What does this mean?

A flux control coefficient of 0.2 means that increasing the enzyme activity of the step by 1% will increase the steady state flux through the pathway by 0.2%.

In expression (3.1), J is the flux through the pathway and e_i the enzyme concentration of the ith step. Operationally, an individual C_{e_i} is measured by making a small change to E_i, waiting for the system to reach a new steady state, and then taking the ratio of the change. Before moving on to another step, the level of E_i must be restored back to its original value.

From a practical standpoint we see that the control coefficients can also be approximated by the ratio of **percentage changes** which is a useful interpretation for measurement purposes. The other point to note is that like elasticities, we can express the control coefficients in log form (See section 2.2).

The flux control coefficient measures the fractional change in flux brought about by a given fractional change in enzyme concentration. The concentration control coefficients measure the fractional change in species concentration given a fractional change in enzyme concentration. Control coefficients are useful because they tell us how much influence each enzyme or protein has in a biochemical reaction network.

It is important to note however, that knowing the values of the control coefficients does not tell us why certain enzymes or proteins have more influence than others. To answer the 'why' question we must consider the theorems associated with how control is distributed, and the relationship of the control coefficients to the elasticities of the network.

3.3 Distribution of Control

Flux control coefficients are a useful measure to judge the degree to which a particular step influences the steady state flux. One key question is how the influences are distributed across a pathway. Consider the simple two-step pathway:

$$X_o \xrightarrow{v_1} S \xrightarrow{v_2} X_1$$

There is a simple graphical technique we can use to study how the enzyme concentrations, e_1 and e_2, control the steady state concentration, s, and the steady state flux, J. In this system, the steady state flux, J, will be numerically equal to the reaction rates v_1 and v_2:

$$J = v_1 = v_2$$

It is important to recall that for an isolated enzyme where all reactants, products and effectors are held constant, the reaction rate v is proportional to the concentration of enzyme, E, i.e $v \propto e$.

Let us plot both reaction rates, v_1 and v_2, against the substrate concentration, s, Figure 3.4.

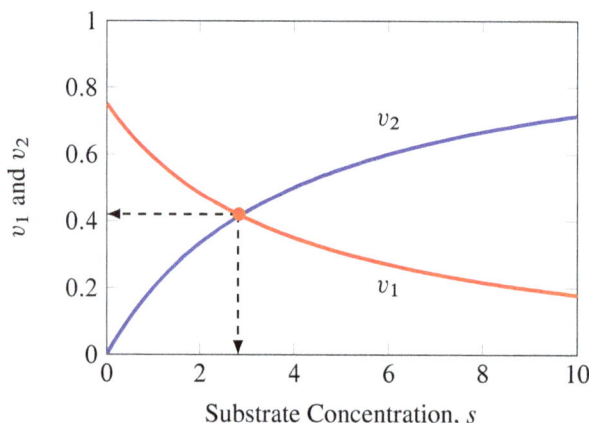

Figure 3.4 Plot of v_1 and v_2 versus the concentration of S for a simple two-step pathway. The intersection of the two curves marks the point when $v_1 = v_2$, that is steady state. A perpendicular dropped from this point gives the steady state concentration of S.

Note the response of v_1 to changes in s. v_1 falls as s increases due to product inhibition by S. The intersection point of the two curves marks the point when $v_1 = v_2$, that is, the steady state. A line dropped perpendicular from the intersection point marks the steady state concentration of S.

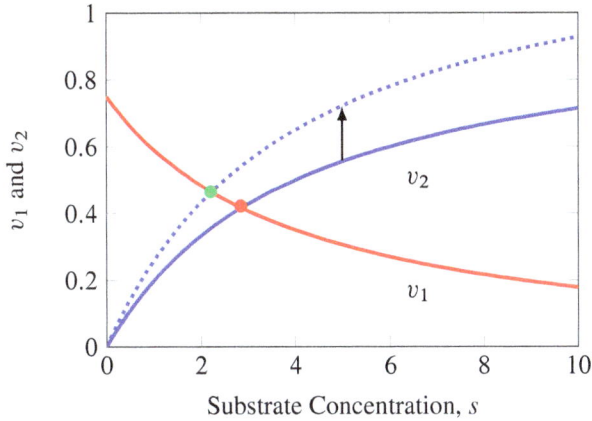

Figure 3.5 v_2 has been increased by 30% (dotted line) by increasing the enzyme concentration on v_2. This results in a displacement of the intersection point to the left, leading to a decrease in the steady state concentration of S.

Let us increase the concentration of E_2 by 30% by adding more enzyme (Figure 3.5). Because the reaction rate is proportional to E_2, the curve is scaled upwards although its general shape stays the same. Note how the intersection point moves to the left, indicating that the steady state concentration of S **decreases** relative to the reference state. This is understandable because with a higher v_2, more S is consumed, therefore S decreases.

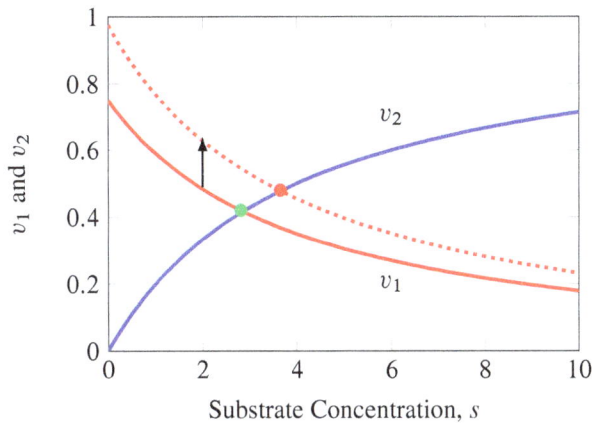

Figure 3.6 v_1 has been increased by 30% (dotted line) by increasing the enzyme concentration on v_1. This results in a displacement of the steady state curve to the right, leading to an increase in the steady state concentration of S.

In the next experiment, restore E_2 back to its original level and instead increase the amount of E_1

by 30% (Figure 3.6). Again, changing E_1 scales the v_1 curve but because of the negative curvature of v_1, the intersection point moves to the right, indicating that the steady state concentration of S **increases** relative to the reference state.

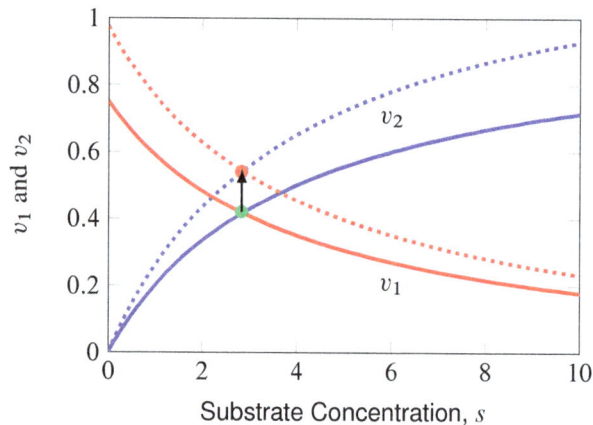

Figure 3.7 In this experiment, both E_1 and E_2 are increased by 30 % (dotted lines). Because both rates are increased by the same amount, the rate of change of S does not change. This means that there is no resulting change to the steady state concentration of S. The net flux through the pathway has however increased by 30 %.

In the final experiment we change the activity of **both** E_1 and E_2 by 30% (Figure 3.7). Note that the curves for v_1 and v_2 are both equally scaled upwards, moving the intersection point vertically upwards and therefore **does not** change the steady state concentration of S. This happens because both curves move vertically by the same fraction so that the intersection point can only move vertically.

This experiment highlights an important result, when all enzyme concentrations are increased by the same fraction, the flux increases by that same fraction but the species or metabolite levels remain **unchanged**. We can summarize this with the following statement:

If all E_i are increased by a factor α, sat 30%, then the steady state change in J and s_i is: αJ and 0 respectively.

> If all enzymes are increased by αe_i then
>
> $$\delta J = \alpha J \quad \text{and all} \quad \delta s_j = 0$$

This is such an important result that it will be repeated again:

> Given a pathway with n steps where every enzyme concentration is raised by the same factor, α, the species concentrations will remain unchanged but the flux increases by a factor α. This observation is true no matter how complex the pathway topology.

A way to understand why J increases by αJ is as follows. Since $\delta s = 0$, the only change that could possibly effect the flux is the change in enzyme concentration, since the enzyme concentration has increased by a given proportion (α), the flux must also have increased by the same proportion since the rate is proportional to the enzyme concentration (i.e $v_i \propto e_i$), hence $J \rightarrow \alpha J$.

Example 3.2

In the following pathway, an increase in S_2 by 20% results in a 5% increase in the steady state flux. Estimate the value of $C_{e_2}^J$.

$$\rightarrow S_1 \xrightarrow{E_2} S_2 \rightarrow S_3 \rightarrow$$

$C_{e_2}^J$ is the ratio of the fractional change in flux divided by the fraction change in the enzyme concentration. Therefore, an estimate for the control coefficient is given by:

$$C_{e_2}^J = \frac{0.05}{0.2} = 0.25$$

3.4 Predicting Flux and Concentration Changes

We can also rearrange equations (3.1), and (3.2) into the following form:

$$\frac{dJ}{J} = C_{e_i}^J \frac{de_i}{e_i}$$

$$\frac{ds_j}{s_j} = C_{e_i}^{s_j} \frac{de_i}{e_i}$$

These simple relations allow us to compute the change in flux or concentration given a change in enzyme concentration. These relations only hold true if the changes in enzyme concentration are infinitesimal. For practical purposes the relationships will approximately hold provided the changes in E_i are small. Of more interest is if we make changes to multiple enzymatic steps, the overall change will be the sum of the individual changes. The technical reason for this is that small changes in E_i mean that the response is linear so that multiple responses can be summed to obtain the total response. In general, if we make changes to n reaction steps, then the overall change in flux and species concentrations is given by:

$$\frac{dJ}{J} = \sum_{i=1}^{n} C_{e_i}^J \frac{de_i}{e_i} \tag{3.3}$$

$$\frac{ds}{s} = \sum_{i=1}^{n} C_{e_i}^s \frac{de_i}{e_i} \tag{3.4}$$

See also Box 7.1.

Box 7.1 Total Derivative. Proof of (3.3) **and** (3.4)**.**

If:

$$J = J(e_1, e_2, \ldots)$$

then the total derivative is given by:

$$dJ = \frac{\partial J}{\partial e_1} de_1 + \frac{\partial J}{\partial e_2} de_2 + \ldots$$

Dividing both sides by J, and for each term on the right, multiply top and bottom by the appropriate e_i, yields:

$$\frac{dJ}{J} = \frac{\partial J}{\partial e_1} \frac{e_1}{J} \frac{de_1}{e_1} + \frac{\partial J}{\partial e_2} \frac{e_2}{J} \frac{de_2}{e_2} + \ldots$$

$$\frac{dJ}{J} = C_{e_1}^J \frac{de_1}{e_1} + C_{e_2}^J \frac{de_2}{e_2} + \ldots = \sum_{i=1}^{n} C_{e_i}^J \frac{de_i}{e_i}$$

The same applies to $s = s(e_1, e_2, \ldots)$.

Example 3.3

In the following pathway, the numbers refer to the flux control coefficients for the respective reaction step:

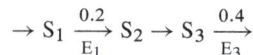

$$\rightarrow S_1 \xrightarrow[E_1]{0.2} S_2 \rightarrow S_3 \xrightarrow[E_3]{0.4}$$

What is the percentage change in flux if we increase E_1 by 10% and E_3 by 20%?

To calculate this we use equation (3.3):

$$\frac{\delta J}{J} = 0.1 \times 0.2 + 0.2 \times 0.4 = 0.1 \text{ or } 10\%$$

Example 3.4

In the following pathway, the numbers refer to the flux control coefficients for the respective reaction steps:

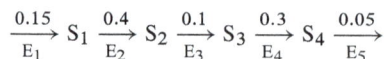

$$\xrightarrow[E_1]{0.15} S_1 \xrightarrow[E_2]{0.4} S_2 \xrightarrow[E_3]{0.1} S_3 \xrightarrow[E_4]{0.3} S_4 \xrightarrow[E_5]{0.05}$$

Given no other information, if you could increase enzyme concentrations by 20%, which two steps would you engineer to increase the flux the most?

Engineering the second and fourth enzymes will have the most effect on the steady state flux since they have the highest flux control coefficients. If we increased the second and fourth step by 20% the percentage change in flux will be:

$$\frac{\delta J}{J} = 0.2 \times 0.4 + 0.2 \times 0.3 = 0.14 \text{ or } 14\%$$

3.5 Summation Theorems

In this section we will introduce the concept of an **operational proof**. These proofs rely on carrying out thought experiments on a system and then casting the experiments in algebraic form from which new results (or theorems) can be derived. Although perhaps not as rigorous as a purely algebraic approach, operational proofs offer insight into the underlying biology and dynamics of the system, and are therefore very useful exercises in their own right.

One step pathway

Although perhaps somewhat trivial, let us first consider a one step pathway:

$$X_o \xrightarrow{E_1} X_1$$

The pathway is a single reaction catalyzed by a single enzyme, E_1. There are no floating metabolites, just two fixed external pools, X_o and X_1. There is a flux from X_o to X_1, denoted J, and is equal to the rate through the enzyme, E_1. If we make a change to the concentration of E_1, there will be a change in flux, given by δJ, which is equal to the change in rate, δv.

In order to work out the change in flux (although trivial in this case) we will use a standard procedure that will be employed in subsequent examples. This procedure involves the use of equation (2.16) given towards the end of the last chapter. This equation is repeated here:

$$\frac{dv}{v} = \sum_{j=1}^{m} \varepsilon_{s_j}^v \frac{ds_j}{s_j} + \varepsilon_e^v \frac{de}{e} \tag{3.5}$$

The equation enables us to compute the total change in rate through a step given changes to all effectors that might affect the rate, including the enzyme catalyzing the reaction. In the one step pathway, the only thing that is changing is the concentration of E_1 which means there are no δs_j terms to consider. Inserting the change in E_1 into the above equation and dropping the δs_j terms, gives us the following relation:

$$\frac{dv}{v} = \varepsilon_{e_1}^v \frac{de_1}{e_1} \tag{3.6}$$

Recall that the elasticity, $\varepsilon_{e_1}^v$, equals one. Although described before, this can be shown again as follows. Assume that the reaction rate can be described using a standard Michaelis-Menten type rate law:

$$v = \frac{V_m x_o}{K_m + x_o}$$

where V_m is the maximal velocity, and K_m the concentration of substrate, X_o, that yields half the maximal rate. The maximal velocity can be further described in terms of the catalytic rate constant, k_{cat}, and the concentration of enzyme, e_1. We can therefore rewrite the rate law as:

$$v = \frac{e_1 k_{cat} x_o}{K_m + x_o}$$

To compute the enzyme elasticity, $\varepsilon_{e_1}^v$, the rate law can be differentiated with respect to e_1 and scaled:

$$\varepsilon_{e_1}^v = \frac{k_{cat} x_o}{(K_m + x_o)} \frac{(K_m + x_o)}{e_1 k_{cat} x_o} e_1 = 1 \tag{3.7}$$

This yields an elasticity of one. Another way to look at this is to realize that $v \propto e_1$ which is a first-order reaction response, and as we saw in the last chapter, first-order responses yield elasticities of one. Setting the enzyme elasticity to one allows us to simplify equation (3.6) to:

$$\frac{dv}{v} = \frac{de_1}{e_1}$$

The change in rate on the left-hand side takes into account all changes that have occurred. The left-hand side must therefore equal the change in the pathway's flux, and so:

$$\frac{dJ}{J} = \frac{de_1}{e_1}$$

Dividing both sides by de_1/e_1 yields the flux control coefficient:

$$C_{e_1}^J = 1$$

> We can conclude that the enzyme of a single step pathway has complete proportional control over the flux. The total amount of control is equal to one.

Two-step pathway

Consider next a two-step pathway:

$$X_o \xrightarrow{v_1} S \xrightarrow{v_2} X_1$$

where X_o and X_1 are fixed. Let the pathway be at steady state and imagine increasing the concentration of enzyme, E_1, catalyzing the first step by an amount, δe_1. The effect of this is to increase the steady state levels of S and flux, J. Let us now increase the level of E_2 by δe_2 **such that** the change in S is restored to the original value it had at steady state.

The net effect of these two changes is by definition, $\delta s = 0$.

There are two ways to look at this thought experiment, from the perspective of the **system** and from the perspective of **local changes**. For the system we can compute the overall change in flux or species concentration by adding the two control coefficient terms using equation (3.3), thus:

$$\frac{\delta J}{J} = C_{e_1}^J \frac{\delta e_1}{e_1} + C_{e_2}^J \frac{\delta e_2}{e_2}$$

$$\frac{\delta s}{s} = C_{e_1}^s \frac{\delta e_1}{e_1} + C_{e_2}^s \frac{\delta e_2}{e_2} = 0$$

(3.8)

We can also look at what is happening locally at every reaction step using equation (3.5) for which there will be two: one for v_1, and another for v_2. Since the thought experiment guarantees that $\delta s = 0$, the local equations are quite simple:

$$\frac{\delta e_1}{e_1} = \frac{\delta v_1}{v_1}$$

$$\frac{\delta e_2}{e_2} = \frac{\delta v_2}{v_2}$$

Because the pathway is linear, at steady state, $v_1 = v_2 = J$. We can substitute these expressions into (3.8) and rewrite the system equations as:

$$\frac{\delta J}{J} = C_{e_1}^J \frac{\delta v_1}{v_1} + C_{e_2}^J \frac{\delta v_2}{v_2}$$

$$\frac{\delta s}{s} = C_{e_1}^s \frac{\delta v_1}{v_1} + C_{e_2}^s \frac{\delta v_2}{v_2} = 0$$

Note that at steady state the change in v_1 and v_2 must be the same, therefore $\delta v_1/v_1 = \delta v_2/v_2$. Set $\alpha = \delta J/J = \delta v_1/v_1 = \delta v_2/v_2$, and rewrite the above equations as:

$$\alpha = C_{e_1}^J \alpha + C_{e_2}^J \alpha = \alpha(C_{e_1}^J + C_{e_2}^J)$$

$$0 = C_{e_1}^s \alpha + C_{e_2}^s \alpha = \alpha(C_{e_1}^s + C_{e_2}^s)$$

We then conclude through cancelation of α since $\alpha \neq 0$, that:

$$1 = C_{e_1}^J + C_{e_2}^J$$

$$0 = C_{e_1}^s + C_{e_2}^s$$

It can be shown that these relations can be extended to longer pathways so that for a linear pathway of n enzyme catalyzed reactions, the summations will include all n steps, that is:

$$\sum_{i=1}^n C_{e_i}^J = 1$$

$$\sum_{i=1}^n C_{e_i}^{s_j} = 0$$

Without proof, the summation theorems apply to pathways of any shape or size with any number of regulatory loops. To help justify this sweeping statement, let us consider one more example, a simple branched pathway. A full justification would require the use of matrix algebra which is beyond the scope of the current edition, but see [44].

Simple branched pathway

Consider the branched pathway shown in Figure 3.8. At steady state the following statement must be true:

$$v_1 = v_2 + v_3$$

As before, let us make a positive perturbation in the concentration of E_1, that is δe_1 to the reaction step v_1. This will cause the steady state level of S and all reactions rates downstream to increase. We now make changes to the concentrations of E_2 and E_3 **such that** $\delta s = 0$. This can be done by decreasing the levels of E_2 and E_3 until the steady state concentration of S is restored back to where it was before the initial perturbation in E_1. We now ask what are the magnitudes of the perturbations in e_1, e_2, and e_3, needed to ensure that $\delta s = 0$?

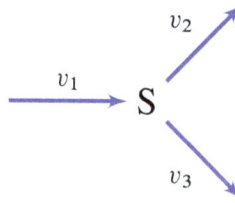

Figure 3.8 Simple Branched Pathway.

The local equations for each reaction step can be written as follows, note there are no δs terms because $\delta s = 0$ and again assuming that $v_i \propto e_i$:

$$\frac{\delta e_1}{e_1} = \frac{\delta v_1}{v_1}, \qquad \frac{\delta e_2}{e_2} = \frac{\delta v_2}{v_2}, \qquad \frac{\delta e_3}{e_3} = \frac{\delta v_3}{v_3} \tag{3.9}$$

It is also true that to satisfy the steady state condition, the sum of the changes in δv_2 and δv_3 must equal the change δv_1, that is:

$$\delta v_1 = \delta v_2 + \delta v_3$$

We divide both sides of this equation by v_1 and adjusting the denominators of v_2 and v_3 to obtain:

$$\frac{\delta v_1}{v_1} = \frac{\delta v_2}{v_2}\frac{v_2}{v_1} + \frac{\delta v_3}{v_3}\frac{v_3}{v_1}$$

To constrain the possible solutions, we impose the condition $\delta v_1/v_1 = \delta v_2/v_2$, allowing us to rewrite the above equation as:

$$\frac{\delta v_1}{v_1} - \frac{\delta v_1}{v_1}\frac{v_2}{v_1} = \frac{\delta v_3}{v_3}\frac{v_3}{v_1}$$

That is:

$$\frac{\delta v_1}{v_1}\left(1 - \frac{v_2}{v_1}\right) = \frac{\delta v_3}{v_3}\frac{v_3}{v_1}$$

where v_2/v_1 is the fraction of flux going down the upper branch, which we will call, α, and v_3/v_1 the fraction of flux going down the lower arm, that is $1 - \alpha$. Therefore with respect to the first step:

$$\frac{\delta v_1}{v_1}(1 - \alpha) = \frac{\delta v_3}{v_3}(1 - \alpha)$$

$$\text{and so } \frac{\delta v_1}{v_1} = \frac{\delta v_3}{v_3}$$

In other words $\delta v_1/v_1 = \delta v_2/v_2 = \delta v_3/v_3$. We now write out the system equations for J:

$$\frac{\delta J}{J} = C_{e_1}^J \frac{\delta e_1}{e_1} + C_{e_2}^J \frac{\delta e_2}{e_2} + C_{e_3}^J \frac{\delta e_3}{e_3}$$

$$\frac{\delta s}{s} = C_{e_1}^s \frac{\delta e_1}{e_1} + C_{e_2}^s \frac{\delta e_2}{e_2} + C_{e_3}^s \frac{\delta e_3}{e_3}$$

Using the previous result that $\delta v_1/v_1 = \delta v_2/v_2 = \delta v_3/v_3$ and the local equation (3.9) we can write:

$$\frac{\delta J}{J} = C_{e_1}^J \frac{\delta v_1}{v_1} + C_{e_2}^J \frac{\delta v_1}{v_1} + C_{e_3}^J \frac{\delta v_1}{v_1}$$

$$\frac{\delta s}{s} = C^s_{e_1} \frac{\delta v_1}{v_1} + C^s_{e_2} \frac{\delta v_1}{v_1} + C^s_{e_3} \frac{\delta v_1}{v_1}$$

Note that the flux change δJ through v_1 is non-zero while $\delta s = 0$:

$$\frac{\delta J}{J} = \frac{\delta v_1}{v_1} \left(C^J_{e_1} + C^J_{e_2} + C^J_{e_3} \right)$$

$$\frac{\delta s}{s} = \frac{\delta v_1}{v_1} \left(C^s_{e_1} + C^s_{e_2} + C^s_{e_3} \right) = 0$$

Because $\delta J/J = \delta v_1/v_1$ and $\delta v_1/v_1 \neq 0$ we can finally write:

$$1 = C^J_{e_1} + C^J_{e_2} + C^J_{e_3}$$
$$0 = C^s_{e_1} + C^s_{e_2} + C^s_{e_3}$$

Once again the flux control coefficients sum to one, and the concentration control coefficients sum to zero. Such operational proofs can be extended to other pathway configurations and a more formal approach using a combination of implicit differentiation together with the application of linear algebra techniques will show that the summation theorems apply to networks of arbitrary complexity. Given a pathway of arbitrary complexity, the following relationships (3.10) are true:

Summation Theorems:

$$\sum_{i=1}^{n} C^J_{e_i} = 1$$

$$\sum_{i=1}^{n} C^{s_j}_{e_i} = 0$$

(3.10)

where n equals the number of reaction steps in the pathway.

The one caveat is that if the control coefficients are expressed in terms of changes in enzyme concentrations, then there is the implicit assumption that $v_i \propto E_i$ and changes in a particular E_i has no effect on other enzyme concentrations. Later on we will see that even this assumption can be relaxed by using an alternative definition for the control coefficient.

Interpreting the summation theorems

How can we interpret the summation theorems described in the last section? Consider the flux summation theorem, repeated below:

$$\sum_{i=1}^{n} C^J_{e_i} = 1$$

What biological insight can be obtained from this result? The flux summation theorem implies that if any one of the enzyme steps has a high control coefficient, then the remaining steps must have small control coefficients. This assumes that there are no negative control coefficients (a situation that is not guaranteed for branched pathways). For a linear chain of n steps, the average control coefficient will be $1/n$. For a pathway with many steps, either the control coefficient for each is small, or one or two steps have significant control and the remaining steps very little. It has been suggested that this property is the molecular basis for the existence of metabolic dominant and recessive genes. Heterozygotes which carry one normal copy and one mutant copy of a gene are found on average to have only 50 % of the enzyme activity of the homozygous individual. Yet heterozygous individuals are generally 'normal' in the sense that they appear indistinguishable from the homozygote. In such situations, the mutant gene is termed 'recessive' and the normal gene dominant. Such behavior can be explained if we assume that many of the control coefficients in a pathway are small, so that even a 50 % reduction in activity has little or no effect.

Another important aspect of the summation theorem is that the flux summation implies a 'total' or unit amount of control available in a pathway, the total being one. The total control is distributed amongst the different enzymes according to each enzyme's control coefficient. Since there is a total amount of control to distribute in a pathway, an effect which causes the control coefficient of one enzyme to change will mean that the control coefficients of other steps **must change** to compensate and maintain the total control at unity. This implies that the distribution of control in a pathway is a dynamic process, changing as the conditions of the pathway change.

Example

Consider the three-step pathway shown in Figure 3.9. A simple mathematical model for this pathway is given in the Tellurium script listing 3.1.

```
import tellurium as te

r = te.loada ('''
  # Assume the kcat values equal 1
  J1: $Xo -> S1;  e1/Km1*(Xo-S1/Keq1)/(1 + Xo/Km1 + S1/Km2);
  J2:  S1 -> S2;  e1/Km2*(S1-S2/Keq2)/(1 + S1/Km3 + S2/Km4);
  J3:  S2 -> $X1; e1*S2/(Km5 + S2);

  Xo = 2;   X1 = 0;
  Keq1 = 1.2; Keq2 = 2.5;
  e1 = 3.4; e2 = 8.2; e3 = 2.3;
  Km1 = 0.6; Km2 = 0.78;
  Km3 = 0.9; Km4 = 1.2;
  Km5 = 0.5;
  S1 = 0; S2 = 0;
''')

r.steadyState()
print (" C1 = ", r.getCC ("J1", "e1"),
       " C2 = ", r.getCC ("J1", "e2"),
       " C3 = ", r.getCC ("J1", "e3"))
```

Listing 3.1 Three step pathway model.

The command getCC() is used to compute the control coefficient. The first argument is the variable we wish to observe, in this case the flux through step one. The second argument is the parameter we which to perturb, in this case one of the enzymes. Table 3.1 shows the values for the flux control coefficients computed from this model. Flux control is distributed across all three reaction steps.

$$C^J_{e_1} = 0.3677$$

$$C^J_{e_2} = 0.1349$$

$$C^J_{e_3} = 0.4989$$

Table 3.1 Flux Control Coefficients. Note that that last step has most of the control.

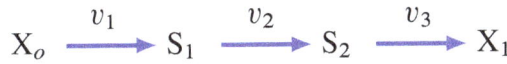

$$X_o \xrightarrow{v_1} S_1 \xrightarrow{v_2} S_2 \xrightarrow{v_3} X_1$$

Figure 3.9 Three-step linear pathway.

Almost 50% of control is located on the last step. This shows that in a linear pathway the committed step (i.e. the first step) is not necessarily the step with the most control. By varying the values of the various kinetic parameters, it is possible to obtain almost any pattern of control.

What determines the value of a flux control coefficient?

Many of the subsequent chapters will focus on the question of what determines the value of a particular flux or concentration control coefficient. In this section a brief description will be given.

Consider the three step pathway shown in Figure 3.9. Assume the pathway is at steady state. What happens to the steady state flux when the concentration of the enzyme that catalyzes the second step is increased? A change δe_2 is made resulting in new steady state concentrations to S_1 and S_2, as well as a new steady state flux, J.

The change in reaction rate, v_2, can be described using the local equation introduced in equation (2.16) and repeated here using the δ notation:

$$\frac{\delta v}{v} = \sum_{j=1}^{m} \varepsilon^v_{s_j} \frac{\delta s_j}{s_j} + \varepsilon^v_e \frac{\delta e}{e}$$

The local equation for v_2 (3.11) will have three terms, one involving e_2, and two other terms involving S_1 and S_2:

$$\frac{\delta v_2}{v_2} = \varepsilon^{v_2}_{s_1} \frac{\delta s_1}{s_1} + \varepsilon^{v_2}_{s_2} \frac{\delta s_2}{s_2} + \varepsilon^{v_2}_{e_2} \frac{\delta e_2}{e_2} \tag{3.11}$$

Collectively the three terms determine how the rate through v_2 changes. The enzyme, e_2, contributes a positive change in rate given by $\varepsilon^{v_2}_{e_2} \delta e_2 / e_2$. Since the enzyme elasticity $\varepsilon^{v_2}_{e_2}$ is one, the change in v_2 is in proportion to the change in e_2. For the sake of argument assume that e_2 is increased by 2% and the resulting net change in the final steady state flux increases by 1%. The above equation can be written as:

$$0.01 = \varepsilon^{v_2}_{s_1} \frac{\delta s_1}{s_1} + \varepsilon^{v_2}_{s_2} \frac{\delta s_2}{s_2} + 0.02 \tag{3.12}$$

It should be clear from the equation that the terms that include S_1 and S_2 must make a negative contribution in order to reduce the contribution from e_2 so that the overall sum equals 0.01.

As a result of the increase in v_2, S_1 will be consumed faster and S_2 made faster. S_1 will therefore **decrease** by an amount, δs_1, and S_2 will **increase** by an amount, δs_2. The drop in S_1 will result in a reduction in the reaction rate, and the increase in S_2 will likewise reduce the rate via product inhibition. Both changes **oppose** the increase in rate caused by e_2.

The degree to which changes in S_1 and S_2 change the rate will be determined by the two elasticities, $\varepsilon_{s_1}^{v_2}$ and $\varepsilon_{s_2}^{v_2}$. If both elasticities are large, then the reduction in rate caused by the change in S_1 and S_2 will be equally great, meaning that any change brought about by changes in e_2 will be effectively zeroed out. The net effect is that the flux control coefficient for this step will be small.

If both $\varepsilon_{s_1}^{v_2}$ and $\varepsilon_{s_2}^{v_2}$ are small then the reduction in rate caused by the change in S_1 and S_2 is likely also to be small. In this case most of the contribution will be from e_2 and therefore the flux control coefficient will be higher.

To summarize:

1. Control coefficients describe the influence a given reaction step has on a flux or species concentration at steady state.

2. Control coefficients are dimensionless quantities.

3. Flux control is shared throughout a pathway; that is, the degree to which flux is limited in a pathway is shared. It is unlikely that a single step has exclusive control over the flux.

4. If one step gains flux control, one or more other steps must lose control.

5. Control coefficients are system properties; they can only be computed or measured in an intact system. Inspection of a single enzymatic step will not reveal its degree of control (or influence) on the pathway.

Rate-limiting steps

In much of the metabolic literature and many contemporary biochemistry textbooks, one will often find a brief discussion of an idea called the rate-limiting step. In text books we will find statements such as:

'It is of course a truism to say that every metabolic pathway can and must have only one rate-liming step", Denton and Pogson, 1976.

'In order to exert control on the flux of metabolites through a metabolic pathway, it is necessary to regulate its rate-limiting step' Voet and Voet Biochemistry (p522), 3rd edition, 2004.

'The rate-limiting step occurs near the beginning of the pathway and is regulated by feedback inhibition...' Wikipedia (April 16, 2018), https://en.wikipedia.org/wiki/Metabolic_pathway

The implication is that a given metabolic pathway has one and only one rate-limiting step, often assigned to either the so-called committed step or regulated step. However this is inconsistent with the previously presented view that control, or rate-limitingness can be shared among all steps and not confined to just one. Other than a few books such as the 4th and later editions of Lehninger Biochemistry, and the latest Voet and Voet, most textbooks and unfortunately many research articles continue to cling to the idea of a single rate-limiting step. The 4th edition or later of Lehninger states more correctly:

'Metabolic control analysis shows that control of the rate of metabolite flux through a pathway is distributed among several of the enzymes in that path'.

The literature also uses an overabundance of similar terms such as rate-determining, pacemaker, choke point, bottleneck, master reaction or key enzyme. These terms are meant to convey a similar idea to the rate-limiting step. One of the earliest references to the concept of the rate-limiting step is a quote from Blackman [5]:

'When a process is conditioned as to its rapidity by a number of separate factors, the rate of the process is limited by the pace of the slowest factors."

This sentence started a century long love affair with the idea of the rate-limiting step in biochemistry, a concept that has lasted to this very day. As described in the introductory chapter, from the 1930s to the 1950s there were however a number of published papers which were highly critical of the concept, most notably Burton [14], Morales [69] and in particular Hearon [40]. Unfortunately much of this work did not find its way into the rapidly expanding fields of biochemistry and molecular biology after the second World War. Instead, the intuitive idea first pronounced by Blackman remains today as one of the cornerstones in understanding cellular regulation. The concept drives much of metabolic engineering and drug targeting of metabolism. What is most surprising however, is that a simple quantitative analysis shows that it cannot be true, and there is ample experimental evidence [41, 13] to support the alternative notion of shared control. The concept of the rate-limiting step is both inconsistent with logic and more importantly, experimental evidence.

The confusion over the existence of rate-limiting steps stems from a failure to realize that rates in cellular networks are governed by the law of mass-action. That is, if a concentration changes, then so does its rate of reaction. Some researchers try to draw analogies between cellular pathways and human experiences such as traffic congestion on roads or customer lines at shopping store checkouts. In each of these analogies, the rate of traffic and the rate of customer checkouts does not depend on how many cars are in the traffic line or how many customers are waiting in line. In a sense, these systems are not governed by 'mass-action' kinetics. In these analogies the use of the phrase rate-limiting step is reasonable. Traffic congestion and the customer line are rate-limiting because the only way to increase the flow is to either widen the road or increase the number of cash tills, i.e. there is a single factor that determines the rate of flow. In reaction networks the flow is governed by many factors including substrate/product/effector concentrations as well as the capacity of the reaction (V_{max}) itself. In biological pathways, rate-limiting steps are therefore the exception rather than the rule. It is highly unlikely for a single reaction to be fully rate limiting because it can be influenced by many factors. Many hundreds of measurements of control coefficients have been born out of this prediction.

Most biochemistry and molecular biology literature interpret the rate-limiting step to be the single step in a pathway which limits the flux. In terms of our control coefficients we can interpret the rate-limiting step as the step with a flux control coefficient of one. This means, by the summation theorem, that all other steps (at least in a linear chain) must have flux control coefficients of zero. However, when we consider branched and cyclic systems it is possible to have flux control coefficients much greater than one (other control coefficient must then be negative to satisfy the summation theorem). In these cases, what adjective should we use: hyper-rate-limiting steps, super-bottleneck, extreme-choke points? In the final analysis, it is better to assign a **value** to the rate-limitingness of a particular step in a pathway rather than designate a given reaction step as either rate-limiting or not.

Further Reading

1. Rafael Moreno-Sanchez, Emma Saavedra, Sara Rodrguez-Enriquez, and Viridiana Olin-Sandoval (2008) Metabolic Control Analysis: A Tool for Designing Strategies to Manipulate Metabolic Pathways, Journal of Biomedicine and Biotechnology, Volume 2008, Article ID 597913, doi:10.1155/2008/597913

Exercises

1. At steady state, all reaction rates are equal in a linear chain of reactions. Explain this statement.

2. The control coefficients are defined in terms of infinitesimal relative changes. An alternative would be to define them using large finite changes, that is $\Delta J / \Delta E$ which could be more easily measured. What is the main disadvantage to defining control coefficients in terms of large finite changes?

3. List three properties of control coefficients.

4. In a given reaction step E_i, the enzyme concentration is increased by 15%. The steady state change in flux was found to be 5% and the change in a species, S_j, changed by -3%. Estimate the values for the flux control coefficient, $C_{e_i}^J$ and the concentration control coefficient, $C_{e_i}^{s_j}$.

5. In the last question you were asked to find estimates for the control coefficients. Why were you asked to estimate the control coefficients and not their precise values?

6. A given reaction step has a flux control coefficient of 0.6. If the enzyme concentration is increased by 40%, what is the approximate change in the steady state flux?

7. Two reactions have flux control coefficients of 0.2 and 0.3, respectively. The concentration of the first enzyme is changed by 10%, and the second enzyme by 30%. What is the approximate change in the steady state flux if both changes are made?

8. What assumption(s) are made in the derivation of the summation theorems?

9. In a linear pathway, the concentration control coefficient for a given species, S, is found to be negative with respect to one enzyme but positive with respect to every other enzymatic step. Where is the species, S, located in the pathway. Explain your answer.

10. The last four steps in a five step pathway are found to have concentration control coefficients for a species, S, of -0.1, -0.2, -0.5 and -0.05. What is the concentration control coefficient with respect to the first step?

11. Locate four biochemistry and molecular cell biology textbooks and explain how the books describe regulation in pathways with respect to flux control. If they mention rate-limiting steps or rate-determining steps, describe how they justify these statements, if at all.

12. Why are rate-limiting steps unlikely to be found in natural pathways?

3.6 Appendix

Python/Tellurium Scripts

```
import tellurium as te
import numpy as np
r = te.loada ('''
   J0: $Xo -> S1; (E1/0.3)*(0.5*Xo-100*S1)/(1+Xo+S1);
   J1: S1 -> S2; E2*(3*S1-0.2*S2)/(1+S1+S2);
   J2: S2 -> S3; E3*(500*S2-10*S3)/(1+S2+S3);
   J3: S3 -> S4; E4*(200*S3-2*S4)/(1+S3+S4);
   J4: S4 -> $X1; E5*(200*S4-2*X1)/(1+S4+X1);

   Xo = 10;  X1 = 0;
   E1 = 3.4; E2 = 8.2;
   E3 = 2.3; E4 = 1.8;
   E5 = 4.5;
   S1 = 0; S2 = 0; S3 = 0; S4 = 0;
''')

m1 = r.simulate (0, 0.2, 100, ['time', 'J0']);
r.E2 = r.E2*4;
m2 = r.simulate (0.2, 0.5, 200, ['time', 'J0']);
alldata = np.vstack ((m1, m2));

r.plot (alldata)
```

Listing 3.2 Script for Figure 3.2.

```
import tellurium as te
import numpy as np
import pylab
r = te.loada ('''
   J0: $Xo -> S1; E1*(10*Xo-2*S1)/(1+Xo+S1);
   J1:  S1 -> S2; E2*(10*S1-2*S2)/(1+S1+S2);
   J2:  S2 -> S3; E3*(10*S2-2*S3)/(1+S2+S3);
   J3:  S3 -> S4; E4*(10*S3-2*S4)/(1+S3+S4);
   J34: S4 -> $X1; E5*(10*S4-2*X1)/(1+S4+X1);

   Xo = 10;  X1 = 0;
   E1 = 3.4; E2 = 8.2;
   E3 = 2.3; E4 = 1.8;
   E5 = 4.5;
   S1 = 8.359; S2 = 17.68; S3 = 6.938; S4 = 0.4816;
''')

# -------------- E1 ---------------
r.simulate (0, 80, 500);
m1 = r.simulate (0, 10, 100, ['time', 'J0']);
```

```
r.E1 = r.E1*1.2;
m2 = r.simulate (10, 30, 100, ['time', 'J0']);
alldata = np.vstack((m1, m2)); # Combine the two segments
pylab.plot (alldata[:,0], alldata[:,1], linewidth=2, label='E1')

# --------------- E2 ---------------
r.resetAll()
r.simulate (0, 80, 500);
m1 = r.simulate (0, 10, 100, ['time', 'J0']);
r.E2 = r.E2*1.2;
m2 = r.simulate (10, 30, 100, ['time', 'J0']);
alldata = np.vstack((m1, m2)); # Combine the two segments
pylab.plot (alldata[:,0], alldata[:,1], linewidth=2, label='E2')

# --------------- E3 ---------------
r.resetAll()
r.simulate (0, 80, 500);
m1 = r.simulate (0, 10, 100, ['time', 'J0']);
r.E3 = r.E3*1.2;
m2 = r.simulate (10, 30, 100, ['time', 'J0']);
alldata = np.vstack((m1, m2)); # Combine the two segments
pylab.plot (alldata[:,0], alldata[:,1], linewidth=2, label='E3')

# --------------- E4 ---------------
r.resetAll()
r.simulate (0, 80, 500);
m1 = r.simulate (0, 10, 100, ['time', 'J0']);
r.E4 = r.E4 *1.2;
m2 = r.simulate (10, 30, 100, ['time', 'J0']);
alldata = np.vstack((m1, m2)); # Combine the two segments
pylab.plot (alldata[:,0], alldata[:,1], linewidth=2, label='E4')

# --------------- E5 ---------------
r.resetAll()
r.simulate (0, 80, 500);
m1 = r.simulate (0, 10, 100, ['time', 'J0']);
r.E5 = r.E5 *1.2;
m2 = r.simulate (10, 30, 100, ['time', 'J0']);
alldata = np.vstack((m1, m2)); # Combine the two segments
pylab.plot (alldata[:,0], alldata[:,1], linewidth=2, label='E4')

pylab.legend(loc='upper left')
pylab.show()
```

Listing 3.3 Script for Figure 3.3.

F

4

Linking the Parts to the Whole

A deeper understanding of metabolism (or any other network) requires us to relate the parts, that is the enzymes, to the whole pathway. We need to understand how the properties of the parts contribute to the behavior we see in intact systems. To put it more grandly, we seek to understand phenotype from genotype.

So far two types of measures have been introduced: (i) the elasticities which describe how individual reactions respond to changes in their participating reactants and other modifiers, and (ii) control coefficients which describe how much influence individual reactions have on the response of whole pathways. In this chapter the aim is to bridge these two descriptions, and in order to do that, the relationship between the elasticities and control coefficients must be described.

4.1 Control Coefficients in Terms of Elasticities

The relationship between the control coefficients and elasticities is best introduced by way of example. Consider the pathway:

$$X_o \xrightarrow{v_1} S_1 \xrightarrow{v_2} X_1$$

Assume v_1 is catalyzed by an enzyme E_1, and v_2 by an enzyme E_2. X_o and X_1 are fixed species. Assume the pathway is at steady state with a steady state flux of J and concentration for the intermediate s. An experiment is carried out where E_1 is increased by an amount δe_1. This will result in a change to the steady state where the flux will change by an amount, δJ, and the concentration of S changed by δs. The experiment can be described using two local (2.16) and two system equations (3.1) and (3.2). The local equations are concerned with changes at v_1 and v_2, while the system equations describe how the change δe_1 changes the steady state flux and concentration of S. In this and subsequent chapters the notation for the elasticities will be simplified. In previous chapters elasticities have been expressed using notation such as $\varepsilon_{s_1}^{v_2}$. From now on we will drop the v and s lettering in the sub and superscript and instead refer to the elasticity using the simpler notation

ε_1^2. This will reduce the amount of clutter in the equations. For example, ε_3^5 refers the elasticity of reaction v_5 with respect to species s_3. Below are shown the two local and two system equations:

$$
\left.
\begin{aligned}
\frac{\delta v_1}{v_1} &= \frac{\delta e_1}{e_1} + \varepsilon_1^1 \frac{\delta s}{s} \\[2ex]
\frac{\delta v_2}{v_2} &= \varepsilon_1^2 \frac{\delta s}{s}
\end{aligned}
\right\} \text{Local equations}
$$

$$
\left.
\begin{aligned}
\frac{\delta s}{s} &= C_{e_1}^s \frac{\delta e_1}{e_1} \\[2ex]
\frac{\delta J}{J} &= C_{e_1}^J \frac{\delta e_1}{e_1}
\end{aligned}
\right\} \text{System equations}
$$

Because $\delta v_1/v_1 = \delta v_2/v_2$ when the system changes to the new steady state, the two local equations can be set equal to each other:

$$
\frac{\delta e_1}{e_1} + \varepsilon_1^1 \frac{\delta s}{s} = \varepsilon_1^2 \frac{\delta s}{s}
$$

Rearranging, and replacing $\delta s/s$ with the system equation, $C_{e_1}^s \delta e_1/e_1$ yields:

$$
\frac{\delta e_1}{e_1} + \varepsilon_1^1 C_{e_1}^s \frac{\delta e_1}{e_1} = \varepsilon_1^2 C_{e_1}^s \frac{\delta e_1}{e_1}
$$

Canceling and rearranging gives:

$$
C_{e_1}^s = \frac{1}{\varepsilon_1^2 - \varepsilon_1^1} \tag{4.1}
$$

This is an important result because we have related a control coefficient to the pathway elasticities. The steady state change in $\delta v_1/v_1$ is the same as the change in pathway flux, $\delta J/J$, therefore we can equate the first system equation to the first local equation:

$$
C_{e_1}^J \frac{\delta e_1}{e_1} = \frac{\delta e_1}{e_1} + \varepsilon_1^1 \frac{\delta s}{s} \tag{4.2}
$$

Dividing both sides of the equation by $\delta e_1/e_1$ gives:

$$
C_{e_1}^J = 1 + \varepsilon_1^1 \frac{\delta s/s}{\delta e_1/e_1}
$$

Noting that in the limit $(\delta s/s)/(\delta e_1/e_1)$ is equal to $C_{e_1}^s$ yields:

$$
C_{e_1}^J = 1 + \varepsilon_1^1 C_{e_1}^s
$$

But from (4.1) $C_{e_1}^s = 1/(\varepsilon_1^2 - \varepsilon_1^1)$, so that:

$$
C_{e_1}^J = 1 + \varepsilon_1^1 \frac{1}{\varepsilon_1^2 - \varepsilon_1^1}
$$

Rearranging yields:

$$C_{e_1}^J = 1 + \frac{\varepsilon_1^1}{\varepsilon_1^2 - \varepsilon_1^1} = \frac{\varepsilon_1^2}{\varepsilon_1^2 - \varepsilon_1^1} \tag{4.3}$$

Given the flux and concentration summation theorems, it is easy to obtain the other two control coefficients by subtraction ($C_{e_2}^J = 1 - C_{e_1}^J$ and $C_{e_2}^s = -C_{e_1}^s$):

$$C_{e_2}^s = -C_{e_1}^s = -\frac{1}{\varepsilon_1^2 - \varepsilon_1^1}$$
$$C_{e_2}^J = 1 - C_{e_1}^J = -\frac{\varepsilon_1^1}{\varepsilon_1^2 - \varepsilon_1^1} \tag{4.4}$$

The resulting equations (4.1), (4.3), and (4.4) describe how the control coefficients are determined by the elasticities. Using these equations it is possible to understand why particular steps are more or less rate limiting than others, and why some reactions may have more or less influence over metabolite levels. One interesting result is that the ratio of the flux control coefficients (4.1) and (4.3) is related to the ratio of the flanking elasticities:

$$\frac{C_{e_1}^J}{C_{e_2}^J} = -\frac{\varepsilon_1^2}{\varepsilon_1^1}$$

Similarly, the ratio of the concentration control coefficients is minus one:

$$\frac{C_{e_1}^s}{C_{e_2}^s} = -1$$

These properties are described by a set of additional theorems that relate the control coefficients with the elasticities, and will be covered in the next section.

4.2 Connectivity Theorems

In this section an additional set of theorems are introduced that relate the control coefficients to the substrate, product and effector elasticities. The theorems are called the **connectivity theorems** and represent one of the most important results in metabolic control analysis.

In deriving the summation theorems in the previous chapter, certain operations were performed on the pathway such that the flux changed value but the concentrations of the species were unchanged, that is, $\delta J / J \neq 0$ and $\delta s / s = 0$.

The constraints on the flux and concentration variables when deriving the summation theorems suggest a complementary set of constraints. That is, one or more operations can be performed on the enzymes such that the opposite is true, $\delta J / J = 0$ and $\delta s / s \neq 0$. A set of operations which preserve the flux but change the species concentrations lead to the second set of theorems, called the **connectivity theorems**.

Consider the following pathway fragment:

$$\xrightarrow{v_1} S_1 \xrightarrow{v_2} S_2 \xrightarrow{v_3} S_3 \xrightarrow{v_4}$$

Let us make a change to the rate through v_2 by increasing the concentration of enzyme E_2. Assume E_2 is increased by an amount, δe_2. This will result in a change to the steady state of the pathway. The concentration of S_2, S_3, and the flux through the pathway will rise, and the concentration of S_1 will decrease because it is upstream of the disturbance.

Let us now impose a second change to the pathway such that the flux is **restored** to what it was before the original change. Since the flux increased when E_2 was changed, the flux can be decreased by decreasing one of the other enzyme levels. If the concentration of E_3 is decreased, this will reduce the flux. Decreasing E_3 will also cause the concentration of S_2 to further increase. However, S_1 and S_3 will change in the opposite direction compared to when E_2 was increased.

When E_3 is sufficiently changed so that the flux is restored to its original value, the concentrations of S_1 and S_3 will also be restored to their original values. It is only S_2 that will differ. This is true because the flux through v_1 is now the same as it was originally (since we've restored the flux), and E_1 has not been manipulated in anyway. This means that the concentration of S_1 and all species upstream of S_1 must be the same as they were before the modulations occurred. The same arguments apply to S_3 and all species downstream of v_4.

To summarize: E_2 has been **increased** by δe_2, resulting in a change, δJ, to the flux. The concentration of E_3 is **decreased** such that the flux is restored to its original value. In the process, S_2 has changed by δs_2 and neither S_1 nor S_3 have changed. In fact *no other* species in the entire system has changed other than S_2.

If a particular species is made and consumed by many steps, it would still be possible to perform the necessary manipulations on all the adjacent enzymes such that only the shared species changed in concentration and the flux was unaltered.

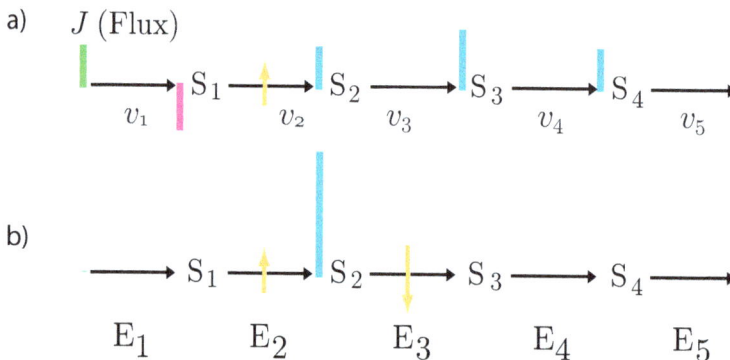

Figure 4.1 Connectivity Theorem: a) A change is made to E_2 that catalyzes v_2. This causes changes to species concentrations upstream and downstream including a change in flux indicated by the vertical bars. Note that S_1 decreases. b) Another change is made to E_3 that catalyzes v_3 to oppose the change in flux. This results in only three net changes, a change in E_2, E_3 and S_2; no other changes occur. The colored bars represent concentration changes. The bar on the extreme left is the change in pathway flux.

Flux Connectivity Theorem

Using the thought experiment described in the last section, two sets of equations can be derived which apply simultaneously to the pathway, a local equation and a system equation. The system equation will describe the effect of the enzyme changes on the flux. Since the net change in flux is zero, and the fact that only E_2 and E_3 were changed, the system flux can be written using the following system equation:

$$\frac{\delta J}{J} = 0 = C_{e_2}^J \frac{\delta e_2}{e_2} + C_{e_3}^J \frac{\delta e_3}{e_3} \tag{4.5}$$

To determine the local equations, we must focus on what is happening at the reaction steps v_2 and v_3. As a result of making changes to E_2 and E_3, the change in rate at v_2 is given by:

$$0 = \frac{\delta v_2}{v_2} = \frac{\delta e_2}{e_2} + \varepsilon_2^2 \frac{\delta s_2}{s_2}$$

and at v_3:

$$0 = \frac{\delta v_3}{v_3} = \frac{\delta e_3}{e_3} + \varepsilon_2^3 \frac{\delta s_2}{s_2}$$

Note that $\delta e_2/e_2$ will not necessarily equal $\delta e_3/e_3$. No other changes took place so these are the only local equations to consider. We can rearrange the local equations so that:

$$\frac{\delta e_2}{e_2} = -\varepsilon_2^2 \frac{\delta s_2}{s_2} \tag{4.6}$$

$$\frac{\delta e_3}{e_3} = -\varepsilon_2^3 \frac{\delta s_2}{s_2} \tag{4.7}$$

$\delta e_2/e_2$ and $\delta e_3/e_3$ can be inserted from the local equations into the system equations (4.5) to obtain:

$$0 = \frac{\delta J}{J} = -\left(C_{e_2}^J \varepsilon_2^2 \frac{\delta s_2}{s_2} + C_{e_3}^J \varepsilon_2^3 \frac{\delta s_2}{s_2} \right)$$

and therefore:

$$0 = \frac{\delta s_2}{s_2} \left(C_{e_2}^J \varepsilon_2^2 + C_{e_3}^J \varepsilon_2^3 \right)$$

Since $\delta s_2/s_2$ is not equal to zero it must be true that:

$$0 = C_{e_2}^J \varepsilon_2^2 + C_{e_3}^J \varepsilon_2^3$$

This relationship is called a **connectivity theorem**. The derivation can be applied to a species that interacts with any number of steps and not just the two in the previous example. For example, consider the pathway fragment in Figure 4.2.

S interacts with its production rate, v_1, a consumption rate, v_2, and an inhibitory interaction with v_3. In the case with three interactions, the connectivity may therefore be written as:

$$C_{e_1}^J \varepsilon_s^1 + C_{e_2}^J \varepsilon_s^2 + C_{e_3}^J \varepsilon_s^3 = 0$$

In general, the number of terms in the connectivity theorem will equal the number of interactions a species makes. For a species S that interacts with r other steps, the flux connectivity theorem is written as:

$$0 = \sum_{i=1}^{r} C_{e_i}^J \varepsilon_s^i$$

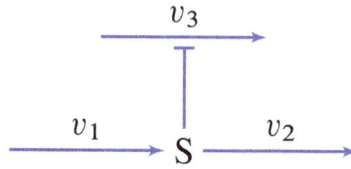

Figure 4.2 Pathway fragment showing three interactions via S.

Concentration Connectivity Theorem

To derive the flux connectivity theorem we used the system equation that was related to the flux. It is however possible to use a different set of system equations, those with respect to the species concentrations. In the case of the species, there will be two distinct system equations. One of these will describe the effect the modulation has on the common species (S_2 in the example), and a second will describe the effect on any other species (S_1, S_3, etc.). Consider first the system equation involving the common species. For the pathway under consideration, this equation is given by:

$$\frac{\delta s_2}{s_2} = C_{e_2}^{s_2}\frac{\delta e_2}{e_2} + C_{e_3}^{s_2}\frac{\delta e_3}{e_3}$$

Given that the change in the common species, $\delta s_2/s_2$, is non-zero (Figure 4.1), substituting in the local equations, equation (4.7), leads to:

$$\frac{\delta s_2}{s_2} = -C_{e_2}^{s_2}\varepsilon_2^2\frac{\delta s_2}{s_2} - C_{e_3}^{s_2}\varepsilon_2^3\frac{\delta s_2}{s_2}$$

Since $\delta s_2/s_2 \neq 0$, the term $\delta s_2/s_2$ can be canceled which leads to the **first** concentration connectivity theorem:

$$-1 = C_{e_2}^{s_2}\varepsilon_2^2 + C_{e_3}^{s_2}\varepsilon_2^3$$

A **second** theorem can be derived by considering the effect of the modulation on a distant species, for example S_3. In this case the system equation with respect to S_3 becomes:

$$0 = \frac{\delta s_3}{s_3} = C_{e_2}^{s_3}\frac{\delta e_2}{e_2} + C_{e_3}^{s_3}\frac{\delta e_3}{e_3}$$

Note that the distance species did not change so that the equation equals zero because the operations we made ensure that species other than the common species, S_2, do not change in concentration.

Substituting once again the local equations into the above system equation leads to:

$$0 = \frac{\delta s_3}{s_3} = -C_{e_2}^{s_3}\varepsilon_2^2\frac{\delta s_2}{s_2} - C_{e_3}^{s_3}\varepsilon_2^3\frac{\delta s_2}{s_2}$$

or:

$$0 = -\frac{\delta s_2}{s_2}\left(C_{e_2}^{s_3}\varepsilon_2^2 + C_{e_3}^{s_3}\varepsilon_2^3\right)$$

However, $\delta s_2/s_2$ is not zero, therefore it must be the case that:

$$0 = C_{e_2}^{s_3}\varepsilon_2^2 + C_{e_3}^{s_3}\varepsilon_2^3$$

That completes the proof for the concentration connectivity theorems. As with the flux connectivity theorems, the concentration connectivity theorems can be generalized to any number of steps that a species might interact with.

To summarize:

Flux Connectivity Theorem

With respect to a common metabolite, S_k, where r is the number of interactions S_k makes with neighboring reaction steps:

$$\sum_{i=1}^{r} C_{e_i}^{J} \varepsilon_{s_k}^{v_i} = 0 \qquad (4.8)$$

Concentration Connectivity Theorem

With respect to the common metabolite, S_k, where r is the number of interactions S_k makes with neighboring reaction steps:

$$\sum_{i=1}^{r} C_{e_i}^{s_k} \varepsilon_{s_k}^{v_i} = -1 \qquad (4.9)$$

Concentration Connectivity Theorem

With respect to the metabolite, S_k, and a distant metabolite, S_m, where r is the number of interactions S_k makes with neighboring reaction steps. **Note:** $k \neq m$:

$$\sum_{i=1}^{r} C_{e_i}^{s_m} \varepsilon_{s_k}^{v_i} = 0 \qquad (4.10)$$

$$X_o \xrightarrow{v_1} S_1 \xrightarrow{v_2} S_2 \xrightarrow{v_3} X_1$$

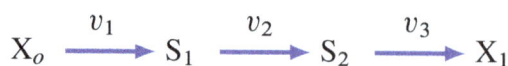

Figure 4.3 Three step linear pathway.

Interpretation

Why are the connectivity theorems important? First and foremost, the theorems link local effects, in terms of the elasticities, to global effects, in terms of the control coefficients. Consider for example the following linear pathway:

$$\xrightarrow{v_1} S_1 \xrightarrow{v_2}$$

The flux connectivity can be written in the form:

$$\frac{C_{e_1}^J}{C_{e_2}^J} = -\frac{\varepsilon_1^2}{\varepsilon_1^1}$$

The ratio of two adjacent flux control coefficients is inversely proportional to the ratio of the corresponding elasticities. This means that high flux control coefficients tend to be associated with small elasticities, and small flux control coefficients with large elasticities.

This effect is explained by species opposing changes in rates. Since species with high elasticities are able to oppose rate changes more effectively than small elasticities, it follows that large elasticities are associated with small flux control coefficients and *vice versa*.

A good example of this is the case of a reaction operating near equilibrium where the elasticities are very high relative to adjacent elasticities on neighboring enzymes. In such situations the flux control coefficients of near equilibrium enzymes are *likely* to be small. However, one must bear in mind that it is the ratio of elasticities which is important and not their absolute values. Simply examining the elasticity of a single reaction is not sufficient to draw a firm conclusion. Even more so, one must also consider all the ratios of the elasticities along a pathway. Even though one elasticity ratio may suggest a high or low flux control coefficient on a particular enzyme, the other ratios coupled to the flux summation theorem must be considered because they will give an absolute value to a particular flux control coefficient. This reinforces the point that control coefficients are system-wide properties and cannot be reliably estimated from the properties of the step alone. The following statement is one of the most important in MCA.

> The examination of a single enzyme will not give an indication of the ability of that enzyme to control the flux or species concentrations.

A second application of the connectivity theorems is for making simple statements about the properties of specific pathway configurations. Chapter 6 will delve more deeply into the properties of linear pathways, but here we can make some simple statements using the connectivity theorems. Consider the three-step pathway in Figure 4.3.

Two flux and four concentration connectivity theorems can be written for this pathway:

$$C_1^J \varepsilon_1^1 + C_2^J \varepsilon_1^2 = 0 \qquad C_2^J \varepsilon_2^2 + C_3^J \varepsilon_2^3 = 0 \qquad (4.11)$$

$$C_1^{S_1} \varepsilon_1^1 + C_2^{S_1} \varepsilon_1^2 = -1 \qquad C_2^{S_2} \varepsilon_2^2 + C_3^{S_2} \varepsilon_2^3 = -1 \qquad (4.12)$$

$$C_1^{S_2} \varepsilon_1^1 + C_2^{S_2} \varepsilon_1^2 = 0 \qquad C_2^{S_1} \varepsilon_2^2 + C_3^{S_1} \varepsilon_2^3 = 0 \qquad (4.13)$$

Consider the situation where there is no product inhibition on the first step, i.e. $\varepsilon_1^1 = 0$. Looking at the flux connectivity theorems and inserting zero for ε_1^1, the flux theorems can be rewritten as:

$$C_2^J \varepsilon_1^2 = 0$$
$$C_2^J \varepsilon_2^2 + C_3^J \varepsilon_2^3 = 0$$

Assuming that ε_1^2 is not zero, it must then be true that $C_2^J = 0$. Since $C_2^J = 0$, the second theorem is reduced to:

$$C_3^J \varepsilon_2^3 = 0$$

Again, ε_2^3 is assumed to be non-zero, therefore it must be true that $C_3^J = 0$. From this simple analysis one can deduce that since $C_2^J = C_3^J = 0$, then by the flux summation theorem, $C_1^J = 1$.

If $\varepsilon_2^2 = 0$, one can similarly show that $C_3^J = 0$ and $C_1^J + C_2^J = 1$. From this it can be concluded that if there is no product inhibition in the first reaction then the first step fully controls the flux. Likewise if the second step has no product inhibition then the flux control coefficient downstream must be zero.

The concentration connectivity theorems can also be used to derive properties of the pathway. Assume as before that $\varepsilon_1^1 = 0$. Using the two connectivity theorems that include ε_1^1, and removing the term containing ε_1^1 from each yields:

$$C_2^{S_2} \varepsilon_1^2 = 0$$
$$C_2^{S_1} \varepsilon_1^2 = -1 \quad \Rightarrow \quad C_2^{S_1} = -\frac{1}{\varepsilon_1^2}$$

The first equation indicates that without product inhibition on the first step, the second step has no influence on S_2, $C_2^{S_2} = 0$. Logically this makes sense because without product inhibition it is impossible to change the steady state flux. Since the flux through v_3 must be unchanged it must be the case the S_2 is also unchanged. The second equation is more interesting, it states that without product inhibition, the ability of the second step to influence S_1 is inversely proportional to ε_1^2. For example, if ε_1^2 is small, then the influence of the second step on S_1 will be large. Note that the relationship also indicates via the sign, that an increase in activity of step two **reduces** the concentration of S_1, which makes logical sense.

Other patterns can be found in behavior by setting some of the other elasticities to zero. For example, if ε_2^3 is set to zero (reaction step three is saturated), then $C_2^J = 0$ and $C_1^J = 0$, meaning that $C_3^J = 1$ and $C_2^1 = 0$.

4.3 Response Coefficients

Control coefficients measure the response of a pathway to changes in enzyme activity. What about the effect of external factors such as inhibitors, pharmaceutical drugs or boundary species? Such effects are measured by another coefficient called the **response coefficient**. The **flux response coefficient** is defined by:

$$R_x^J = \frac{dJ}{dx}\frac{x}{J}$$

and the **concentration response coefficient** by:

$$R_x^s = \frac{ds}{dx}\frac{x}{s}$$

where x is the concentration of the external factor. The response coefficient measures how sensitive a pathway is to changes in external factors other than enzyme activities. What is the relationship of the response coefficients with respect to the control coefficients and elasticities of a pathway?

Like many of the proofs in this chapter, we can carry out a thought experiment to investigate the response coefficients more closely. Consider the pathway fragment below:

$$X \xrightarrow{v_1 E_1} S \xrightarrow{v_2, E_2}$$

where X is the fixed boundary species. Let us increase the concentration of E_1 by an amount δe_1. This will cause the steady state flux and concentration of S and in fact all downstream species beyond v_2 to increase. Now decrease the concentration of X such that we restore the flux and steady state concentration of S back to their original values. From this thought experiment we can write the operations in terms of the local response equation and a system response equation as follows:

$$\frac{\delta v_1}{v_1} = \varepsilon_x^1 \frac{\delta x}{x} + \varepsilon_{e_1}^1 \frac{\delta e_1}{e_1} = 0 \quad \Big\}\, \text{Local equation}$$

$$\frac{\delta J}{J} = R_x^J \frac{\delta x}{x} + C_{e_1}^J \frac{\delta e_1}{e_1} = 0 \quad \Big\}\, \text{System equation}$$

Note that the right-hand sides are zero because the thought experiment guarantees that the flux has not changed. We can eliminate the $\delta e_1/e_1$ term in the system response equation by substituting the term from the local response equation. In addition, if we assume that the reaction rate for an enzyme catalyzed reaction is proportional to the enzyme concentration, then we know from previous considerations that $\varepsilon_{e_1}^1 = 1$, see (3.7).

Therefore:

$$0 = R_x^J \frac{\delta x}{x} - C_{e_1}^J \varepsilon_x^1 \frac{\delta x}{x}$$

Since $\delta x/x \neq 0$, we can cancel $\delta x/x$ to yield:

$$R_x^J = C_{e_1}^J \varepsilon_x^1 \qquad\qquad (4.14)$$

This gives use a useful relationship. It can be generalized for multiple external factors acting simultaneously at multiple site, by summing up individual responses:

$$R_x^J = \sum_{i=1}^{n} C_{e_i}^J \varepsilon_x^{v_i}$$

Likewise, the response of a species, S, to an external factor, X, is given by:

$$R_x^s = \sum_{i=1}^{n} C_{e_i}^s \varepsilon_x^{v_i} \qquad (4.15)$$

These relationships apply to any external factor that can affect one or more reaction rates. The relationship between the response coefficient, control and elasticity coefficients carries an important message. The response of an external factor, X, is a function of two things:

1) The effect the factor has on the step it acts upon.
2) The effect that the step itself has on changing the system.

This means than an effective external factor, such as a pharmaceutical drug, must not only be able to bind and inhibit the enzyme or protein being targeted, but the step itself must be able to transmit the disturbance to the rest of the pathway and ultimately affect the phenotype.

The ability of an external factor to influence a given species or flux depends on:

1. The ability of the external factor to influence its immediate target.

2. The ability of the target to influence the network it is connected to.

4.4 Canonical Control Coefficients

Let us write the response coefficient equation in a different way. Consider:

$$C_{e_i}^J = \frac{R_{e_i}^J}{\varepsilon_{e_i}^{v_i}}$$

Expand the terms and replace e_i with a general parameter p to yield:

$$C_{v_i}^J = \frac{\dfrac{dJ}{dp}\dfrac{p}{J}}{\dfrac{\partial v_i}{\partial p}\dfrac{p}{v_i}} = \frac{dJ}{dv_i}\frac{v_i}{J} \qquad (4.16)$$

Equation (4.16) defines a control coefficient as a **parameterless control coefficient**, $C_{v_i}^J$. In this text such control coefficients will be called **canonical control coefficients** to distinguish them from enzyme based control coefficients. These coefficients describe the effect of a change in the local reaction rate on the steady state level of the pathway flux. This may seem like an odd definition; isn't the change in reaction rate the same as the change in steady state pathway flux? In this case, no. We have to be clear what the derivative, dv_i, in the denominator actually means. Operationally, the change indicated by dv_i refers to a change in v_i by some unspecified means under the conditions where the reactants, products and any other effectors remain constant. dv_i in this context is also sometimes referred to as the local rate, i.e. the change in the reaction rate we could impose if the reaction were not connected to the rest of the network. The way dv_i is changed is unspecified, but it must be done via a parameter of the system, not by a variable quantity such as one of the reactants or products. The identity of the parameter will depend on the constraints imposed by the system, but common parameters could be the concentration of expressed enzyme, the catalytic constant of the enzyme, or an external inhibitor.

For example, suppose that the change to alter v_i is via a change in the enzyme concentration, e_i. Such a change will cause an immediate change, dv_i, in the reaction rate. The system is now allowed to evolve to its new steady state. The dv_i will cause changes in the immediate environment of the reaction, causing the enzyme's substrate to decrease and its product to increase. These changes in turn will propagate throughout the system. Once the system has settled to the new steady state, an inspection of v_i will reveal that the final change in rate does not equal the original dv_i (because the local environment has now changed). We refer to the final change in v_i as dJ, that is the change in flux through the system. Taking the ratio of dJ and dv_i and scaling, we obtain the canonical control coefficient. In many situations, the enzyme elasticity, $\varepsilon_{e_i}^{v_i} = 1$, which means that:

$$C_{e_i}^J = C_{v_i}^J$$

In other words, the control coefficients that are measured, C_e^J, are often identical to the canonical control coefficients. Strictly speaking, the summation and connectivity theorems only apply to the canonical control coefficients, but because $\varepsilon_{e_i}^{v_i} = 1$, we can often safely express the theorems using the control coefficients with respect to enzyme concentration.

4.5 How to Derive Control Equations

A key aspect of MCA is being able to derive the control equations, that is the equations that relate the elasticities to the control coefficients. Section 4.1 illustrated one way that used the local and system equations. In this section a number of other approaches will be described, some of which can be automated by computer.

Derivation using the Theorems

One way to derive the control equations is to combine the summation and connectivity theorems. For example, consider a two-step pathway such as:

$$X_o \xrightarrow{v_1} S \xrightarrow{v_2} X_1$$

where X_o and X_1 are fixed species. There is only one flux connectivity theorem with respect to the intermediate species S, that is:

$$C_{e_1}^J \varepsilon_s^1 + C_{e_2}^J \varepsilon_s^2 = 0$$

In addition, there will be a single flux summation theorem:

$$C_{e_1}^J + C_{e_2}^J = 1$$

These two equations can be combined and solved for $C_{e_1}^J$ and $C_{e_2}^J$ in terms of the elasticities, thus:

$$C_{e_1}^J = \frac{\varepsilon_s^2}{\varepsilon_s^2 - \varepsilon_s^1}$$

$$C_{e_2}^J = -\frac{\varepsilon_s^1}{\varepsilon_s^2 - \varepsilon_s^1}$$

(4.17)

The concentration control coefficients can be derived in similar fashion by using the concentration control coefficient theorems. For a two-step pathway these would be:

$$C_{e_1}^s + C_{e_2}^s = 0$$
$$C_{e_1}^s \varepsilon_s^1 + C_{e_2}^s \varepsilon_s^2 = -1$$

Combining both equations and solving for $C_{e_1}^s$ and $C_{e_2}^s$ yields the following control equations:

$$C_{e_1}^s = \frac{1}{\varepsilon_s^2 - \varepsilon_s^1}$$

$$C_{e_2}^s = -\frac{1}{\varepsilon_s^2 - \varepsilon_s^1}$$

These are the same equations that were derived at the start of the chapter. For more complex pathways that include branches and cycles, additional theorems are required to solve the equations.

As a convenience, the theorems can be rendered in matrix form and solved using standard matrix methods. Both the flux and concentration control coefficient theorems can be recast in matrix form as follows:

$$
\begin{bmatrix} 1 & 1 & 1 \\ \varepsilon_1^1 & \varepsilon_1^2 & 0 \\ 0 & \varepsilon_2^2 & \varepsilon_2^3 \end{bmatrix}
\begin{bmatrix} C_1^J & C_1^{S_1} & C_1^{S_2} \\ C_2^J & C_2^{S_1} & C_2^{S_2} \\ C_3^J & C_3^{S_1} & C_3^{S_2} \end{bmatrix}
=
\begin{bmatrix} 1 & 0 & 0 \\ 0 & -1 & 0 \\ 0 & 0 & -1 \end{bmatrix}
\tag{4.18}
$$

For example, the first row of the first matrix multiplied by the first column of the elasticity matrix yields the flux summation theorem. It is a simple matter to extend the matrix to any size linear pathway by following the pattern. For example, a four-step pathway is represented by:

$$
\begin{bmatrix} 1 & 1 & 1 & 1 \\ \varepsilon_1^1 & \varepsilon_1^2 & 0 & 0 \\ 0 & \varepsilon_2^2 & \varepsilon_2^3 & 0 \\ 0 & 0 & \varepsilon_3^3 & \varepsilon_3^4 \end{bmatrix}
\begin{bmatrix} C_1^J & C_1^{S_1} & C_1^{S_2} & C_1^{S_3} \\ C_2^J & C_2^{S_1} & C_2^{S_2} & C_2^{S_3} \\ C_3^J & C_3^{S_1} & C_3^{S_2} & C_3^{S_3} \\ C_4^J & C_4^{S_1} & C_4^{S_2} & C_4^{S_3} \end{bmatrix}
=
\begin{bmatrix} 1 & 0 & 0 & 0 \\ 0 & -1 & 0 & 0 \\ 0 & 0 & -1 & 0 \\ 0 & 0 & 0 & -1 \end{bmatrix}
$$

Multiplying both sides by the inverse of the matrix containing the elasticities yields:

$$
\begin{bmatrix} C_1^J & C_1^{S_1} & C_1^{S_2} & C_1^{S_3} \\ C_2^J & C_2^{S_1} & C_2^{S_2} & C_2^{S_3} \\ C_3^J & C_3^{S_1} & C_3^{S_2} & C_3^{S_3} \\ C_4^J & C_4^{S_1} & C_4^{S_2} & C_4^{S_3} \end{bmatrix}
=
\begin{bmatrix} 1 & 1 & 1 & 1 \\ \varepsilon_1^1 & \varepsilon_1^2 & 0 & 0 \\ 0 & \varepsilon_2^2 & \varepsilon_2^3 & 0 \\ 0 & 0 & \varepsilon_3^3 & \varepsilon_3^4 \end{bmatrix}^{-1}
\begin{bmatrix} 1 & 0 & 0 & 0 \\ 0 & -1 & 0 & 0 \\ 0 & 0 & -1 & 0 \\ 0 & 0 & 0 & -1 \end{bmatrix}
$$

One advantage of writing the equations in matrix form is that it makes it easy to numerically evaluate the control coefficients by inverting the elasticity matrix. For example, if a three-step linear pathway has the following elasticities:

$$
\begin{bmatrix} 1 & 1 & 1 \\ -0.6 & 1.2 & 0 \\ 0 & -0.2 & 0.5 \end{bmatrix}
$$

Then the control coefficient matrix is given by:

$$\begin{bmatrix} 1 & 1 & 1 \\ -0.6 & 1.2 & 0 \\ 0 & -0.2 & 0.5 \end{bmatrix}^{-1} \begin{bmatrix} 1 & 0 & 0 \\ 0 & -1 & 0 \\ 0 & 0 & -1 \end{bmatrix} = \begin{bmatrix} 0.192 & 1.347 & 0.385 \\ 0.577 & -0.962 & 1.154 \\ 0.231 & -0.385 & -1.538 \end{bmatrix}$$

Note how the left column of values sum to one, reflecting the flux summation theorem. The second and third columns sum to zero, corresponding to the concentration summation theorem.

One advantage of the matrix formalism is that it is easy to enter such matrices into tools such as Mathematica and carry out a symbolic inversion to obtain the control coefficient equations.

It is possible to rearrange the control matrix so that the flux control coefficients are positioned along the top row, and concentration control coefficients along the subsequent rows. Also, one can move the negative signs on the right-hand side of equation (4.18) to the elasticity terms so that the right-hand side becomes an identity matrix. See equation below, (4.19):

$$\begin{bmatrix} C_1^J & C_2^J & C_3^J \\ C_1^{s_1} & C_2^{s_1} & C_3^{s_1} \\ C_1^{s_2} & C_2^{s_2} & C_3^{s_2} \end{bmatrix} \begin{bmatrix} 1 & -\varepsilon_1^1 & 0 \\ 1 & -\varepsilon_1^2 & -\varepsilon_2^2 \\ 1 & 0 & -\varepsilon_2^3 \end{bmatrix} = \begin{bmatrix} 1 & 0 & 0 \\ 0 & 1 & 0 \\ 0 & 0 & 1 \end{bmatrix} \qquad (4.19)$$

For more complex pathways that include branches and cycles, this arrangement becomes more convenient.

The following shows how Mathematica can be used to invert the elasticity matrix to generate the control coefficient matrix.

```
ee = {{1, -e1, 0}, {1, -e2, -e3}, {1, 0, -e4}}
Inverse[ee]
{{(e2 e4)/(e1 e3-e1 e4+e2 e4),-((e1 e4)/(e1 e3-e1 e4+e2 e4)),(e1 e3)/(e1 e3-e1 e4+e2 e4)},
{(-e3+e4)/(e1 e3-e1 e4+e2 e4),-(e4/(e1 e3-e1 e4+e2 e4)),e3/(e1 e3-e1 e4+e2 e4)},
{e2/(e1 e3-e1 e4+e2 e4),-(e1/(e1 e3-e1 e4+e2 e4)),(e1-e2)/(e1 e3-e1 e4+e2 e4)}}
```

Pure Algebraic Method - Advanced Topic

The pure algebraic method relies on the use of implicit differentiation of the system equation and is a formalization of the thought experiments presented at the start of the chapter. Consider again the simple two-step pathway:

$$X_o \xrightarrow{v_1} S \xrightarrow{v_2} X_1$$

At steady state the rate of change of S is given by:

$$\frac{ds}{dt} = v_1 - v_2 = 0$$

Assuming that v_1 can be changed by perturbations to the concentration of catalyzing enzyme, E_1, we can write the rate of change as:

$$\frac{ds}{dt} = v_1(s(e_1), e_1) - v_2(s(e_1)) = 0$$

In this equation each reaction rate, v_i, is a function of both the steady state species concentration, and the perturbing parameter, e_1. Note that v_1 is both a function of s and e_1, and s in turn is a function of e_1. v_2 is only a function of s and not *directly* a function of e_1 but indirectly via s. We can implicitly differentiate this equation with respect to e_1 to yield:

$$0 = \frac{\partial v_1}{\partial s}\frac{ds}{de_1} + \frac{\partial v_1}{\partial e_1} - \frac{\partial v_2}{\partial s}\frac{ds}{de_1}$$

We can scale each of the derivatives by multiplying by the appropriate factors, that is:

$$0 = \frac{\partial v_1}{\partial s}\frac{s}{v_1}\frac{ds}{de_1}\frac{e_1}{s} + \frac{\partial v_1}{\partial e_1}\frac{e_1}{v_1} - \frac{\partial v_2}{\partial s}\frac{s}{v_1}\frac{ds}{de_1}\frac{e_1}{s}$$

which can be simplified to:

$$0 = C_{e_1}^s \varepsilon_s^1 + \varepsilon_{e_1}^1 - C_{e_1}^s \varepsilon_s^2$$

Solving for $C_{e_1}^s$ and assuming v_1 is first-order with respect to e_1 so that $\varepsilon_{e_1}^1 = 1$, yields:

$$C_{e_1}^s = \frac{1}{\varepsilon_s^2 - \varepsilon_s^1} \qquad (4.20)$$

We can derive $C_{e_2}^s$ in the same way by implicitly differentiating:

$$\frac{ds}{dt} = v_1(s(e_2)) - v_2(s(e_2), e_2) = 0$$

The flux control coefficients can be computed in a similar way. For example, to find $C_{e_1}^J$ we can implicitly differentiate:

$$J = v_1(s(e_1), e_1)$$

$$\frac{dJ}{de_1} = \frac{\partial v_1}{\partial s}\frac{ds}{de_1} + \frac{\partial v_1}{\partial e_1}$$

Scaling yields:

$$C_{e_1}^J = C_{e_1}^s \varepsilon_s^1 + 1$$

Substituting $C_{e_1}^s$ gives:

$$C_{e_1}^J = -\frac{1}{\varepsilon_s^1 - \varepsilon_s^2}\varepsilon_s^1 + 1 = \frac{\varepsilon_s^1}{\varepsilon_s^2 - \varepsilon_S^1}$$

The use of implicit differentiation can also be cast into matrix form. We can write the system equation in matrix form [87]:

$$\frac{d\mathbf{s}}{dt} = \mathbf{N}v(\mathbf{s}(\mathbf{e}), \mathbf{e}) \qquad (4.21)$$

where $\mathbf{N}$ is the stoichiometry matrix, v the rate vector, and $d\mathbf{s}/dt$ the rate of change vector. The rate vector has been made an explicit function of the concentrations, $\mathbf{s}$, and the vector of enzyme concentrations, $\mathbf{e}$. At steady state this equation is equal to zero and differentiation with respect to $\mathbf{e}$ yields the following:

$$0 = \mathbf{N}\frac{\partial v}{\partial \mathbf{s}}\frac{d\mathbf{s}}{d\mathbf{e}} + \mathbf{N}\frac{\partial v}{\partial \mathbf{e}}$$

Rearrangement gives:

$$\frac{d\mathbf{s}}{d\mathbf{e}} = \left(\mathbf{N}\frac{\partial v}{\partial \mathbf{s}}\right)^{-1}\mathbf{N}\frac{\partial v}{\partial \mathbf{e}} \qquad (4.22)$$

The left-hand side represents the unscaled concentration control coefficient. $\partial v/\partial \mathbf{s}$ is a n by m matrix of unscaled elasticity coefficients, where n is the number of reactions and m the number of floating species. $\partial v/\partial \mathbf{e}$ is a n by n matrix of elasticities with respect to each enzyme, e. To illustrate how this equation can be used, consider the two-step pathway:

$$X_o \overset{v_1}{\to} S \overset{v_2}{\to} X_1$$

where X_o and X_1 are fixed species. In this model, $n = 2$, $m = 1$. With two steps the $\mathbf{e}$ vector will have two entries, e_1 and e_2. The elasticity matrix will be a 2 by 1 matrix:

$$\frac{\partial \mathbf{v}}{\partial \mathbf{s}} = \begin{bmatrix} \dfrac{\partial v_1}{\partial s} \\[2ex] \dfrac{\partial v_2}{\partial s} \end{bmatrix}$$

and $\partial \mathbf{v}/\partial \mathbf{e}$ is a 2 by 2 matrix:

$$\frac{\partial \mathbf{v}}{\partial \mathbf{e}} = \begin{bmatrix} \dfrac{\partial v_1}{\partial e_1} & 0 \\[2ex] 0 & \dfrac{\partial v_2}{\partial e_2} \end{bmatrix}$$

Note that e_1 has no direct affect on v_2 and e_2 has no direct effect on v_1, hence the entries $\partial v_2/\partial e_1$ and $\partial v_1/\partial e_2$ are zero. Combined together using equation (4.22):

$$\begin{bmatrix} \dfrac{ds}{de_1} \\[2ex] \dfrac{ds}{de_2} \end{bmatrix} = \left(\begin{bmatrix} 1 & -1 \end{bmatrix} \begin{bmatrix} \dfrac{\partial v_1}{\partial s} \\[2ex] \dfrac{\partial v_2}{\partial s} \end{bmatrix} \right)^{-1} \begin{bmatrix} 1 & -1 \end{bmatrix} \begin{bmatrix} \dfrac{\partial v_1}{\partial e_1} & 0 \\[2ex] 0 & \dfrac{\partial v_2}{\partial e_2} \end{bmatrix}$$

Multiplied out we obtain:

$$\begin{bmatrix} \dfrac{ds}{de_1} \\[2ex] \dfrac{ds}{de_2} \end{bmatrix} = \begin{bmatrix} \dfrac{\partial v_1/\partial e_1}{\partial v_1/\partial s - \partial v_2/\partial s} \\[3ex] \dfrac{\partial v_1/\partial e_2}{\partial v_1/\partial s - \partial v_2/\partial s} \end{bmatrix}$$

Each entry can be scaled by multiplying both sides by e_1 and dividing by s, for example, to scale ds/de_1:

$$\frac{ds}{de_1}\frac{e_1}{s} = \frac{e_1}{v_1}\frac{\partial v_1}{\partial e_1} \frac{1}{\dfrac{\partial v_1}{\partial s}\dfrac{s}{v_1} - \dfrac{\partial v_2}{\partial s}\dfrac{s}{v_2}}$$

Note the trick for scaling the denominator terms on the right-side where top and bottom are multiplied by v_1, noting that $v_1 = v_2$ at steady state. Rearranging yields:

$$C_{e_1}^s = \varepsilon_{e_1}^1 \frac{1}{\varepsilon_s^1 - \varepsilon_s^2}$$

As before, $\varepsilon_{e_1}^1 = 1$ so that:

$$C_{e_1}^s = \frac{1}{\varepsilon_s^1 - \varepsilon_s^2}$$

This is the same equation as the one previously derived (4.20). The one limitation of this approach is that it cannot, without a minor modification, be used on pathways that include conserved moieties. This is because the inverse in equation (4.22) cannot be evaluated. This issue will be revisited in a later chapter when moiety cycles are discussed in some depth.

The flux control coefficients can be similarly derived. The system equation for fluxes is shown below:

$$J = \mathbf{v}(\mathbf{s}(\mathbf{e}), \mathbf{e})$$

Differentiating with respect to $\mathbf{e}$ gives:

$$\frac{dJ}{d\mathbf{e}} = \frac{\partial \mathbf{v}}{\partial \mathbf{s}}\frac{d\mathbf{s}}{d\mathbf{e}} + \frac{\partial \mathbf{v}}{\partial \mathbf{e}}$$

Scaling with $\boldsymbol{J}$, $\mathbf{v}$ and $\mathbf{e}$ yields:

$$C_e^J = \varepsilon_s^v C_e^s + \varepsilon_e^v$$

If we assume $v \propto e_i$ then $\varepsilon_e^v = \boldsymbol{I}$, where $\boldsymbol{I}$ is the identity matrix, so that:

$$C_e^J = \varepsilon_s^v C_e^s + \boldsymbol{I}$$

This approach is is useful because it can be encoded using symbolic algebra tools such as Mathematica, Maxima or in the Python based tool, PySCeSToolbox [?].

4.6 Relationship to S-Systems

An alternative approach to MCA is Biochemical Systems Theory, or BST that was developed by Savageau and colleagues over many years. There is considerable overlap between BST and MCA. The main difference is the emphasis that BST places on developing approximate simulation models based on power laws:

$$\alpha x^g$$

A particular type of power law formalism, called S-systems, consists of two terms. The first term aggregates all inputs to a species as a single power law term. The second term aggregates all consumption terms as a single power law and is subtracted from the first term. A single S-systems equation that describes the rate of change of a species X, would look like:

$$\frac{dx}{dt} = \alpha \prod_{j=1}^{n} x_j^{g_{ij}} - \beta \prod_{j=1}^{n} x_j^{h_{ij}}$$

α and β are called the rate constants, and g and h the kinetic orders. The product terms include all species that might influence production in the first term, and all species that might influence consumption in the second term.

For example, consider a branch with a single reaction, v_1, producing a species, x, and two reactions, v_2 and v_3 consuming x. The differential equation for this system is:

$$\frac{dx}{dt} = v_1 - (v_2 + v_3)$$

The S-system equation for the rate of change in x is given by:

$$\frac{dx}{dt} = \alpha x^g - \beta x^h$$

Note that the two consumption steps have been merged into a single term βx^h. In the process of aggregating, some information is lost. For example, information on the relative magnitudes of the consumption rates, v_2 and v_3, is missing and cannot be recovered. The chief advantage in aggregating is that the equation can be analytically solved for the steady state solution, something that is invariably impossible to do otherwise. The analysis of S-system models however leads to results that are identical to those of MCA, and many of the terms can be interchanged between the two approaches.

Instead of elasticities, BST uses the term kinetic orders. Instead of control coefficients, BST uses the term logarithmic gains. The main difference is that the analysis of branched systems or systems with moiety conserved cycles (Chapter 12) can be more difficult when using BST. The approaches used to derive the logarithmic gains are quite different in BST and MCA but the outcomes are identical. BST differentiates the analytical steady state solutions from the S-system. MCA can derive the same results using operational proofs, implicit differentiation or a variety of matrix methods. The stability of steady state solutions – see Chapter 9 – is automatically obtained from BST, whereas with MCA, stability must be determined by computing the Jacobian separately. Finally, the theorems that are highlighted by MCA are of secondary importance to BST and are rarely mentioned. Thus there are pros and cons to both approaches. In this textbook the focus is obviously MCA.

BST can be illustrated with a simple two-step pathway:

$$X_o \rightarrow S \rightarrow$$

The S-system equation can be written as:

$$\frac{ds}{dt} = \alpha x_o^{g_1} s^{g_2} - \beta s^h$$

Note that the first reaction $X_o \rightarrow s$ has two terms raised to a power because the reaction is influenced by both X_o and S. To compute the steady state level of S, ds/dt can be set to zero such that:

$$\alpha x_o^{g_1} s^{g_2} = \beta s^h$$

Further rearrangement leads to:

$$\frac{\beta}{\alpha} = x_o^{g_1} s^{g_2} s^{-h}$$

Taking logs on both sides:

$$\log\left(\frac{\beta}{\alpha}\right) = g_1 \log(x_o) + (g_2 - h) \log(s)$$

And solving for $\log(s)$:

$$\log(s) = \frac{\log\left(\frac{\beta}{\alpha}\right) - g_1 \log(x_o)}{(g_2 - h)} \tag{4.23}$$

Recall that the definition (3.1) of the control coefficients was given in terms of the logarithmic derivative:

$$C = \frac{\partial \log V}{\partial \log p}$$

where V is the variable and p the parameter. Given that in the previous example the steady state solution is in terms of $\log(s)$, it is straightforward to compute the control coefficients (or logarithmic gains) by differentiating equation (4.23) with respect to one of the parameters. For example, the logarithmic gain for the steady state concentration of S with respect to the rate constants α and β can be computed as follows. To make things clearer, equation (4.23) is reexpressed as: Therefore, differentiating with respect to $\log(\alpha)$ and $\log(\beta)$, yields:

$$\frac{\partial \log(s)}{\partial \log(\alpha)} = \frac{1}{h - g_2} \qquad \frac{\partial \log(s)}{\partial \log(\beta)} = -\frac{1}{h - g_2} \tag{4.24}$$

Note that the signs have been swapped in the numerators and denominators. Similarly, the logarithmic gain with respect to x_o can also be derived:

$$\frac{\partial \log(s)}{\partial \log(x_o)} = \frac{g_1}{h - g_2}$$

The logarithmic gains $\partial \log(s)/\partial \log(\alpha)$ and $\partial \log(s)/\partial \log(\beta)$ correspond to the control coefficients, $C_{e_1}^s$ and $C_{e_2}^s$. Since the kinetic orders g_1 and h are equivalent to the elasticities ε_s^1 and ε_s^2 respectively, the analysis using BST and MCA yields identical results:

$$C_{e_1}^s = \frac{\partial \log(s)}{\partial \log(\alpha)} = \frac{1}{\varepsilon_s^2 - \varepsilon_s^1} = \frac{1}{h - g_2}$$

$$C_{e_2}^s = \frac{\partial \log(s)}{\partial \log(\beta)} = -\frac{1}{\varepsilon_s^2 - \varepsilon_s^1} = -\frac{1}{h - g_2}$$

MCA Symbol			BST Symbol	
Control Coefficient	$C_{e_1}^s$		Logarithmic Gain	$\dfrac{\partial \log(s)}{\partial \log(\alpha)}$
Control Coefficient	$C_{e_2}^s$		Logarithmic Gain	$\dfrac{\partial \log(s)}{\partial \log(\beta)}$
Response Coefficient	$R_{x_o}^s$		Logarithmic Gain	$\dfrac{\partial \log(s)}{\partial \log(x_o)}$
Elasticity	ε_s^1		Kinetic Order	g_2
Elasticity	ε_s^2		Kinetic Order	h
Elasticity	$\varepsilon_{x_o}^1$		Kinetic Order	g_1

Table 4.1 Summary of equivalent terms used in MCA and BST.

The logarithmic gain $\partial \log(s)/\partial \log(x_o)$ corresponds to the response coefficient, $R_{x_o}^s$. Recall that the response coefficient is equal to the elasticity, $\varepsilon_{x_o}^1$ times the concentration control coefficient $C_{e_1}^s$, that is:

$$R_{x_o}^s = \varepsilon_{x_o}^1 C_{e_1}^s$$

In the BST formalism, $\varepsilon_{x_o}^1 = g_1$, therefore we can make the following equivalence:

$$R_{x_o}^s = \varepsilon_{x_o}^1 C_{e_1}^s = \varepsilon_{x_o}^1 \frac{1}{\varepsilon_s^2 - \varepsilon_s^1} = \frac{\partial \log(s)}{\partial \log(x_o)} = g_1 \frac{1}{h - g_2}$$

Table 4.1 summarizes some of the equivalent terms between MCA and BST.

Using this small example it is evident that MCA and BST are very similar. Perhaps by historical accident MCA is more developed given its European origins where theory tends to get more attention. Both approaches ask some of the same questions and yield the same answers. The use of power laws in deriving the steady state solutions and logarithmic gains is certainly elegant, but the aggregation that S-systems require poses challenges for more complex pathways.

MCA derives theorems which BST does not emphasize. However the theorems allow additional insight to be gained. MCA also focuses much more on fluxes through pathways which BST tends to be more silent on. Steady state stability is dealt with explicitly in BST and has resulted in a variety of very useful results (see later chapters), especially for oscillating systems. However with recent developments in the theory [52], MCA can accomplish the same.

MCA has been integrated much more fully into stoichiometric analysis and offers an elegant mathematical framework for dealing with conserved moieties and branched systems (see later chapters). The use of generalized power laws where aggregation is not required may however offer a useful approach to building approximate models when data is limited.

Further Reading

1. Kacser H and Burns JA, (1973). In: Davies, D.D (ed.), Rate Control of Biological Processes, vol. 27 of Symp. Soc. Exp. Biol. p. 65-104. Cambridge University Press.

2. Kacser H and Burns J (1979) Molecular Democracy: Who Shares the Controls? Biochem Soc Trans, 7, 1149-1160.

3. Heinrich R and TA Rapoport (1974) A linear steady-state treatment of enzymatic chains. General properties, control and effector strength. Eur. J. Biochem. 42:89-95.

4. Heinrich R, Rapoport SM, Rapoport TA (1977) Metabolic regulation and mathematical models. Prog Biophys Mol Biol. 1977;32(1):1-82.

5. Fell D A (1996) Understanding the Control of Metabolism, Frontiers in Metabolism, ISBN-10: 185578047X

6. Savageau M (1976) Biochemical systems analysis: a study of function and design in molecular biology, Addison-Wesley. Note: The text was republished in 2010 and is available on Amazon and is an excellent source for insight into a variety of pathway structures.

Exercises

1. A given species S has a single production step and a single consumption step. The elasticity of the production step with respect to S was found to be -1.6, and for the consumption step, 0.12.

   a) Explain why the production step elasticity is negative.

   b) From the information, what can you say about the flux control coefficients of the production and consumption steps?

2. Explain why examination of a single enzyme in a pathway will not necessarily give a good indication of how flux limiting the enzyme is.

3. The response coefficient relationship has two important lessons for those looking to develop new therapeutic drugs. What are they?

4. What is a canonical control coefficient?

5. Given the model described by the Tellurium script (4.1), ignore everything in the script after line 14. Find the flux control coefficient for each step and then confirm numerically that the flux summation theorem holds. Use perturbations to estimate the flux control coefficients.

6. Use BST to derive the logarithmic gains for S_1 and S_2 with respect to the rate constants in all three steps for the three-step pathway:

$$X_o \rightarrow S_1 \rightarrow S_2 \rightarrow X_1$$

   where X_o and X_1 are fixed (boundary) species. Confirm that the results you get correspond to those obtained using MCA.

4.A   Python/Tellurium Scripts

Calculating control coefficients and elasticities using Python and Tellurium.

```
import tellurium as te
import numpy as np
r = te.loada ('''
    J1: $Xo -> S1; E1*(k1*Xo - k2*S1);
    J2: S1 -> S2; E2*(k3*S1 - k4*S2);
    J3: S2 -> $X1; E3*(k5*S2);

    Xo = 5
    k1 = 0.2; k2 = 0.04;
```

```
    k2 = 0.7; k3 = 0.34;
    k4 = 0.65;
    E1 = 1; E2 = 1; E3 = 1;
''')

print r.steadyState()
print "Flux Control Coefficients:"
print r.getCC ("J1", "E1"),  r.getCC ("J1", "E2")
print r.getCC ("J1", "E3")

print "S1 Concentration Control Coefficients:"
print r.getCC ("S1", "E1"),  r.getCC ("S1", "E2")
print r.getCC ("S1", "E3")

print "S2 Concentration Control Coefficients:"
print r.getCC ("S2", "E1"),  r.getCC ("S2", "E2")
print r.getCC ("S2", "E3")

print "Elasticities:"
print r.getEE ("J1", "Xo"), r.getEE ("J2", "S1")
print r.getEE ("J2", "S1"), r.getEE ("J2", "S2")
print r.getEE ("J3", "S2"),
```

Listing 4.1 Computing Control Coefficients.

5

Experimental Methods

5.1 Introduction

Up to this point, we've largely discussed the theory of metabolic control analysis. However, without the ability to make measurements and test predictions on real biological systems, the approach outlined in the previous chapters will find little practical use. In this chapter we will outline various methods that have been developed to measure control coefficients and some examples of applications used in real metabolic pathways. At the heart of MCA is the use of perturbations to illicite a response. Perturbations can be made to external factors such as nutrients or internally by either changing gene expression or use of specific enzymatic inhibitors. When metabolic control analysis was first developed, genetic engineering of DNA was in its infancy and in order to perturb enzyme activity researchers had to devise creative approaches to manipulating gene dosage. With modern developments in genetic engineering, changing the expression level of a given gene can now even be done by amateurs and most certainly by undergraduates. The measurement of control coefficients is therefore now relatively straightforward.

Measuring Control Coefficients

Various approaches have been used in the past to measure control coefficients experimentally. All revolve around the need to change either the concentration or activity of an enzyme or protein. The different methods can be categorized into six general approaches as shown in Table 5.1. We've also included computer modeling even though its not really an experimental method. However experimentally determined values for the various rate constants can be used to build models that are reliable enough to estimate control coefficients [100, 72]. We will cover this topic in a later section towards the end of the chapter.

5.2   Using Classical Genetics

There is an interesting story related to the origins of metabolic control analysis from the Kacser lab[1]. They noticed that a mutation in one of the amino acid biosynthesis enzymes in the fungus *Neurospora crassa* yielded a loss of 95% enzyme activity, and yet this hardly affected the observed phenotype. How was this possible? Instinctively, one might try to answer this question by studying the enzyme in question, perhaps studying its kinetics in detail or understanding its catalytic activity by determining its protein structure. However such information could not explain why enzyme loss had so little effect on the phenotype. In an insightful move, the Kacser lab decided that the answer must lie in the enzyme's cellular context. That is, the network within which the enzyme operated. It was this observation that began the study of how networks behaved and how phenotype was related to phenotype, not in terms of individual proteins or enzymes, but in terms of networks. Similar conclusions were made by two other groups who also developed MCA independently, namely Heinrich and Rapoport in Berlin and Savageau in Michigan, USA.

1. Use of classical genetics to manipulate gene expression(s)
2. Titration of enzymes with specific inhibitors
3. Double modulation method
4. *in vitro* reconstitution and enzyme titration
5. Gene engineering to change enzyme levels *in vivo*
6. Computer modeling (not really an experimental method)

Table 5.1 Different approaches to measuring control coefficients.

The Kacser lab investigated the amino acid biosynthesis pathway by changing the copy number of a gene using classical genetics [34]. They assumed that the activity or concentration of the enzyme or protein in question was proportional to the copy number. This work was carried out using the fungus *Neurospora crassa* where arginine biosynthesis was studied. This fungus forms multinucleated mycelia that generate polyploid spores. By mixing different ratios of spores containing genes encoding wild and mutant enzymes, it was possible to generate mycelia with different activities of the arginine pathway enzymes. From such experiments it was determined that four enzymes, acetyl-ornithine aminotransferase, ornithine transcarbamoylase, arginine succinate synthetase, and arginine-succinate lyase all had flux control in the range of 0.02 to 0.2. This indicated that none of the enzymes exerted significant control over arginine synthesis.

Another study by the same group investigated the flux control of alcohol dehydrogenase (ADH) in *Drosophila melanogaster*. ADH is present in three alleles that encode isoforms with different maximal activities. When mixing the various combinations (including mutations of the isoforms), it was possible to change the total activity of ADH and measure the ethanol production. From this study it was concluded that ADH has a flux control coefficient of zero.

5.3   Genetic Engineering

Neurospora crassa and *Drosophila melanogaster* were special cases where gene dosage could be engineered by classical genetic methods. However most systems are not so easily altered. As a result, alterative methods based on inhibitors were developed. In particular oxidative phosphorylation is susceptible to a large repertoire of inhibitors. Control coefficients were estimated by titrating a given inhibitor and measuring the effect on a flux or species concentration [39]. By taking into account how the inhibitor acted, it is possible to obtain estimates for the control coefficient of the inhibited step. We will come back to this in the next section.

---

[1] personal communication

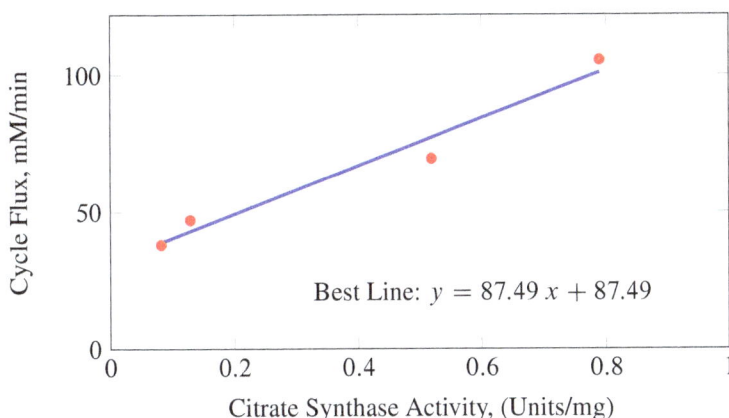

Figure 5.1 Data from Walsh and Koshland [116] replotted. The cycle flux represents the sum of the Krebs and Glyoxylate fluxes under acetate medium conditions. Wildtype level of citrate synthase is at 0.675 units/mg. Original error bars are omitted from this plot.

With modern developments in genetic engineering and molecular biology, it is now relatively straightforward to estimate control coefficients. For example, inducible or repressor operator sites can be added to a gene of interest and the effect of up regulating or down regulating the wild-type activity can be measured. In addition, protein levels can be knocked down by using a variety of method such as RNA antisense or CRISPR. RNA antisense was used to estimate the flux control coefficient for ribulose-bisphosphate-carboxylase (rubisco) which is responsible for fixing carbon dioxide in plants [102]. The traditional view has been that Rubisco catalyzes a rate-limiting step (and still is by many), that is a step with a high flux control coefficient. However, studies using RNA antisense in tobacco plants to reduce the level of Rubisco show that during high illumination, flux control was estimated in the range of 0.69 to 0.83. In moderate illumination or high carbon dioxide levels, the flux control fell dramatically to 0.05 to 0.2. This study reconfirms two important points that were highlighted in the theoretical analysis, the first is that control is not fixed but depends on external conditions. Secondly, rate limitingness cannot be determined by simple inspection of the enzyme but instead must be actively measured.

One of the earliest uses of an inducible promotor to control gene expression and thereby estimate rate control was the work by Walsh and Koshland [116]. In their paper they describe using the inducer IPTG (isopropyl-β-D-thiogalactopyranoside) to control the Krebs cycle enzyme citrate sythase in *E. coli* via a *tac* promoter[2]. The flux through citrate synthase was measured indirectly by measuring radiolabelled substrate incorporation into carbon dioxide, fatty acids and cellular mass. Figure 5.1 shows the data replotted from their paper (note errors bars were displayed in the original data).

Using this data, Walsh and Koshland suggested that under an acetate medium, citrate synthase was rate-controlling. However, the extent of control was not considered. Using their data, it is relatively easy to compute the flux control coefficient for citrate synthase. Given the straight line fit through the points (Figure 5.1), the slope of the line is 87.49 flux units per citrate synthase activity. In order to relate this to the flux control coefficient, the slope must be multiplied by the enzyme activity and divided by the flux. This is easily done using the regression parameters from the straight line fit. Figure 5.2 shows the flux control coefficient plotted as a function of citrate synthase activity. The first thing to note is that at the wildtype point, citrate synthase is not completely rate limiting, but has a flux control coefficient of 0.65. This means that a 1% increase in citrate synthase activity will lead to a 0.65% change in cycle flux. Because the coefficient is less than one, flux control must also reside elsewhere.

---

[2]The *tac* promotor is a synthetic promoter made from a combination of the lac and trp promoters.

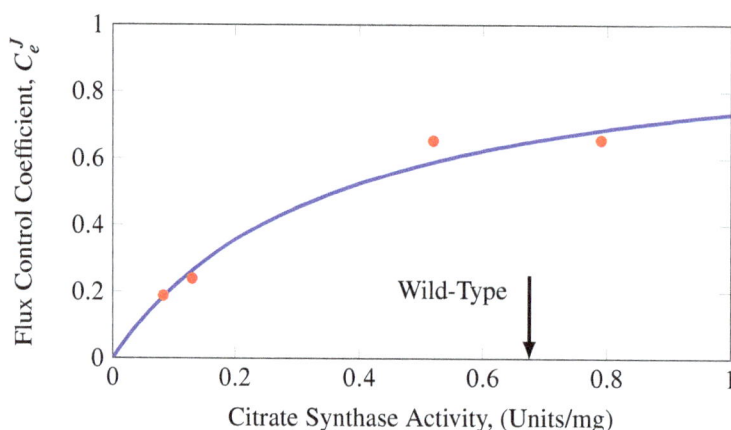

Figure 5.2 Data from Walsh and Koshland [116]. Flux control coefficient computed using a best line fit from Figure 5.1. At the wildtype level of citrate synthase, the flux control coefficient has a value of 0.65 indicated by the arrow. The response is completely counter intuitive to what one would expect and suggests that either the data is incomplete or that there are other physiological changes taking place.

The second and somewhat unusual observation is that the flux control coefficient decreases as the citrate synthase activity decreases. Under normal circumstances the expectation is that flux control *increases* as enzyme activity decreases. There are a number of possible reasons for this: i) the data could be wrong; ii) the straight line fit through the data is inappropriate. If the flux had been measured at lower and higher levels of citrate synthase, the data would show more of a hyperbolic like response, meaning that flux control would be high at low enzyme activities, and low at higher enzyme activities. The straight line fit assumes that the data does not go through zero flux at zero enzyme activity which one would imagine it would in practice; iii) there are unrecorded changes to other enzymes in the pathway as a result of changes to the citrate synthase level. For example, it is possible that other enzymes are expressed less, such that any flux control is transferred from these other enzymatic steps to citrate synthase so that the flux control at citrate synthase appears to rise as citrate synthase is decreased. At the end of the day there isn't enough data to make any firm conclusion about how much citrate synthase influences flux.

In the original paper Walsh and Koshland did not compute the flux control coefficient. Instead they assumed that the linear relationship meant that flux control at citrate synthase was fixed and high. A closer investigation suggests the opposite, and also raises some additional and interesting questions.

5.4   Titration by Inhibitors

The first approach devised to estimate control coefficients was the use of classical genetic to manipulate gene copy number. This however is difficult to do and requires considerable expertise in genetic manipulation. In the late 1970s and early 1980s, researchers realized that inhibitors could also be used to change protein and enzyme activity. One of the most well studied pathways is oxidative phosphorylation, due in part to the wide variety of available inhibitors. Inhibitors include irreversible types, for example cyanide can bind to cytochrome c oxidase (Site 3), noncompetitive inhibitors such as Rotenone that can bind to NADH-CoQ-oxidoreductase (Site I) or Antimycin that can bind to CoA-cytochrome c oxidoreductase and a competitive inhibitor such as malate that binds to dicarboxylate transporter.

By titrating the inhibitor and measuring the steady state response, it is possible to estimate the control co-efficient at the inhibited site by extrapolating the response curve back to zero inhibitor. Each inhibitor type however must be treated differently [39]. In a previous chapter the response coefficient was introduced:

$$R_x^J = \frac{dJ}{dx}\frac{x}{J}$$

The response coefficient describes the effect of an external signal, X, on a pathway. The response coefficient can be expressed in terms of the control coefficient and the external signal elasticity using the relationship:

$$R_x^J = C_e^J \varepsilon_x^v$$

Expanding this relationship yields:

$$\frac{dJ}{dx}\frac{x}{J} = C_e^J \frac{\partial v}{\partial x}\frac{x}{v} \tag{5.1}$$

We can cancel the x terms on both sides and rearranging so that C_e^J is on one side we get:

$$C_e^J = \frac{dJ}{dx}\frac{1}{J} \bigg/ \frac{\partial v}{\partial x}\frac{1}{v}$$

The term $\frac{dJ}{dxJ}$ can be derived from the initial slop of the inhibition curve, and $v\partial v/\partial x$ from the inhibition characteristics of the inhibited enzyme. Since we seek the value of C_e^J when there is no inhibitor present, we should measure both terms when $x = 0$. We can illustrate this approach with a number of examples. Consider first a non-competitive inhibitor. In the simplest case the rate of an enzymatic reaction subject to non-competitive inhibition is given by the well known equation:

$$v = \frac{V_m\, s}{(K_m + s)(1 + i/K_i)}$$

The elasticity for this rate laws with respect to inhibitor concentration, I can be shown to be equal to:

$$\varepsilon_i^v = -\frac{i}{K_i + i}$$

Substituting this into equation (5.1) and setting the inhibitor concentration to zero, ($i = 0$) we arrive at an expression for the control coefficient when the inhibitor concentration is extrapolated to zero:

$$C_x^J = -\frac{K_i}{J}\frac{dJ}{dx}$$

This approach requires an estimate for the inhibition constant, K_i and the slope of the response flux versus inhibitor at zero inhibitor. The same analysis can be done for competitive inhibitors. The irreversible rate law for a competitive inhibitor is given by the well known equation:

$$v = \frac{V_m\, s}{s + K_m\left(1 + \frac{i}{K_i}\right)}$$

The elasticity for this rate law with respect to the competitive inhibitor, I, can be shown to be equal to:

$$\varepsilon_i^v = -\frac{i/K_i}{1 + s/K_m}$$

Inserting this into equation (5.1) and setting the inhibitor concentration to zero ($i = 0$) we arrive at an expression for the control coefficient based on titrating an enzymatic step with a competitive inhibitor:

$$C_e^J = -\frac{K_i}{J}\left(1 + \frac{s}{K_m}\right)\frac{dJ}{di}$$

This can be generalized if need be to the reversible case:

$$C_e^J = -\frac{K_i}{J}\left(1 + \frac{s}{K_s} + \frac{p}{K_p}\right)\frac{dJ}{di}$$

Where s is the concentration of substrate and p the concentration of product. K_s and K_p are the K_m constants for the substrate and product respectively.

Example 5.1

The following data (constructed from a simulated pathway with added noise) was collected from a pathway and measures the flux through the pathway at various concentrations of an irreversible inhibitor. Use the data to estimate the flux control coefficient through the pathway.

Inhibitor Concentration	Pathway Flux
0	1.5
0.2	1.4
0.3	1.5
0.4	1.2
0.5	1.0
0.6	0.8
0.7	0.7
0.8	0.3
0.9	0.2
0.94	0.0

A plot of the data is shown below:

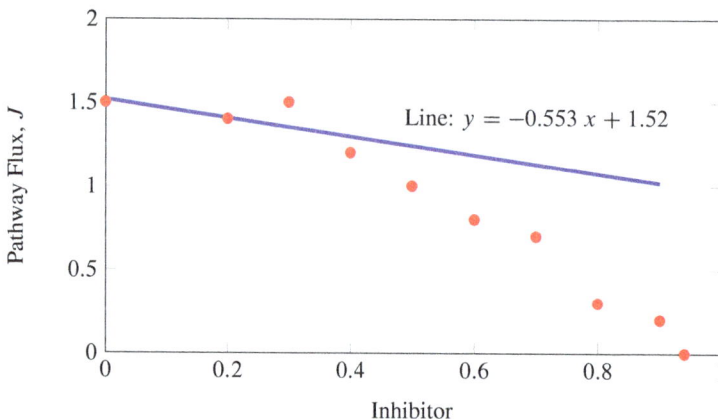

The curve that follows the points is not known, which means it is difficult to select a suitable nonlinear function to fit the points. Instead we will take the first four data points and plot a straight line through them. This is shown by the blue continuous line on the plot. The slope of the fitted line was found to be -0.553. This corresponds to the dJ/dx term in equation (5.2). The flux at zero inhibitor is 1.5 and the I_{max} is 0.94, the concentration of inhibitor that yields zero flux. Taking these together and inserting the values into equation (5.2), we obtain the flux control coefficient:

$$C_e^J = -0.553\frac{0.94}{1.5} = 0.35$$

The simulation model that was used to obtain the data gave a control coefficient of 0.367 which is close to the estimated value. The key to obtaining a reasonable estimate is to secure sufficient points at low inhibitor concentration in order to compute a best line fit through the first few points. Attempts to fit polynomials, logistic curves, or hyperbolic curves will likely yield poor estimates.

Irreversible Inhibitors

Irreversible inhibitors are a special case; the amount of inhibitor required to completely inhibit the enzyme should be equal to the amount of enzyme. This assumes that one molecule of inhibitor binds to a single enzyme and completely inhibits the enzyme's activity. That is, $x_{max} = e$. The equation to compute the control coefficient using an irreversible enzyme is given by:

$$C_e^J = -\frac{x_{max}}{J}\frac{dJ}{dx} \tag{5.2}$$

The negative sign is included because the slope of dJ/dx is negative.

5.5 Double Modulation Technique

The double modulation method, first proposed by Kacser and Burns in 1979 is an elegant method for estimating elasticities *in vivo* but is difficult to execute experimentally. Consider the pathway:

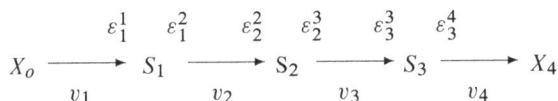

$$X_o \xrightarrow[v_1]{\varepsilon_1^1 \quad \varepsilon_1^2} S_1 \xrightarrow[v_2]{\varepsilon_2^2 \quad \varepsilon_2^3} S_2 \xrightarrow[v_3]{\varepsilon_3^3 \quad \varepsilon_3^4} S_3 \xrightarrow{} X_4$$

Let us focus on reaction v_3 flanked by species S_2 and S_3, respectively. Let us make a perturbation in the upstream source metabolite, X_o. Changes will propagate through the pathway resulting in changes in S_2 and S_3. If the changes are sufficiently small, we can write down the change in flux using the following relation:

$$\frac{\delta J^\dagger}{J^\dagger} = \varepsilon_2^3 \frac{\delta s_2^\dagger}{s_2^\dagger} + \varepsilon_3^3 \frac{\delta s_3^\dagger}{s_3^\dagger} \tag{5.3}$$

The quantities have a superscript $\dagger$ to indicate this is the set of changes that accompanied the change due to a perturbation in X_o. We can now carry out a separate experiment where we perturb the downstream sink pool X_4, this time using the superscript $\sharp$. Again, we will observe propagations in the pathway resulting in changes to S_2 and S_3. Note that because the disturbance is from another source, the changes in S_2, S_3 and flux J will be different. That is:

$$\frac{\delta J^\sharp}{J^\sharp} = \varepsilon_2^3 \frac{\delta s_2^\sharp}{s_2^\sharp} + \varepsilon_3^3 \frac{\delta s_3^\sharp}{s_3^\sharp}$$

We now have two equations in two unknowns, ε_2^3 and ε_3^3. Assuming changes in metabolite concentration and fluxes can be measured, the two equations can be used to solve for the elasticities. In principle, if all the metabolite changes were measured in the entire pathway, we could estimate all the elasticities. Once we have the elasticities the control coefficients can be estimated using the methods described in the last chapter. Another point worth making, it doesn't matter what changes are made to illicit the perturbation. In this case we used X_o and X_4 but changes in enzyme levels or addition of inhibitors are equally valid ways to perturb the system. The method has been generalized by Acerenza and Cornish-Bowden [1] and an experimental application was published by Giersch [35].

Example 5.2

Using the same system from the previous example, the following perturbation data were obtained by carrying out two perturbations, one upstream and one downstream of the reaction under observation. Estimate the values for the elasticities ε_1^2 and ε_2^2.

Before any perturbations were made, the following reference flux and concentrations were recorded:

$$J = 1.5$$
$$s_1 = 0.74$$
$$s_2 = 0.92$$

A perturbation that involved increasing the input pool by 30% was applied and a new steady state flux and concentrations were recorded:

$$J = 1.6$$
$$s_1 = 0.91$$
$$s_2 = 1.15$$

Note that it was not possible to perturb the output metabolite from the pathway, instead an inhibitor was applied to one of the downstream steps. The degree of inhibition at the inhibited site is not known (and is irrelevant), but the new steady state flux and concentrations were recorded:

$$J = 1.28$$
$$s_1 = 0.905$$
$$s_2 = 1.32$$

From this data, estimate the two elasticities, ε_1^2 and ε_2^2 with respect to S_1 and S_2. Formulate two equations, corresponding to each perturbation, of the form:

$$\frac{\delta J}{J} = \varepsilon_1^2 \frac{\delta s_1}{s_1} + \varepsilon_2^2 \frac{\delta s_2}{s_2}$$

From the data these two equations are:

$$\frac{0.1}{1.5} = \varepsilon_1^2 \frac{0.17}{0.74} + \varepsilon_2^2 \frac{0.23}{0.92}$$

$$\frac{-0.22}{1.5} = \varepsilon_1^2 \frac{-0.165}{0.74} + \varepsilon_2^2 \frac{0.4}{0.92}$$

Evaluating the ratios we obtain:

$$0.066 = \varepsilon_1^2\, 0.23 + \varepsilon_2^2\, 0.25$$

$$-0.14 = -\varepsilon_1^2\, 0.23 + \varepsilon_2^2\, 0.43$$

Solving for ε_1^2 and ε_2^2 yields:

$$\varepsilon_1^2 = 1.531 \quad \varepsilon_2^2 = -1.144$$

The true values for the elasticities obtained form the original model were $\varepsilon_1^2 = 1.69$ $\varepsilon_2^2 = -1.3$. Compared to the 'experimentally' determined values there is a discrepancy. This is due to two factors: the first is that all values were rounded down to two decimal places, secondly and more importantly, relatively large perturbations were made. The double modulation method depends on making small enough changes such that the relationship described by equation (5.3) remains true. If the perturbations are too high, equation (5.3) is only an approximation. Ideally one might make multiple perturbation of different strengths, plotting the resulting changes and extrapolating the plotted response back to the zero axis. The method however is a difficult one to execute experientially.

5.6 Reconstituted Systems

There has been a long history of using reconstituted systems as a way to understand cellular function. In the 1940s, actomyosin threads from muscle were reconstituted by Szent-Gyoergyi and later Straub, which led to the understanding of how muscles contract [105]. Perhaps more well known is the work by Arthur Kornberg [60] in the late 1950s who studied DNA replication. Liu and Fletcher [63] provide a more detailed view of reconstituted systems such as mitotic spindles, cell motility and membrane dynamics. The Liu and Fletcher article makes no mention of reconstituting metabolic or signaling pathways however. It is therefore worth mentioning a remarkable set of papers published in the late 1970s and early 1980s by a group of East German scientists. Their work involved the partial reconstitution of glycolysis using purified enzymes: Pyruvate Kinase, Adenylate Kinase, Phosphofructokinase, Fructose 1,6-bisphosphatase and Glucose-6-phosphatase. The five enzymes were studied in a continuously fed stirred tank reactor where enzymes and metabolites were pumped into a 1 ml chamber. Effluent was collected and metabolite levels measured. A number of variants of this setup were also explored including trapping the enzymes in a polyacrylamide gel and varying the enzymes in the mix. In a series of papers, this system (including variants) was used to study bistablity, oscillatory and other behaviors [95, 27, 96, 28, 29].

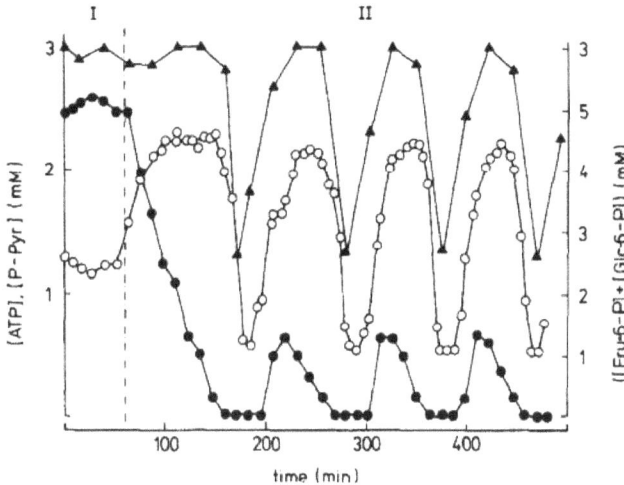

Figure 5.3 This plot [28] shows sustained oscillations in a reconstituted system containing five enzymes Pyruvate Kinase, Adenylate Kinase, Phosphofructokinase, Fructose 1,6-bisphosphatase and Glucose-6-phosphatase. Oscillations are shown for ATP, Phosphoenolpyruvate and Fructose-6-P-Glucose-6-P.

Very little additional work has been done in reconstituted systems until recently. It is worth mentioning the work by Chassagnole et al [17] on reconstituting the threonine pathway, branch-point studies by Curien [21], and Kouril [61] on reconstituting gluconeogenesis in sulfolobus solfataricus. More interesting from the perspective of this book, was the work carried out by Torres et al [108], who applied metabolic control analysis to the reconstituted system using rat liver extract containing six enzymes. By titrating the extract with additional enzyme, the authors where able to calculate the flux control coefficients. More recently, Panke and his group [12] have put together a remarkable experimental setup for studying a reconstituted system of glycolysis. What was most interesting about this study was the use of online mass-spectrometry to obtain real-time measurements of fourteen metabolites. In addition, they titrated the system with enzymes to determine the flux control coefficients. Given experimental setups such as those developed by Panke, it is possible to obtain highly detailed measurements of the dynamics of metabolic pathways. Of all the experimental approaches, this probably offers the most exciting potential to understand the dynamics of metabolic pathways.

5.7 By Computer Simulation

There are many instances in the literature where control coefficients have been computed by simulation. This is achieved by building a kinetic model of the pathway and using the computer to carry out the perturbations and thus estimate the control coefficients. Computer models can be developed to investigate the properties of particular pathway motifs or as a means to build predictive models of actual pathways. Some of the earliest uses of computer simulation to compute control coefficients was work done by Reinhart Heinrich and colleagues in developing a series of models of red blood cell metabolism [80].[3] In doing so they used computer simulation to estimate the flux and concentration control coefficients.

However one of the impediments to using computer simulation has been the availability of suitable data and more significantly, the lack of validation of the models once built. In recent years this has begun to change and there now exist a number of well validated and hence reliable computer models of metabolism.

Of particular note is the work by Smallbone, Mendes and coworkers [99] who developed a validated computer model of yeast glycolysis. The authors developed an initial model based on existing literature data. From this model estimates for the control coefficients were calculated using the simulation model. Those enzymatic steps with the largest flux control where isolated and their kinetics remeasured experimentally under physiological conditions. The new kinetic data was used to improve the original simulation model; the method was then repeated until all the enzymes had been experimentally characterized. This resulted in the development of a much more reliable model of glycolysis than was hitherto possible. A similar approach was very successfully applied to developing a reliable kinetic model of glycolysis in the malaria parasite, *Plasmodium falciparum* [72, 77] by Penlker, Snoep and coworkers.

The experience gained by these studies has highlighted the fact that one of the most important aspects when building a kinetic model is to have the kinetic properties of the individual enzyme measured under physiological conditions. This may seem like an obvious thing to do but until recently it was an aspect that was largely ignored. One possible reason for this is that previously, model building relied on global fitting of the model parameters to experimental data. That is, all model parameters were fitted simultaneously. This appears to have resulted in models that fitted the experimental data but failed when an attempt was made to use the model to predict behaviour beyond the scope of the fitted model, indicating that the model had failed to generalize. The approaches used by Mendes and Snoep indicate that it is possible to build reliable computer models of metabolism and to use computer simulation to estimate the control coefficients. Further commentary on this approach to modeling can be found in the article by van Eunen and Bakker [113].

5.8 By Calculation - Serine Pathway

Fell and Shell [31] (see also [101]) considered the serine biosynthesis pathway in rabbit liver. Serine is made by a linear pathway of three steps that branches off glycolysis at 3-phosphoglycerate.

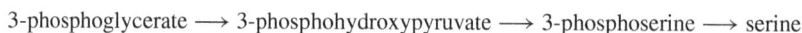

$$\text{3-phosphoglycerate} \longrightarrow \text{3-phosphohydroxypyruvate} \longrightarrow \text{3-phosphoserine} \longrightarrow \text{serine}$$

The first step, catalyzed by 3-phosphoglycerate dehydrogenase, requires the reduction of NAD to NADH. The second step is catalyzed by a 3-phosphoserine aminotransferase and requires glutamate as a cofactor (producing 2-oxoglutarate), and the third step catalyzed by 3-phosphoserine phosphatase results in the release of free phosphate. The question that the authors set out to answer was what determines the flux control coefficients for the three steps? Based on traditional views of metabolic regulation, it would be expected that the committed

---

[3]Even early, researchers such as Higgins and Burns used analog computers to carry out their simulations and in some cases estimate control coefficients (personal communication).

step (3-phosphoglycerate dehydrogenase) would be the rate-limiting step in the pathway. The question is whether this can be shown by computing the flux control coefficients.

In order to make the analysis simpler, the authors assumed that the flux through the serine pathway was small compared to the main glycolytic pathway. This means that whatever changes occurred in the serine pathway, there were negligible effects on the glycolytic intermediate 3-phosphoglycerate. Likewise, it was assumed that the cofactor concentrations for NAD/NADH and glutamate/2-oxoglutarate would be held relatively constant by other cellular processes. An additional assumption had to be made regarding the first two enzymes. Because the concentration of phosphohydroxypyruvate has been found to be too low to measure, the authors decided to merge the two first steps into a single grouped step. As a result, the final calculation would yield a single flux control coefficient for the group, and another flux control coefficient for the last step.

In order to compute the flux control coefficients, the authors needed to estimate the elasticities. The first elasticity to compute is the elasticity for the first two grouped enzymes with respect to the intermediate 3-phosphoserine. The maximal activities for the first two enzymes are high in rabbit liver which allowed the authors to assume that the degree of saturation of the enzymes by substrate was low. Note that the intermediate 3-phosphohydroxypyruvate is already very low, and is an additional factor that justifies this assumption. If the saturation levels are very low then the elasticities can be computed using equation (2.9) and repeated here:

$$\varepsilon_s^v = \frac{1}{1 - \Gamma/K_{eq}} = \frac{1}{1 - \rho}$$

$$\varepsilon_p^v = -\frac{\Gamma/K_{eq}}{1 - \Gamma/K_{eq}} = -\frac{\rho}{1 - \rho}$$

To use these equations the mass-action ratio, Γ, is required together with the equilibrium constant for the two step group. The concentrations for 3-phosphoglycerate, 3-phosphoserine, NAD/NADH and glutamate/2-oxoglutarate were obtained from the literature and used to compute the mass-action ratio. The joint equilibrium constant was obtained by multiplying the two individual equilibrium constants, again obtained from the literature. From this information; the elasticity for 3-phosphoserine with respect to the group was computed to be -1.43.

The next step was to compute the elasticity of 3-phosphoserine phosphatase with respect to 3-phosphoserine (PSer). It was determined by kinetic fitting and that 3-phosphoserine phosphatase is uncompetitively inhibited by serine. This resulted in formulating the following rate law for 3-phosphoserine phosphatase:

$$v = \frac{V_m \text{PSer}\, a}{\text{PSer} + K_m\, a} \quad \text{where} \quad a = \frac{1 + \text{Ser}/K_1}{1 + \text{Ser}/K_2}$$

where the various kinetic constants were determined from non-linear fitting of literature data - $K_m = 0.089$ mM, $K_1 = 16.5\ \mu$M and $K_2 = 0.6$ mM. The elasticity computed in this way was determined to be 0.041. The low value is attributed to the significant inhibition from serine. Combining the summation and connectivity theorems allows the flux control coefficients to be computed:

$$C_{1,2}^J = 0.03 \qquad C_3^J = 0.97 \tag{5.4}$$

What we see is that the last step has most of the control, the first two steps have very little control. This tells us that the committed step is not necessarily the rate determining step in this case. The authors go on to make further observations depending on the feeding state of the animals. For example, under a glucose/ethanol feed regime, the control coefficients are computed to be 0.46 and 0.54, respectively. That is, neither step dominates. What is more interesting is that the first two steps under these conditions are closer to equilibrium and yet have a flux control of 0.46. This shows that it is possible for near-equilibrium steps to acquire significant flux control. The paper includes much more detail on other aspects of the study and the reader should refer to the original paper [31].

Further Reading

1. Kacser H and Burns J. (1979) Molecular Democracy: Who Shares the Controls? Biochem Soc Trans, 7, 1149-1160.

2. Rafael Moreno-Sanchez, Emma Saavedra, Sara Rodrguez-Enriquez, and Viridiana Olin-Sandoval (2008) Metabolic Control Analysis: A Tool for Designing Strategies to Manipulate Metabolic Pathways, Journal of Biomedicine and Biotechnology, Volume 2008, Article ID 597913, doi:10.1155/2008/597913

Exercises

1. Write an essay on the use of physiologically measured enzyme kinetics to build reliable simulation models of metabolism.

Linear Pathways

6.1 Basic Properties

Simulations

Linear pathways represent the simplest network architecture and are a good starting point to gain insight into how cellular networks operate. However before examining the properties of linear pathways algebraically it is worth considering some simple properties we can obtain from simulations. Consider a five step pathway:

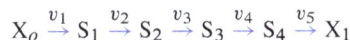

$$X_o \xrightarrow{v_1} S_1 \xrightarrow{v_2} S_2 \xrightarrow{v_3} S_3 \xrightarrow{v_4} S_4 \xrightarrow{v_5} X_1$$

where X_o and X_1 are fixed species. The simplest case to consider is where the kinetics of each step follows simple reversible mass-action. We will also assume throughout this chapter that there are no feedback loops; each step is only affected by its immediate reactant or product. For any given reaction, an increase in reactant will cause the reaction rate to increase and an increase in product will cause the reaction rate to decrease. These properties result in specific changes to species levels depending on the location of a perturbation. For example if a perturbation is made to the source species, X_o, then all species concentrations downstream will increase. Figure 6.1 show six plots, showing how a pulse in X_o travels down the pathway. Note that all the species increase in response to an increase in X_o.

Consider a perturbation near the center of the pathway. Figure 6.2 shows the effect of increasing the level of E_3 on the concentrations of the species. All species upstream decrease and all species downstream increase.

Algebraic Analysis

There is only so much that simulations can do to help us understand pathway dynamics. Instead we can resort to an algebraic approach to obtain deeper insights. Consider the following linear pathway:

$$X_o \xrightarrow{v_1} S_1 \xrightarrow{v_2} S_2 \xrightarrow{v_3} \ldots S_m \to X_1 \tag{6.1}$$

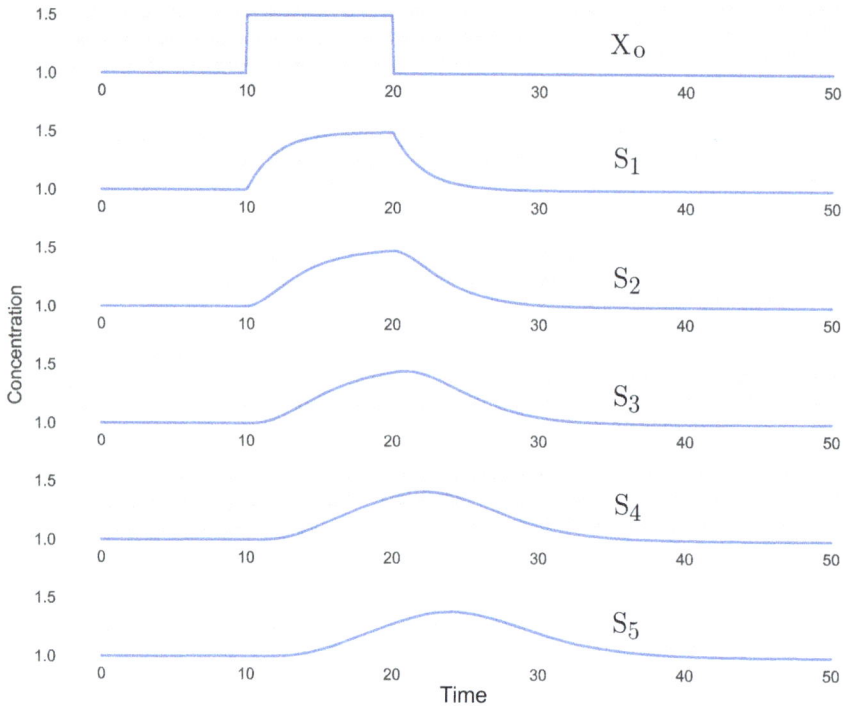

Figure 6.1 Propagation of a perturbation in species concentrations along a five step pathway as a result of a pulse input at X_o. Note how the initial square pulse is smoothed out and that species further downstream experience the perturbation later in time. Tellurium script can be found at 6.1

This pathway has m floating species and n reactions ($n = m + 1$). X_o and X_1 are fixed species representing the source and sink pools, respectively. We can assume that each reaction obeys the following simple reversible mass-action kinetic law:

$$v_i = k_i s_{i-1} - k_{-i} s_i \qquad (6.2)$$

where k_i and k_{-1} are the forward and reverse rate constants, respectively. S_{i-1} is the substrate and S_i the product. Recall that the equilibrium constant for such as simple reaction is given by:

$$K_{eq} = q_i = \frac{k_i}{k_{-i}} = \frac{s_i}{s_{i-1}}$$

which means we can replace the reverse rate constant and rewrite the rate law as:

$$v_i = k_i \left(s_{i-1} - \frac{s_i}{q_i} \right) \qquad (6.3)$$

This model is simple enough that we can derive the analytical equation for the steady state flux through the pathway. One way to do this is to first start with a two-step pathway:

$$X_o \xrightarrow{v_1} S_1 \xrightarrow{v_2} X_1$$

The rates for the two steps are given by:

$$v_1 = k_1 \left(x_0 - \frac{s_1}{q_1} \right) \qquad v_2 = k_2 \left(s_1 - \frac{x_1}{q_2} \right)$$

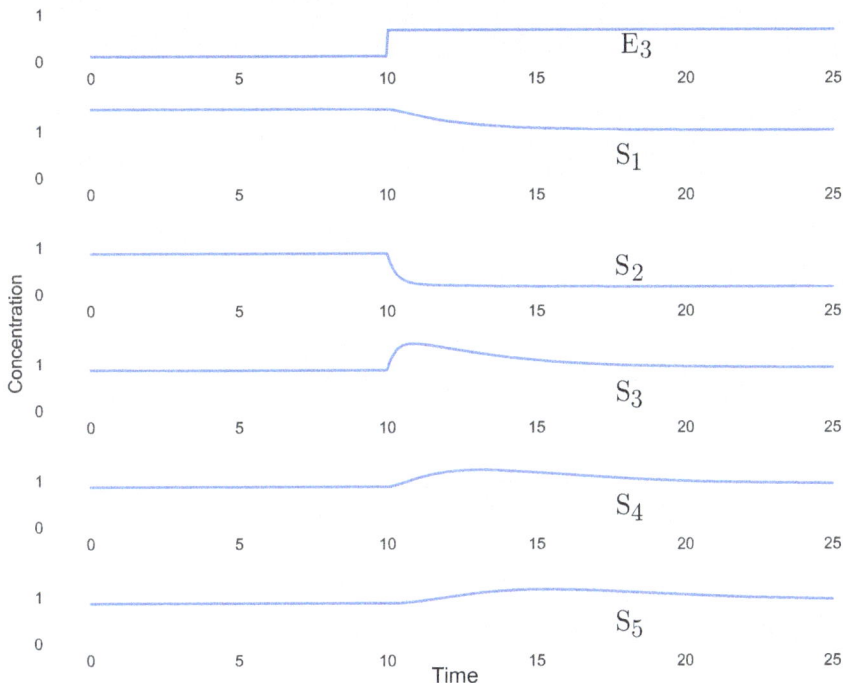

Figure 6.2 Propagation of a perturbation in species concentrations along a five step pathway as a result of a step increase in E_3. Species S_1 and S_2 decrease, all other species increase. Tellurium script can be found at 6.2

By setting $v_1 = v_2$, we can solve for the steady state concentration of S_1:

$$s_1 = \frac{q_1}{q_2} \frac{k_2 x_1 + k_1 q_2 x_o}{k_1 + k_2 q_1}$$

Inserting this solution into one of the rate laws leads to the steady state flux:

$$J = \frac{x_o q_1 q_2 - x_1}{\dfrac{1}{k_2} q_1 q_2 + \dfrac{1}{k_1} q_2} \tag{6.4}$$

We can also derive the flux equation for a three-step pathway

$$X_o \xrightarrow{v_1} S_1 \xrightarrow{v_2} S_2 \xrightarrow{v_3} X_1$$

Setting $v_1 = v_2 = v_3$, we can solve for steady state concentration of S_1 and S_2:

$$s_1 = \frac{q_1}{q_2} \frac{k_2 k_3 x_1 + k_1 k_2 x_o + k_1 k_3 q_2 q_3 x_o}{k_1 k_2 + k_1 k_3 q_2 + k_2 k_3 q_1 q_2}$$

$$s_2 = \frac{q_2}{q_3} \frac{k_1 k_3 x_1 + k_2 k_3 q_1 x_1 + k_1 k_2 q_1 q_3 x_o}{k_1 k_2 + k_1 k_3 q_2 + k_2 k_3 q_1 q_2}$$

Inserting either s_1 or s_2 into one of the reaction rate laws will produce the pathway steady state flux:

$$J = \frac{x_o \, q_1 q_2 q_3 - x_1}{\frac{1}{k_1} q_1 q_2 q_3 + \frac{1}{k_2} q_2 q_3 + \frac{1}{k_3} q_3}$$

If we compare the solutions to the two step and three step pathway we see a pattern forming from which we can deduce that the flux for a pathway of arbitrary length will be:

$$J = \frac{x_o \prod_{i=1}^{n} q_i - x_1}{\sum_{i=1}^{n} \frac{1}{k_i} \left(\prod_{j=i}^{n} q_j \right)} \tag{6.5}$$

where n is the number of steps in the linear chain of reactions. For example, if the pathway has four steps, then the steady state flux is given by:

$$J = \frac{x_o \, q_1 q_2 q_3 q_4 - x_1}{\frac{1}{k_1} q_1 q_2 q_3 q_4 + \frac{1}{k_2} q_2 q_3 q_4 + \frac{1}{k_3} q_3 q_4 + \frac{1}{k_4} q_4}$$

and so on. A pattern in the steady state concentrations equations is also evident but it is more subtle. The Chapter appendix gives the solutions to steady state species concentrations for a four-step pathway where the pattern is clearer.

The first thing to note about the flux relationship is that the flux is a function of **all** kinetic and thermodynamic parameters. There is no single parameter that determines the flux completely. This means that for a pathway with randomly assigned parameters, it is extremely unlikely to have the first step as the rate limiting step. It would require a very unlikely set of parameter values for that to occur.

From the flux expression we can compute the corresponding flux control coefficients. For this we need to differentiate the flux equation with respect to an enzyme activity-like parameter. We can use the k_i parameter as a proxy for the enzyme activity. The result of this yields the following expression for the flux control coefficient of the ith step:

$$C_i^J = \frac{\frac{1}{k_i} \prod_{j=i}^{n} q_j}{\sum_{j=1}^{n} \frac{1}{k_j} \prod_{k=j}^{n} q_k} \tag{6.6}$$

Note that the sum, $\sum C_i^J = 1$, is in accordance to the flux summation theorem. The equation also indicates that, at least in this case, the control coefficients are less than one but greater than zero, $0 \leq C_i^J \leq 1$.

For a linear chain where an increase in reactant concentration results in increases in the reaction rate and products cause reaction rates to decrease, then the flux control coefficients are limited in range between 0 and 1.0.

For a three-step pathway the flux control coefficients for each step will be given by:

$$C_1^J = \frac{1}{k_1} q_1 q_2 q_3 / D$$

$$C_2^J = \frac{1}{k_2} q_2 q_3 / D \quad \text{where } D = \frac{1}{k_1} q_1 q_2 q_3 + \frac{1}{k_2} q_2 q_3 + \frac{1}{k_3} q_3$$

$$C_3^J = \frac{1}{k_3} q_3 / D$$

Note that each term in a numerator can be found in the common denominator.

From the flux control coefficient equation (6.6), we can make some general statements. Let us assume for example that each equilibrium constant, q_i, is greater than one, $q_i > 1$, and that all forward rate constants are equal to each other, and all reverse rate constants are equal to each other. This also means that all equilibrium constants are equal. If we now take the ratio of two adjacent steps, for example the ith and $i + 1^{\text{th}}$ step, we find:

$$\frac{C_i^J}{C_{i+1}^J} = \frac{1/k_i \prod_{j=i}^{n} q_j}{1/k_{i+1} \prod_{j=i+1}^{n} q_j} = \frac{k_{i+1}}{k_i} q_i$$

Since $q_i = k_i / k_{-i}$:

$$\frac{C_i^J}{C_{i+1}^J} = \frac{k_{i+1}}{k_i} \frac{k_i}{k_{-i}} = \frac{k_{i+1}}{k_{-i}}$$

Given that we set all the forward rate constants to equal to each other and all the reverse rate constants equal equal to each other, the ratio k_{i+1}/k_{-i} must equal the equilibrium constant, q, therefore:

$$\frac{C_i^J}{C_{i+1}^J} = q \tag{6.7}$$

That is, the ratio of two adjacent control coefficients is equal to the equilibrium constant. Because we assumed that $q > 1$, it must be true that $C_i^J > C_{i+1}^J$, that is **earlier steps** will have **more** flux control. This pattern applies across the entire pathway such that steps near the beginning of a pathway will have more control than steps near the end. We will call this effect **front loading** and gives some credence to the traditional idea that the first or committed step is the most important step in a pathway. However, front loading only applies to unregulated pathways; the moment we add regulation to the pathway, this picture changes. We will consider front loading again in a later section.

As an illustration, consider a five step linear pathway and assume the equilibrium constant for each step is, $q = 2.0$. This means the ratio of adjacent flux control coefficients will be two. Taking into account the summation theorem we arrive at the following flux control coefficient values across the pathway:

Step	1	2	3	4	5
C_i^J	0.52	0.26	0.13	0.06	0.03

Another way to look at a linear pathway is via the mass-action ratio:

$$\Gamma = \frac{s_i}{s_{i-1}}$$

where the species concentrations are measured at steady state. We define the disequilibrium ratio, ρ, to be:

$$\rho = \frac{\Gamma}{K_{eq}}$$

If a step is near equilibrium, then $\rho \simeq 1$, whereas if a step is far from equilibrium, then $\rho \ll 1$.

Consider the following linear pathway where X_o and X_1 are fixed species:

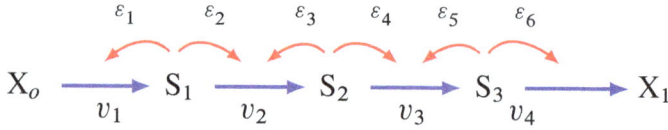

The elasticities have been labeled 1 to 6, for example ε_1 represents ε_1^1, ε_2 represents ε_1^2 etc. Considering the connectivity theorem for each metabolite, the ratios of all the flux control coefficients can be shown to be:

$$C_1^J : C_2^J : C_3^J : C_4^J =$$

$$1 : -\frac{\varepsilon_1}{\varepsilon_2} : -\frac{\varepsilon_1}{\varepsilon_2}\left(-\frac{\varepsilon_3}{\varepsilon_4}\right) : -\frac{\varepsilon_1}{\varepsilon_2}\left(-\frac{\varepsilon_3}{\varepsilon_4}\right)\left(-\frac{\varepsilon_5}{\varepsilon_6}\right)$$

or for a pathway of arbitrary length, the nth term will equal:

$$\prod_{i=1}^{n-1}\left(-\frac{\varepsilon_i}{\varepsilon_{i+1}}\right)$$

If we assume that the enzymes are operating below saturation so that they are governed by the rate law, $v_i = Vm_i/Km_i(s_{i-1} - s_i/Keq_i)$, then we can replace the substrate elasticities by $1/(1 - \rho_i)$ and the product elasticities by $-\rho_i/(1 - \rho_i)$ (See (2.9)). If we apply these substitutions, the ratios of flux control coefficients become:

$$C_1^J : C_2^J : C_3^J : C_4^J =$$

$$(1 - \rho_1) : \rho_1(1 - \rho_2) : \rho_1\rho_2(1 - \rho_3) : \rho_1\rho_2\rho_3(1 - \rho_4) \tag{6.8}$$

or for an arbitrary length pathway, the nth term is equal to:

$$\left(\prod_{i=1}^{n-1}\rho_i\right)(1 - \rho_n) \tag{6.9}$$

This is an important result, because by knowing just the equilibrium constants and the concentrations of the intermediate pools, it is possible to obtain an idea of the relative strengths of the flux control coefficients across the pathway.

> For a linear pathway without regulation and where each catalyzed reaction is operating below its substrate and product K_m, the relative distribution of flux control coefficients can be determined from the metabolite concentrations and equilibrium constants.

6.2 Product Insensitive Steps

We can draw some interesting conclusions from relation (6.9). Let us make one of the steps product insensitive and irreversible[1], say step i, so that the disequilibrium ratio for that step is zero, ($\rho_i = 0$). Since ρ_i appears as

---

[1]Whenever we indicate that a reaction is product insensitive we will also assume it is irreversible unless specified otherwise. It is possible for example for a step to be product irreversible but have a zero-order non-zero reverse rate.

a multiplier in the terms downstream of the product insensitive step, all the flux control coefficients for steps beyond will be zero. Thus steps beyond a product insensitive reaction have no control over the flux. However, steps upstream of the product insensitive step may still have control. Therefore, providing that the product insensitive step is not the first step of the pathway, a product insensitive step will not necessarily carry a control coefficient of one.

> In a linear pathway governed by **linear kinetics** and without the presence of regulatory interactions, all steps downstream of a product insensitive step, ($\rho_i = 0$), have no flux control.

Although this result was derived assuming linear kinetics, the result is more general and applies equally to steps governed by non-linear Michaelis-Menten kinetic laws or steps that show cooperativity. The more general result will be shown in a later section.

It is easy to understand why steps beyond a product insensitive step have no control. Imagine a perturbation in an enzyme activity at a step downstream of an product insensitive step. This perturbation will result in changes in metabolite concentrations upstream of the perturbed step. However, a perturbation in the concentration of the product of the product insensitive step will, by definition, have no effect on the reaction rate of the product insensitive step. This also means that reaction rates of all reaction steps upstream of the product insensitive reaction remain unchanged. It is impossible, therefore, for downstream perturbations to change the overall flux through the pathway. In the extreme case where the first step is product insensitive, the only step that has any influence on the pathway flux is the first step. All other steps have no influence. This means by the flux summation theorem that the flux control coefficient for the first step will be one, and all downstream steps zero.

6.3 Steps Close to Equilibrium

If any of the steps are near equilibrium, then the disequilibrium ratio for that step will be nearly equal to one (i.e. for step i close to equilibrium, $\rho_i \approx 1$). Under these conditions, the term $(1 - \rho_i)$ will equal approximately zero, and therefore the flux control coefficient for that step will also be near zero. In addition, steps other than step i act as if step i is not part of the pathway and the pathway appears shortened.

> In a linear pathway governed by linear kinetics and without regulation, any step that is very close to equilibrium is likely to have a flux control coefficient close to zero.

It is possible to show that the disequilibrium ratio, ρ, is equal to the ratio of the reverse and forward rates for a given reaction:

$$\rho = \frac{v_r}{v_f}$$

Since the forward rate will always be greater than the reverse rate for a pathway showing a positive net rate, the disequilibrium ratio will always be less than one:

$$\rho \leq 1$$

Because ρ is always less than one, flux control tends to be higher near the front of the pathway since downstream steps have greater multiples of ρ values that are less than one (see later section on front loading 6.5). This is another confirmation of the result given in (6.7).

Relaxation Times

For a simple decay process, the half-life is given by $\ln 2/k_1$ where k_1 is the rate constant for the process. The term $1/k_1$ is often called the **relaxation time** and gives an idea of how fast the process changes. For a reversible system such as:

$$A \rightleftharpoons B$$

where the initial concentration of $a = a_o$ and for B is zero, the change in the concentration of A as a function of time is given by:

$$a(t) = a_o e^{-(k_1+k_2)t} + \frac{k_2 T}{k_1 + k_2} \left(1 - e^{-(k_1+k_2)t}\right)$$

If $k_1 + k_2$ is large then the dynamics will be dominated by the first term in the equation and we can approximate $a(t)$ by:

$$a(t) = a_o e^{-t(k_1+k_{-1})}$$

where k_1 and k_{-1} are the forward and reverse rate contacts, respectively. The term $(k_1 + k_{-1})$ is analogous to the half-life for the simple decay process and by analogy, the reciprocal of $(k_1 + k_{-1})$ is called the **relaxation time**, usually denoted by τ:

$$\tau = \frac{1}{k_1 + k_{-1}}$$

Returning to the linear pathway in (6.1), let us assume that all the equilibrium constants are equal to one, $q_i = 1$. This means that the forward and reverse rate constants for each reaction are equal. Applying these assumptions to the flux control equation (6.6) we find that:

$$C_i^J = \frac{\dfrac{1}{k_i}}{\displaystyle\sum_{j=1}^{n} \dfrac{1}{k_j}}$$

and noting that since $k_i = k_{-1}$, then $\tau = 1/(k_i + k_{i-1}) = 1/(2k_i)$, we finally obtain:

$$C_i^J = \frac{\tau_i}{\tau_1 + \ldots + \tau_n}$$

This relation shows how a given flux control coefficient depends on the relaxation time of the particular step relative to the sum of all the relaxation times. That is, the higher the relaxation time, the larger the flux control. This result relates to the previous section where steps close to equilibrium tend to have small flux control coefficients. Steps close to equilibrium will necessarily have small relaxation times.

Although the results shown in this section and the previous section tell us that steps close to equilibrium tend to have small flux control coefficients, we must be very careful with this assertion. In all the equations that predict the values for the flux control coefficients, the one common theme is that no step can be considered in isolation. Thus although a step may be close to equilibrium, this observation must be considered in the context of all others, i.e there may be other steps that are even closer to equilibrium.

> Although a step may be close to equilibrium, the step must be considered in the context of all the others in order to determine the absolute flux control coefficient.

6.4 Saturable Enzyme Kinetics

The previous examples used linear mass-action kinetics for the individual steps. What happens if we replace linear mass-action kinetics with saturable enzymatic rate laws? In such situations we are unable to generate analytical solutions for the flux, as in equation (6.5), and since we cannot derive a flux expression, we also cannot generate control coefficient equations such as (6.6). Instead, we must use the method described in section (4.5) and derive the control coefficients in terms of the elasticities. One way to derive the control

State of v_1	$C_{e_1}^J$	$C_{e_2}^J$	$C_{e_1}^S$	$C_{e_2}^S$
Product insensitive ($\varepsilon_1^1 = 0$)	1	0	$1/\varepsilon_1^2$	$-1/\varepsilon_1^2$
Close to equilibrium ($\varepsilon_1^1 \to -\infty$)	0	1	0	-0

Table 6.1 Values of flux and concentration control coefficients given the state of v_1.

coefficients is to use the summation and connectivity theorems. Section (4.5) derived the equations for a two-step pathway:

$$C_{e_1}^J = \frac{\varepsilon_1^2}{\varepsilon_1^2 - \varepsilon_1^1} \quad C_{e_2}^J = -\frac{\varepsilon_1^1}{\varepsilon_1^2 - \varepsilon_1^1}$$

$$C_{e_1}^S = \frac{1}{\varepsilon_1^2 - \varepsilon_1^1} \quad C_{e_2}^S = -\frac{1}{\varepsilon_1^2 - \varepsilon_1^1}$$

Using these equations we can look at some extreme behaviors. For example, let us assume that the first step is *completely* insensitive to its product, S, such that $\varepsilon_1^1 = 0$. In this case, the control coefficients reduce to:

$$C_{e_1}^J = 1 \quad C_{e_2}^J = 0$$

That is, all the control (or sensitivity) is on the first step. This situation represents the classic rate-limiting step. The flux through the pathway is completely dependent on the first step. Under these conditions, no other step in the pathway can affect the flux. The effect is however dependent on the complete insensitivity of the first step to its product. Such a situation is likely to be rare in real pathways. In fact, the classic rate limiting step has almost never been observed experimentally. Instead, a range of "limitingness" is observed, with some steps having more "limitingness" (control) than others. To shift control off the first step, the strength of product inhibition must be increased.

What happens if the first step is near equilibrium? In this situation, ε_1^1 will approach $-\infty$ (see Figure 2.5) so that the first step hardly has any flux control and all the control is on the second step.

Concentration Control Coefficients for a Two-Step Pathway

A similar analysis can be carried out for the concentration control coefficients, $C_{e_1}^S$ and $C_{e_2}^S$. The following were derived in section (4.5):

$$C_{e_1}^S = \frac{1}{\varepsilon_1^2 - \varepsilon_1^1}$$

$$C_{e_2}^S = -\frac{1}{\varepsilon_1^2 - \varepsilon_1^1}$$

Table 6.1 summarises the effect of product insensitivity on the first step and whether the first reaction is close to equilibrium or not on the values of concentration control coefficients.

When the first step is product insensitivity, the concentration control coefficient become only dependent on the substrate elasticity of the second step, ε_1^2. If for example the concentration of S is roughly in the range of the K_m for second step then $\varepsilon_1^2 \approx 0.5$ and the concentration control coefficients will be 2 and -2 respectively.

If the first reaction is close to equilibrium, then C_1^S and C_2^S both tend to zero. That is both reactions, even the second reaction that is out of equilibrium will have no influence on the concentration of S.

Flux Control Coefficients for a Three-Step Pathway

What about a three-step pathway in Figure 6.3.

$$X_o \xrightarrow{\quad v_1 \quad} S_1 \xrightarrow{\quad v_2 \quad} S_2 \xrightarrow{\quad v_3 \quad} X_1$$

Figure 6.3 Three-step Pathway.

The flux control coefficient summation theorem is given by:

$$C_{e_1}^J + C_{e_2}^J + C_{e_3}^J = 1$$

Given that we have two species concentrations, S_1 and S_2, we have two connectivity theorems:

$$C_{e_1}^J \varepsilon_1^1 + C_{e_2}^J \varepsilon_1^2 = 0$$

$$C_{e_2}^J \varepsilon_2^2 + C_{e_3}^J \varepsilon_2^3 = 0$$

These three equations can be combined to give expressions that relate the flux control coefficients in terms of the elasticities, thus:

$$C_{e_1}^J = \frac{\varepsilon_1^2 \varepsilon_2^3}{\varepsilon_1^2 \varepsilon_2^3 - \varepsilon_1^1 \varepsilon_2^3 + \varepsilon_1^1 \varepsilon_2^2}$$

$$C_{e_2}^J = -\frac{\varepsilon_1^1 \varepsilon_2^3}{\varepsilon_1^2 \varepsilon_2^3 - \varepsilon_1^1 \varepsilon_2^3 + \varepsilon_1^1 \varepsilon_2^2} \qquad (6.10)$$

$$C_{e_3}^J = \frac{\varepsilon_1^1 \varepsilon_2^2}{\varepsilon_1^2 \varepsilon_2^3 - \varepsilon_1^1 \varepsilon_2^3 + \varepsilon_1^1 \varepsilon_2^2}$$

It is probably worth reminding ourselves of the signs of the various elasticities in equations (6.10). All substrate elasticities, ε_1^2, ε_2^3 will be positive, while all product elasticities ε_1^1 and ε_2^2 will be negative.

The first thing to note from these equations is that if the first step is product insensitive, that is $\varepsilon_1^1 = 0$, then $C_{e_1}^J = 1$ and $C_{e_2}^J$ and $C_{e_3}^J$ are zero (confirm this yourself). As with the two-step example, if any of the steps are close to equilibrium (compared to the other two), its flux control coefficient will be close to zero. For example, if the second step is close to equilibrium, that is, $\varepsilon_1^2 \to \infty$ and $\varepsilon_2^2 \to -\infty$, then $C_{e_2}^J \to 0$.

Now assume that each enzyme experiences a small amount of product inhibition, for example each product elasticity, ε_1^1, and ε_2^2 equals -0.1. Also assume that the substrate levels are roughly at the K_m for each enzyme. This means that each substrate elasticity will be 0.5, including ε_1^2 and ε_2^3. With these values, the flux control coefficients can be estimated, as shown in Table 6.2. Flux control is clearly biased towards the start of the pathway but some control is found in steps downstream of the first reaction. As previously mentioned, flux control that is biased towards the front of the pathway is called front loading, a topic we will discuss later in the chapter.

What happens if all three steps are close to equilibrium? At first glance it might seem that no step has flux control. We know from the previous results, steps close to equilibrium have little ability to control flux. However, every system must obey the flux summation theorem where all flux control coefficients sum to one (unless the flux is zero). The division of control in a pathway where all steps are close to equilibrium is instead decided by the relative degree of equilibrium between each step. This is something that has already been mentioned in a previous section but worth repeating here.

Step	Flux Control Coefficient (-0.1)	(-0.2)
J_1	0.806	0.64
J_2	0.161	0.26
J_3	0.032	0.1

Table 6.2 Distribution of flux control assuming weak (-0.1) and moderately weak (-0.2) product inhibition and substrate levels at the enzyme's K_m. Note that when the product inhibition is strengthened to -0.2, there is a significant shift in flux control.

> It is possible for steps close to equilibrium to have significant flux control depending on the context of the reaction.

Concentration Control Coefficients

To compute concentration control coefficients we need a different set of theorems. There are two sets of concentration control coefficients, one with respect to S_1, and another with respect to S_2. For example, if we consider the control coefficients with respect to S_2, we would use the following summation theorem:

$$C_{e_1}^{s_2} + C_{e_2}^{s_2} + C_{e_3}^{s_2} = 0$$

and the two connectivity theorems:

$$C_{e_2}^{s_2} \varepsilon_2^2 + C_{e_3}^{s_2} \varepsilon_2^3 = -1$$

$$C_{e_1}^{s_2} \varepsilon_1^1 + C_{e_2}^{s_2} \varepsilon_1^3 = 0$$

Solving for $C_{e_1}^{s_2}$, $C_{e_2}^{s_2}$ and $C_{e_3}^{s_2}$ yields:

$$C_{e_1}^{s_2} = \frac{\varepsilon_1^2}{\varepsilon_1^2 \varepsilon_2^3 - \varepsilon_1^1 \varepsilon_2^3 + \varepsilon_1^1 \varepsilon_2^2}$$

$$C_{e_2}^{s_2} = \frac{-\varepsilon_1^1}{\varepsilon_1^2 \varepsilon_2^3 - \varepsilon_1^1 \varepsilon_2^3 + \varepsilon_1^1 \varepsilon_2^2}$$

$$C_{e_3}^{s_2} = \frac{\varepsilon_1^1 - \varepsilon_1^2}{\varepsilon_1^2 \varepsilon_2^3 - \varepsilon_1^1 \varepsilon_2^3 + \varepsilon_1^1 \varepsilon_2^2}$$

Note that all the denominators are the same and positive and the numerators for $C_{e_1}^{s_2}$ and $C_{e_2}^{s_2}$ are positive, indicating that increases in e_1 or e_2 result in increases in s_2 which makes logical sense. In contrast, the numerator for $C_{e_3}^{s_2}$ is net negative indicating that increases in e_3 result in decreases in s_2. We can apply similar reasoning to derive the concentration control coefficients with respect to s_1:

$$C_{e_1}^{s_1} = \frac{\varepsilon_2^3 - \varepsilon_2^2}{\varepsilon_1^2 \varepsilon_2^3 - \varepsilon_1^1 \varepsilon_2^3 + \varepsilon_1^1 \varepsilon_2^2}$$

$$C_{e_2}^{s_1} = \frac{-\varepsilon_2^3}{\varepsilon_1^2 \varepsilon_2^3 - \varepsilon_1^1 \varepsilon_2^3 + \varepsilon_1^1 \varepsilon_2^2}$$

$$C_{e_3}^{s_1} = \frac{\varepsilon_2^2}{\varepsilon_1^2 \varepsilon_2^3 - \varepsilon_1^1 \varepsilon_2^3 + \varepsilon_1^1 \varepsilon_2^2}$$

The concentration control coefficients can be estimated using some realistic values for the elasticities. Assume that each enzyme experiences a small amount of product inhibition, such that each product elasticity, ε_1^1 and ε_2^2 are equal to -0.1. Let us also assume that the substrate levels are roughly at the K_m for each enzyme. This means that each substrate elasticity will be 0.5, including ε_1^2 and ε_2^3. Table 6.3 shows the results of the calculations.

Step	$C_i^{s_1}$	$C_i^{s_2}$
e_1	1.982	1.802
e_2	-1.802	0.180
e_3	-0.180	-1.982
Sum	0	0

Table 6.3 Distribution of concentration control assuming weak product inhibition and substrate levels at the enzyme's K_m.

The signs of the coefficients confirm what we already know, changes to enzymes downstream result in metabolites decreasing in concentration (negative coefficient), while changes in enzymes upstream of a metabolite result in increases in the metabolite.

Another important observation is that reactions close to equilibrium have little influence over the species concentrations. Consider the middle reaction, v_2. If v_2 is close to equilibrium, then its substrate elasticity, $\varepsilon_1^2 \gg 0$, and the product elasticity, $\varepsilon_2^2 \ll 0$. Under these conditions $C_{e_2}^{s_1}$ and $C_{e_2}^{s_2}$ tend to zero because the terms ε_1^2 or ε_2^2 only appears in the denominator.

> Reaction steps which are close to equilibrium have little influence over species concentrations in a linear pathway.

Heinrich and Shuster's book 'The Regulation of Cellular System' [44] also showed it is possible to express the concentration control coefficients in terms of the flux control coefficients in a linear pathway. They showed by application of the connectivity and summation theorems that:

For steps at or before i, that is $1 \leq j \leq i$:

$$C_j^{s_i} = \frac{C_j^J}{C_{i+1}^J \varepsilon_i^{i+1}} \sum_{k=i+1}^{n+1} C_k^J$$

For steps downstream of i, that is $i + 1 \leq j \leq n + 1$:

$$C_j^{s_i} = \frac{C_j^J}{C_i^J \varepsilon_i^i} \sum_{k=1}^{i} C_k^J$$

Both equations tell us that the value for a concentration control coefficient at step j is proportional to the flux control coefficient at step j. Therefore if flux control at a particular step is small, then the ability of the same step to control concentration is also diminished. Given that the denominator contains elasticity terms, a low flux control coefficient isn't however a sufficient criterion for low concentration control. The distribution of the concentration control coefficients will be considered in more detail in the next section.

6.5 Front Loading

Consider a linear pathway with linear reversible kinetics on each step where q is the equilibrium constant, $q > 1$. Given two adjacent flux control coefficients, the upstream coefficient will always be equal or larger than the downstream coefficient, that is, for the ith step, the following is true (see equation (6.7)):

$$C_i^J \geq C_{i+1}^J$$

This means that in a linear pathway, control tends to be concentrated upstream. To understand why this is the case, consider the elasticities and control equations for a linear pathway.

Using the flux summation and connectivity theorems, it is straightforward to derive the flux control equations. For example, for the three-step pathway:

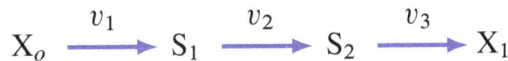

$$X_o \xrightarrow{v_1} S_1 \xrightarrow{v_2} S_2 \xrightarrow{v_3} X_1$$

one can derive the following flux control coefficient equations – see (6.10):

$$C_{e_1}^J = \varepsilon_1^2 \varepsilon_2^3 / D$$

$$C_{e_2}^J = -\varepsilon_1^1 \varepsilon_2^3 / D$$

$$C_{e_3}^J = \varepsilon_1^1 \varepsilon_2^2 / D$$

where D, the denominator, is given by:

$$D = \varepsilon_1^2 \varepsilon_2^3 - \varepsilon_1^1 \varepsilon_2^3 + \varepsilon_1^1 \varepsilon_2^2$$

It is possible to do this for pathways with additional steps, where a clear pattern emerges. For a pathway with n steps where n is even, we have the following equations:

$$C_1^J = \qquad \varepsilon_1^2 \, \varepsilon_2^3 \, \varepsilon_3^4 \, \varepsilon_4^5 \cdots \varepsilon_n^{n+1} / D$$

$$\vdots$$

$$C_m^J = \qquad \prod_{k=m}^{n} \varepsilon_k^{k+1} \prod_{k=m-1}^{1} \varepsilon_k^k / D$$

$$\vdots$$

$$C_n^J = \qquad \varepsilon_1^1 \, \varepsilon_2^2 \, \varepsilon_3^3 \, \varepsilon_4^4 \cdots \varepsilon_{n+1}^{n+1} / D$$

A careful examination of C_1^J reveals that the numerator is the product of all the substrate elasticities. This implies that a perturbation in e_1 'hops' from one enzyme to the next until it reaches the end of the pathway. Conversely, the control coefficient of the last enzyme includes all the product elasticities, that is the perturbation 'hops' from one enzyme to the next until it reaches the beginning of the pathway.

Looking at any intermediate enzyme step we find two groups of elasticities, one group representing the perturbation traveling downstream via the substrate elasticities, and the other representing the perturbation traveling upstream via product elasticities. This implies that the pattern of elasticities in the numerator reflects the path the disturbances takes [18] as it ripples out from the source of the perturbation.

> The pattern of elasticities in the numerator reflects the path taken by the disturbance.

Recall that given a reversible mass-action rate law such as $k_1 s - k_2 p$, the elasticities are given by:

$$\varepsilon_s^v = \frac{1}{1 - \rho}$$

$$\varepsilon_p^v = -\frac{\rho}{1 - \rho}$$

From these equations it follows that $\varepsilon_s^v + \varepsilon_p^v = 1$, that is:

$$\| \varepsilon_s^v \| \geq \| \varepsilon_p^v \|$$

The absolute value of the substrate elasticity is always greater than the product elasticity. Given that an upstream enzyme will have more substrate elasticities than product elasticities in the numerator of its control equation, it follows that the numerator will be larger when compared to an enzyme further downstream, which will have the smaller value product elasticities. This means that perturbations at a downstream enzyme will be attenuated compared to a similar perturbation at an upstream step. Hence the control coefficients upstream will, on average, be larger.

The origins of the asymmetry between the substrate and product elasticities is a thermodynamic one. If the thermodynamic gradient were to be reversed so that the pathway flux traveled 'upstream', the elasticity values exchange so that now the front loading occurs downstream, although 'downstream' is now 'upstream' because the flux has reversed.

The set of perturbations and the signal flow is shown in Figure 6.4.

> In a linear pathway governed by linear kinetics and without regulation, flux control is biased towards the start of the pathway, an effect called **front loading**.

To summarize, it is easier for a disturbance to travel downstream than upstream, meaning that steps upstream have more influence over the flux compared to downstream steps.

Distribution of Concentration Control

In the previous section it was shown that in an unregulated pathway, flux control tends, on average, to be concentrated on the upstream steps. What about the control of concentrations? The equations that describe the concentration control coefficients are more complex however. Consider a four-step linear pathway. A pattern in the concentration control coefficients can be discerned if we look at $C_{e_1}^{s_1}$, $C_{e_1}^{s_2}$, and $C_{e_1}^{s_3}$. For reference, these

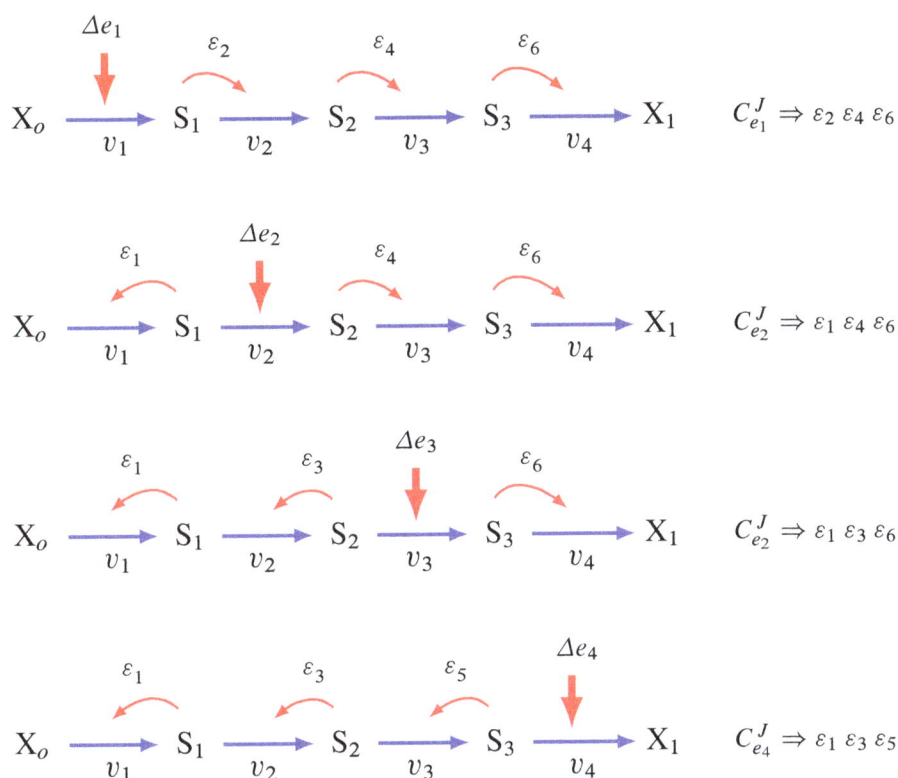

Figure 6.4 Set of four perturbation patterns. In each perturbation, the signal travels outwards up and downstream from the point of disturbance. In each case the elasticities in the numerator of the control equation indicate the strength and path of the signal transmission.

are given below:

$$C_{e_1}^{s_1} = \frac{\varepsilon_2^2 \varepsilon_3^3 - \varepsilon_2^2 \varepsilon_3^4 + \varepsilon_2^3 \varepsilon_3^4}{D}$$

$$C_{e_1}^{s_2} = \frac{\varepsilon_1^2 \varepsilon_3^4 - \varepsilon_1^2 \varepsilon_3^3}{D}$$

$$C_{e_1}^{s_3} = \frac{\varepsilon_1^2 \varepsilon_2^3}{D}$$

D is the common denominator which is a positive quantity. The number of terms in the numerator reduces by one for each species s_i, as we move downstream. Given that the denominator doesn't change, fewer terms in the numerator means smaller control coefficients. In other words, the further away a species is from a perturbation, the smaller the effect it experiences. This is intuitively reasonable because a signal, unless subject to other mechanisms, will attenuate as it propagates from the source of the disturbance. On average, for a linear unregulated pathway, we can state that:

$$C_{e_1}^{s_1} > C_{e_1}^{s_2} > C_{e_1}^{s_3} \ldots > C_{e_1}^{s_n}$$

The pattern also applies to those perturbations at the center of a pathway. Intermediates close to the disturbance will be affected more that distant ones.

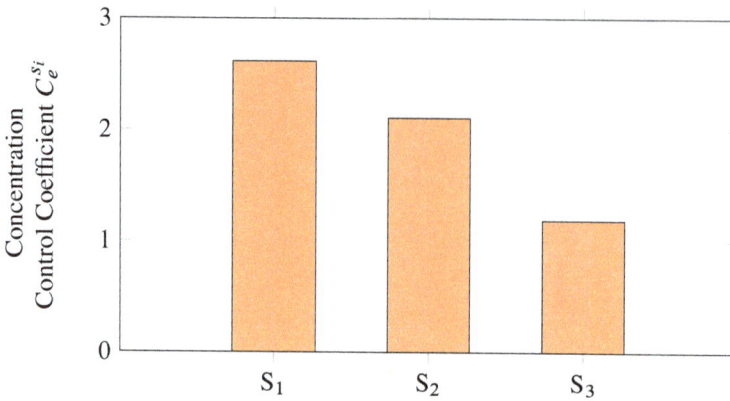

Figure 6.5 Plot showing the sensitivity of changes to downstream species as a function of a perturbation at the first step.

This result also affects the response coefficient, $R_{x_o}^{s_i}$. Because the response coefficient is the product of the x_o elasticity on v_1 and the corresponding concentration control coefficient $C_{e_1}^{s_i}$, the response coefficient of a species s_i with respect to x_o will progressively decrease as we observe species further and further away from x_o.

This effect can easily be illustrated via simulation. The elasticity values can be given random values and the corresponding control coefficients computed for a four-step pathway. This can be repeated 1000 times, each time with a different random sample of elasticities. The product elasticities can be uniformly sampled between 0 and -1, and the substrate elasticities sampled between 0 and 1. The result of the simulation is 1000 concentration control coefficients for $C_{e_1}^{s_1}$, $C_{e_1}^{s_2}$ and $C_{e_1}^{s_3}$. Taking the mean for each group of control coefficients, the following mean values are obtained:

$$C_{e_1}^{s_1} = 2.6128 \quad C_{e_1}^{s_2} = 2.1083 \quad C_{e_1}^{s_3} = 1.1886$$

The results show the influence of e_1 on a species diminishing as the species is further away from e_1.

> In a linear pathway without regulation, concentration control diminishes the further away the species is from the disturbance.

There is one caveat to the above result when considering a linear pathway with linear kinetics. For example, where each reaction is governed by the rate law:

$$v_i = k_i \left(s_{i-1} - \frac{s_i}{q_i} \right)$$

In this situation the elasticities for s_{i-1} and s_i are not free to take on any value, see (2.10), but are constrained by the relation:

$$\varepsilon_{i-1}^v + \varepsilon_i^v = 1$$

Consider the three-step pathway with concentration control coefficients:

$$C_{e_1}^{s_1} = \frac{\varepsilon_2^3 - \varepsilon_2^2}{\varepsilon_1^2 \varepsilon_2^3 - \varepsilon_1^1 \varepsilon_2^3 + \varepsilon_1^1 \varepsilon_2^2}$$

$$C_{e_1}^{s_2} = \frac{\varepsilon_1^2}{\varepsilon_1^2 \varepsilon_2^3 - \varepsilon_1^1 \varepsilon_2^3 + \varepsilon_1^1 \varepsilon_2^2}$$

Assume that the last step in the pathway is irreversible such that $\varepsilon_3^3 = 0$. Given the elasticity constraint, this means that $\varepsilon_2^3 = 1$. Also note that for the middle reaction, $\varepsilon_1^2 = 1 - \varepsilon_2^2$. Inserting these values for the elasticities into $C_{e_1}^{s_1}$ and $C_{e_1}^{s_2}$ we obtain the simpler relations:

$$C_{e_1}^{s_1} = \frac{1}{1 - \varepsilon_1^1}$$

$$C_{e_1}^{s_2} = \frac{1}{1 - \varepsilon_1^1}$$

Note the two control coefficients are identical. This generalizes to any linear pathway length that uses linear kinetics, where all $C_{e_1}^{s_j}$ will be equal to each other. The sensitivity for every species with respect to e_1 is the same. In other words, no matter how far the species is from the disturbance, the species will respond in the same way as the species closest to the disturbance. This counter intuitive behavior is because the transmission of a signal through a set of linear kinetic laws is not attenuated. For steps beyond e_1, the responses downstream of a perturbation point will be the identical. For example, all disturbances in species beyond the 3rd reaction will be identical if the perturbation occurs at the third reaction. This pattern is illustrated in the heat map shown in Figure 6.6 which shows the concentration control coefficients from a six step pathway.

In practical terms this also means that the response coefficients, $R_{x_o}^{s_1}$, $R_{x_o}^{s_2}$, ... will be equal in value. Note however that the magnitude of the disturbances traveling in upstream are different, this is a thermodynamic effect where upstream signal transmission is weaker.

If a pathway uses reversible Michaelis-Menten rate laws where the elasticities are less constrained and saturation effects attenuate the signals then species downstream will experience diminishing changes with respect to a perturbation in e_1.

6.6 Optimal Allocation of Protein

Protein synthesis constitutes a significant drain on resources in a cell [7, 58]. For example, protein synthesis consumes approximately 7.5 ATP equivalents per peptide bond. One glucose molecule yields roughly 36 molecules of ATP. Thus if the average number of peptide bonds in a protein is 300, it takes roughly 62 molecules of glucose to make just one protein molecule, not including the cost of the amino acids. In some cultured mammalian cells, protein synthesis consumes 35% to 50% of all ATP production. In addition to the energetic cost, proteins also occupy a significant proportion of cell volume at around 20 to 30% of the cell. This high level approaches the solubility limit of proteins and also limits the diffusion of other small molecules. These and other issues effectively put an upper limit on the total amount of protein in a cell. It would seem logical to assume that the distribution of a fixed amount of protein is not evenly distributed because some processes may require higher levels of protein compared to others, suggesting competition for protein between different processes. Such distributions are likely to be under evolutionary selection so that there exists an optimal allocation of the fixed amount of protein to cover all processes in the cell. The optimal allocation is also likely to shift as environmental conditions change.

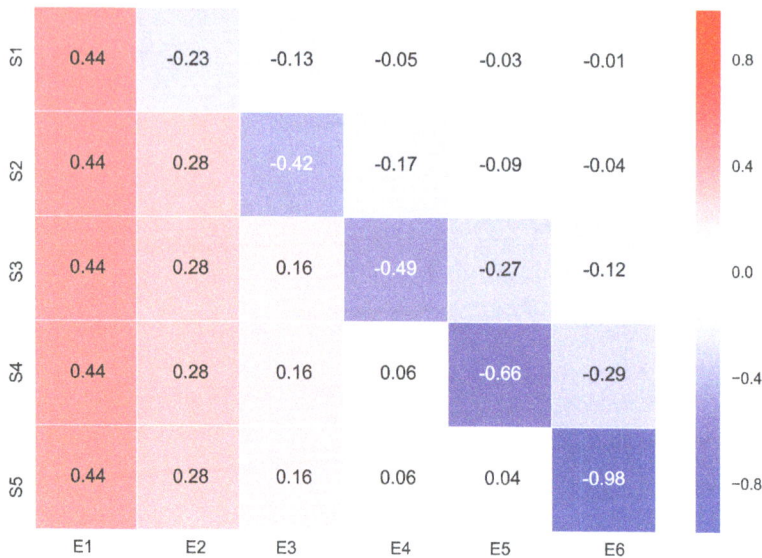

Figure 6.6 Heat map shown the values of the concentration control coefficients for all combinations of species and reaction steps for a six-step pathway using reversible linear kinetics.

In this section we will briefly consider what the optimal allocation of a fixed amount of protein in a metabolic pathways, such that the steady state pathway flux is maximized.

Consider a very simple two-step metabolic scheme shown below:

$$X_o \xrightarrow{v_1} S_1 \xrightarrow{v_2} X_1$$

Assume that the first step is catalyzed by an enzyme E_1, and the second step by enzyme, E_2. Let us reduce the amount of enzyme E_1 by a small amount, δe_1, such that the pathway flux is reduced by an amount, δJ. We can now increase the level of E_2 by δe_2 so that the pathway flux is returned to its original state. The net change in protein is therefore $\delta e_1 + \delta e_2$.

Let us also assume that the levels of E_1 and E_2 had previously been adjusted so that for a given flux, the total $e_1 + e_2$ was at a minimum, that is the distribution of protein was optimal. In other words, it would not be possible to reduce the total amount of protein and at the same time adjust the protein distribution such that the flux is unchanged. Given this, it must be true that:

$$\delta e_1 + \delta e_2 = 0$$

With these changes in E_i and the fact that the flux does not change, we can write the following:

$$C_{e_1}^J \frac{\delta e_1}{e_1} + C_{e_2}^J \frac{\delta e_2}{e_2} = \frac{\delta J}{J} = 0$$

Substituting $\delta e_1 + \delta e_2 = 0$ into the above relation yields:

$$C_{e_1}^J \frac{1}{e_1} = C_{e_2}^J \frac{1}{e_2}$$

We can now invoke the flux summation theorem to eliminate one of the control coefficients to yield:

$$C_{e_1}^J \frac{1}{e_1} = \left(1 - C_{e_1}^J\right) \frac{1}{e_2}$$

Rearranging this to solve for $C_{e_1}^J$ yields:

$$C_{e_1}^J = \frac{e_1}{e_1 + e_2}$$

This result can be generalized to any length pathway so that for a given total amount of protein and a given flux, the optimal allocation of protein at a particular step, i, is given by:

$$C_{e_i}^J = \frac{e_i}{\sum e_i}$$

Further Reading

1. Heinrich R and Rapoport TA (1974) A linear steady-state treatment of enzymatic chains. General properties, control and effector strength. Eur J Biochem. 1974 Feb 15;42(1):89-95.

2. Schuster S, Heinrich R (1996) The Regulation of Cellular Systems, Springer, ISBN 978-1-4613-1161-4 (Unfortunately this book is out of print and second-hand editions can cost as much as 300 to 400 dollars).

3. Brown, GC (1991) Total cell protein concentration as an evolutionary constraint on the metabolic control distribution in cells. J Theor Biol., 153(2), 195-203

4. Klipp E and Heinrich R (1999) Competition for enzymes in metabolic pathways:: Implications for optimal distributions of enzyme concentrations and for the distribution of flux control. Biosystems, 54, 1-14

5. Christensen, CD, Hofmeyr, JHS, and Rohwer, JM (2015). Tracing regulatory routes in metabolism using generalised supply-demand analysis. BMC systems biology, 9(1), 89.

Exercises

1. Show that summing all the C_i^J coefficients in equation (6.6) equals one.

2. Prove equation (6.8) in the main text.

3. In general, if a given enzymatic step is very close to equilibrium, what can be said about the flux control coefficient of that step?

4. Given a six-step linear pathway where the equilibrium constant for each step is 1.5 and the forward rate constants are equal to each other and reverse rate constants are equal to each other, compute the values for the flux control coefficients for each step. Hint: See (6.7).

5. In a four-step linear pathway each step is catalyzed by a reversible Michaelis-Menten rate law. In addition, each step is close to equilibrium. Does this mean that no step in the pathway can control flux? Explain your answer.

6. An unregulated linear pathway is made up of eight enzymatic reaction steps, all steps are product sensitive except for the fifth step. What can you say about the distribution of flux control in this pathway?

7. Show that the ratio of flux control coefficients in a linear pathway, such as in (6.1), where each reaction is governed by equation (6.3), is given by:

$$C_1^J : C_2^J : C_3^J : \ldots =$$

$$(x_o - s_1/q_1) : (s_1 - s_2/q_2)/q_1 : (s_2 - s_3/q_3)/(q_1 q_2) : \ldots$$

8. What is front loading?

9. Metabolic engineers wish to increase the production of an important commodity that is synthesized by a four-step metabolic pathway (6.11).

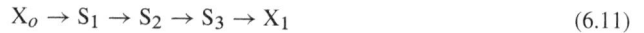

$$X_o \rightarrow S_1 \rightarrow S_2 \rightarrow S_3 \rightarrow X_1 \qquad (6.11)$$

The pathway has no known negative feedback loops. In order to obtain a rough idea of the distribution of control in this pathway, the engineers obtain values for all the standard ΔG^os and estimates for the steady state concentrations of all the metabolite pools in the pathway. ΔG^os were obtained at 25^oC. The table below shows the data they collected:

Step	ΔG^o	Metabolite	Concentration
1	-4.067 kJ mol^{-1}	X_o	5 mM
2	-6.387 kJ mol^{-1}	S_1	13.487 mM
3	11.27 kJ mol^{-1}	S_2	136.75 mM
4	-11.519 kJ mol^{-1}	S_3	0.918 mM
		X_1	0.1 mM

From the data they collected, what advice would you give concerning which step(s) are worth increasing in activity in order to increase the flux through the pathway? Assume that $R = 8.31446 \times 10^{-3}$ J K^{-1} mol^{-1}

10. Using a four-step linear pathway where each reaction uses reversible mass-action kinetics (Equation (6.3)), generate 10,000 variations of this pathway. Do this by setting the equilibrium constants to fixed values of $q_1 = 2; q_2 = 4; q_3 = 8; q_4 = 16$ and then randomizing the forward rate constant between 0 and 1.0. For each pathway variant, compute the flux control coefficients. This can be done by modulating the rate constant for each step and observing the effect on the pathway flux, or by inserting the relevant values into equation (6.6). From the 10,000 variants, compute the distribution of flux control coefficients in the pathway. Explain the distribution of control coefficients you observe.

11. Derive the concentration control coefficient equations for a three-step pathway.

12. It is known that in a given linear pathway the distribution of protein across the enzymes is optimized for flux. In this situation, what it the easiest way to estimate all the flux control coefficients?

13. Prove equation (6.4).

6.7 Appendix

Steady state concentrations for the four step pathway:

$$X_o \xrightarrow{v_1} S_1 \xrightarrow{v_2} S_2 \xrightarrow{v_3} S_3 \xrightarrow{v_4} X_1$$

are given by:

$$s_1 = \frac{q_1}{q_4} \frac{k_1 k_2 k_3 q_4 x_o + k_1 k_2 k_4 q_3 q_4 x_o + k_1 k_3 k_4 q_2 q_3 q_4 x_o + k_2 k_3 k_4 x)}{k_1 k_2 k_3 + k_1 k_2 k_4 q_3 + k_1 k_3 k_4 q_2 q_3 + k_2 k_3 k_4 q_1 q_2 q_3}$$

$$s_2 = \frac{q_2}{q_4} \frac{k_1 k_2 k_3 q_1 q_4 x_o + k_1 k_2 k_4 q_1 q_3 q_4 x_o + k_1 k_3 k_4 x_1 + k_2 k_3 k_4 q_1 x)}{k_1 k_2 k_3 + k_1 k_2 k_4 q_3 + k_1 k_3 k_4 q_2 q_3 + k_2 k_3 k_4 q_1 q_2 q_3}$$

$$s_3 = \frac{q_3}{q_4} \frac{k_1 k_2 k_3 q_1 q_2 q_4 x_o + k_1 k_2 k_4 x_1 + k_1 k_3 k_4 q_2 x_1 + k_2 k_3 k_4 q_1 q_2 x_1}{k_1 k_2 k_3 + k_1 k_2 k_4 q_3 + k_1 k_3 k_4 q_2 q_3 + k_2 k_3 k_4 q_1 q_2 q_3}$$

```
import tellurium as te
import roadrunner
import matplotlib.pyplot as plt

r = te.loada("""
    E1: $Xo -> S1; e1*(k1*Xo - k11*S1*reverse);
    E2: S1 -> S2; e2*(k2*S1 - k21*S2*reverse);
    E3: S2 -> S3; e3*(k3*S2 - k31*S3*reverse);
    E4: S3 -> S4; e4*(k4*S3 - k41*S4*reverse);
    E5: S4 -> S5; e5*(k5*S4 - k51*S5*reverse);
    E6: S5 ->;    e6*k6*S5;

    at ((time > 10) && (flag > 1)): Xo = 1.5;
    at ((time > 20) && (flag > 1)): Xo = 1;

    Xo = 1; flag = 0; reverse = 0;
    e1 = 1; e2 = 1; e3 = 1; e4 = 1; e5 = 1; e6 = 1;
    k1 = 0.5;  k2 = 0.5;  k3 = 0.5;  k4 = 0.5;  k5 = 0.5; k6 = 0.5;

    k11 = 0.8; k21 = 0.7; k31 = 0.5; k41 = 0.7; k51 = 0.55;
    k11 = 0.5; k21 = 0.5; k31 = 0.5; k41 = 0.5; k51 = 0.5;
    S1 = 1; S2 = 1; S3 = 1; S4 = 1; S5 = 1
""")

species = ['time', 'Xo', 'S1', 'S2', 'S3', 'S4', 'S5']

r.reset()
r.flag = 2
print r.dv()
m = r.simulate (0, 50, 500, species)
fig = plt.figure(1)
fig, ax = plt.subplots(figsize=(8, 7))

yposn = 0.86
for i in range (6):
    s = '61' + str (i+1)
    ax = plt.subplot(int (s))
    ax.set_ylim ([1,1.6])
    fig.text(0.7, yposn, species[i+1])
    yposn = yposn - 0.14
    plt.plot (m[:,0], m[:,i+1])
```

```
plt.xlabel('Time')
fig.text(0.02, 0.5, 'Concentration', va='center', rotation='vertical')

plt.tight_layout(rect=[0.02, 0.03, 1, 0.95])
plt.savefig ('figure.pdf')
plt.show()
```

Listing 6.1 Series of plots showing pulse moving downstream.

```
import tellurium as te
import roadrunner
import matplotlib.pyplot as plt

r = te.loada("""
    E1: $Xo -> S1;  e1*(k1*Xo - k11*S1*reverse1);
    E2: S1 -> S2;   e2*(k2*S1 - k21*S2*reverse2);
    E3: S2 -> S3;   e3*(k3*S2 - k31*S3*reverse);
    E4: S3 -> S4;   e4*(k4*S3 - k41*S4*reverse);
    E5: S4 -> S5;   e5*(k5*S4 - k51*S5*reverse);
    E6: S5 ->;      e6*k6*S5;

    at ((time > 10) && (flag > 1)): e3 = 6.5;

    Xo = 1; flag = 0; reverse = 0; reverse1 = 0.1; reverse2 = 0.7
    e1 = 1; e2 = 1; e3 = 1; e4 = 1; e5 = 1; e6 = 1;
    k1 = 0.5;  k2 = 0.5;  k3 = 0.5;  k4 = 0.5;  k5 = 0.5; k6 = 0.5;

    k11 = 0.8; k21 = 0.7; k31 = 0.5; k41 = 0.7; k51 = 0.55;
    k11 = 0.5; k21 = 0.5; k31 = 0.5; k41 = 0.5; k51 = 0.5;
    S1 = 1; S2 = 1; S3 = 1; S4 = 1; S5 = 1
""")

species = ['time', 'e3', 'S1', 'S2', 'S3', 'S4', 'S5']

r.simulate (0, 100, 100)
r.flag = 2
print r.dv()
m = r.simulate (0, 25, 500, species)
fig = plt.figure(1)
#plt.figure(figsize=(8,8))
fig, ax = plt.subplots(figsize=(8, 7))

yposn = 0.855
for i in range (6):
    s = '61' + str (i+1)
    ax = plt.subplot(int (s))
    ax.set_ylim ([0,1.5])
    fig.text(0.7, yposn, species[i+1])
    yposn = yposn - 0.145
    if i == 0:
```

```
      plt.plot (m[:,0], m[:,i+1]/10)
   else:
      plt.plot (m[:,0], m[:,i+1])

plt.xlabel('Time')
fig.text(0.02, 0.5, 'Concentration', va='center', rotation='vertical')

plt.tight_layout(rect=[0.02, 0.03, 1, 0.95])
plt.savefig ('figure.pdf')
plt.show()
```

Listing 6.2 Series of plots showing pulse moving out from a middle perturbation.

7

Branched and Cyclic Systems

In this chapter we will review branched and cyclic systems. Moiety conserved cycles will be treated in a separate chapter.

7.1 Branched Pathways

Branching structures are one of the most common patterns in biochemical networks. Even a pathway such as glycolysis, often depicted as a straight chain in textbooks, is in fact a highly branched pathway.

At any given branch node, where a node is a molecular species, there will be conservation of mass. Given a node species, s_i, with b branches entering the node and d branches leaving, the net rate of change in concentration of s_i is:

$$\sum_{i=1}^{b} v_i - \sum_{j=1}^{d} v_j = \frac{ds_i}{dt}$$

At steady state when $ds_i/dt = 0$, it must therefore be true that:

$$\sum_{i=1}^{b} v_i = \sum_{j=1}^{d} v_j$$

For example consider the simple branched pathway shown in Figure 7.1. J_1, J_2 and J_3 are the steady state fluxes. By the law of conservation of mass, the fluxes in each limb, at steady state will be governed by the relationship:

$$J_1 - (J_2 + J_3) = 0 \qquad \text{or} \qquad J_1 = J_2 + J_3$$

Let E_1, E_2, and e_3 correspond to the concentration of enzyme at each step in the branched pathway. We can therefore define control coefficients, for example $C_{e_1}^{J_1}$ which corresponds to the influence that E_1 has on the flux J_1 or $C_{e_2}^{s}$ which corresponds to the influence enzyme E_2 has on the concentration of the intermediate S. In total there will be three concentration control coefficients $C_{e_1}^{s}$, $C_{e_2}^{s}$, and $C_{e_3}^{s}$. For the flux control coefficients there will be one set of three flux control coefficients for each flux in the branched pathway. Given that there are three fluxes, it must mean there are nine flux control coefficients in total. Table 7.1 lists all twelve control

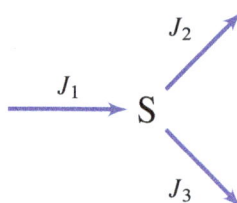

Figure 7.1 Simple branched pathway. This pathway has three different fluxes, J_1, J_2, and J_3, which at steady state are constrained by $J_1 = J_2 + J_3$.

$C_{e_1}^{J_1}$	$C_{e_1}^{J_2}$	$C_{e_1}^{J_3}$	$C_{e_1}^{s}$
$C_{e_2}^{J_1}$	$C_{e_2}^{J_2}$	$C_{e_2}^{J_3}$	$C_{e_2}^{s}$
$C_{e_3}^{J_1}$	$C_{e_3}^{J_2}$	$C_{e_3}^{J_3}$	$C_{e_3}^{s}$

Table 7.1 Set of all control coefficients for the simple branch pathway shown in Figure 7.1.

coefficients. For pathways with more complex branching the number of flux control coefficients increases further.

Each set of three flux control coefficients must obey the flux summation theorem:

$$C_{e_1}^{J_1} + C_{e_2}^{J_1} + C_{e_3}^{J_1} = 1$$

$$C_{e_1}^{J_2} + C_{e_2}^{J_2} + C_{e_3}^{J_2} = 1$$

$$C_{e_1}^{J_3} + C_{e_2}^{J_3} + C_{e_3}^{J_3} = 1$$

Likewise there will also be three connectivity theorems that must be obeyed:

$$C_{e_1}^{J_1} \varepsilon_s^{v_1} + C_{e_2}^{J_1} \varepsilon_s^{v_2} + C_{e_3}^{J_1} \varepsilon_s^{v_3} = 0$$

$$C_{e_1}^{J_2} \varepsilon_s^{v_1} + C_{e_2}^{J_2} \varepsilon_s^{v_2} + C_{e_3}^{J_2} \varepsilon_s^{v_3} = 0$$

$$C_{e_1}^{J_3} \varepsilon_s^{v_1} + C_{e_2}^{J_3} \varepsilon_s^{v_2} + C_{e_3}^{J_3} \varepsilon_s^{v_3} = 0$$

If we consider one set of matching pairs such as:

$$C_{e_1}^{J_1} + C_{e_2}^{J_1} + C_{e_3}^{J_1} = 1$$

$$C_{e_1}^{J_1} \varepsilon_s^{v_1} + C_{e_2}^{J_1} \varepsilon_s^{v_2} + C_{e_3}^{J_1} \varepsilon_s^{v_3} = 0$$

there are three unknowns, $C_{e_1}^{J_1}, C_{e_3}^{J_1}$, and $C_{e_3}^{J_1}$ but only two equations. To solve for $C_{e_i}^{J_1}$, we need another equation.

Let the fraction of flux through J_2 be given by $\alpha = J_2/J_1$, and the fraction of flux through J_3 be given by $1 - \alpha = J_3/J_1$. Let us carry out the following thought experiment:

1. Increase the concentration of E_2 by δe_2. This will cause a decrease in S, an increase in J_1 (relief of product inhibition) and a decrease in J_3.

2. Restore the change in J_1 by decreasing S_3 such that S is restored to its pre-perturbation state. At the end of the thought experiment, $\delta s = 0$.

3. Since we have not changed E_1 and $\delta s = 0$, it must be the case that $\delta J_1 = 0$.

From this experiment we can write down the system and local equations. The system equation is given by:

$$C_{e_2}^{J_1} \frac{\delta e_2}{e_2} + C_{e_3}^{J_1} \frac{\delta e_3}{e_3} = \frac{\delta J_1}{J_1} = 0$$

Note that the system equation only has two terms because we did not change E_1. The local equations are quite simple because $\delta s = 0$ and as before we assume that $\varepsilon_{e_i}^v = 1$, therefore:

$$\frac{\delta v_2}{v_2} = \frac{\delta e_2}{e_2} \quad \text{and} \quad \frac{\delta v_3}{v_3} = \frac{\delta e_3}{e_3}$$

By substitution, the system equation can be written as:

$$C_{e_2}^{J_1} \frac{\delta v_2}{v_2} + C_{e_3}^{J_1} \frac{\delta v_3}{v_3} = 0$$

Since $\delta J_1 = 0$, it must be the case that the net change in flux downstream of S must also be zero. That is, $\delta v_2 + \delta v_3 = 0$, or $\delta v_2 = -\delta v_3$. We can therefore eliminate the δv_3 term:

$$C_{e_2}^{J_1} \frac{\delta v_2}{v_2} - C_{e_3}^{J_1} \frac{\delta v_2}{v_2} \frac{v_2}{v_3} = 0$$

Canceling terms we obtain:

$$C_{e_2}^{J_1} - C_{e_3}^{J_1} \frac{v_2}{v_3} = 1$$

We can substitute the absolute rates, v_2 and v_3 with the fractional rates, α and $1 - \alpha$ to give:

$$C_{e_2}^{J_1} - C_{e_3}^{J_1} \frac{\alpha}{1 - \alpha} = 0$$

One final rearrangement yields:

$$C_{e_2}^{J_1} (1 - \alpha) - C_{e_3}^{J_1} \alpha = 0$$

This result is called the **flux branch point theorem**. We can derive similar theorems with respect to J_2 and J_3. In each case we carry out the same thought experiment such that the reference flux, J_2 or J_3, is unchanged. The two additional theorems are given below with respect to J_2 and J_3.

$$C_{e_1}^{J_2} (1 - \alpha) + C_{e_3}^{J_2} = 0$$

$$C_{e_1}^{J_3} \alpha + C_{e_2}^{J_3} = 0$$

We can also derive, using the same thought experiment, branch point theorems with respect to the species concentration, S if we only perturb E_2 and E_1. This time the system equation is:

$$C_{e_2}^{s} \frac{\delta e_2}{e_2} + C_{e_3}^{s} \frac{\delta e_3}{e_3} = \frac{\delta s}{s} = 0$$

Substituting in the same local equations as before and noting that $\delta v_2 = -\delta v_3$, we obtain after some rearrangement:

$$C_{e_2}^{s} (1 - \alpha) + C_{e_3}^{s} \alpha = 0$$

This result is known as the **concentration branch point theorem** and it is very similar to the flux branch point theorem. There are also a set of variants that correspond to the concentration branch theorems for changes to E_1 and e_3 and E_1 and E_2:

$$C_{e_1}^s(1-\alpha) + C_{e_3}^s = 0$$
$$C_{e_1}^s\alpha + C_{e_2}^s = 0$$

We can write out the theorems in matrix form (See equation (4.18)) using the theorems expressed in terms of J_2; this includes one summation, one connectivity and one branch theorem [32]:

$$
\begin{bmatrix} C_{e_1}^{J_2} & C_{e_2}^{J_2} & C_{e_3}^{J_2} \\ C_{e_1}^s & C_{e_2}^s & C_{e_3}^s \end{bmatrix}
\begin{bmatrix} 1 & -\varepsilon_1^1 & 0 \\ 1 & -\varepsilon_1^2 & 1-\alpha \\ 1 & -\varepsilon_1^3 & 1 \end{bmatrix}
=
\begin{bmatrix} 1 & 0 & 0 \\ 0 & 1 & 0 \end{bmatrix}
$$

Now solve for the control coefficient matrix by rearranging:

$$
\begin{bmatrix} C_1^{J_2} & C_2^{J_2} & C_3^{J_2} \\ C_1^s & C_2^s & C_3^s \end{bmatrix}
=
\begin{bmatrix} 1 & 0 & 0 \\ 0 & 1 & 0 \end{bmatrix}
\begin{bmatrix} 1 & -\varepsilon_1^1 & 0 \\ 1 & -\varepsilon_1^2 & 1-\alpha \\ 1 & -\varepsilon_1^3 & 1 \end{bmatrix}^{-1}
$$

Inverting the matrix will yield the algebraic equations for $C_{e_1}^{J_2}$, $C_{e_2}^{J_2}$, and $C_{e_3}^{J_2}$ (7.1). In the following, the notation has been simplified by setting $\varepsilon_1 = \varepsilon_s^1, \varepsilon_2 = \varepsilon_s^2$, and $\varepsilon_3 = \varepsilon_s^3$.

The first thing to note in equations (7.1) and (7.2) is that the common denominator, $\varepsilon_2\alpha + \varepsilon_3(1-\alpha) - \varepsilon_1$ is **positive**, therefore the following inequalities hold given that $\varepsilon_1 < 0, \varepsilon_2 > 0$ and $\varepsilon_3 > 0$:

$$C_{e_1}^{J_2} = \frac{\varepsilon_2}{\varepsilon_2\alpha + \varepsilon_3(1-\alpha) - \varepsilon_1} > 0$$

$$C_{e_2}^{J_2} = \frac{\varepsilon_3(1-\alpha) - \varepsilon_1}{\varepsilon_2\alpha + \varepsilon_3(1-\alpha) - \varepsilon_1} > 0 \qquad (7.1)$$

$$C_{e_3}^{J_2} = \frac{-\varepsilon_2(1-\alpha)}{\varepsilon_2\alpha + \varepsilon_3(1-\alpha) - \varepsilon_1} < 0$$

And for the concentration control coefficients:

$$C_{e_1}^s = \frac{1}{\varepsilon_2\alpha + \varepsilon_3(1-\alpha) - \varepsilon_1} > 0$$

$$C_{e_2}^s = \frac{-\alpha}{\varepsilon_2\alpha + \varepsilon_3(1-\alpha) - \varepsilon_1} < 0 \qquad (7.2)$$

$$C_{e_3}^s = \frac{-(1-\alpha)}{\varepsilon_2\alpha + \varepsilon_3(1-\alpha) - \varepsilon_1} < 0$$

With respect to the concentration control coefficient, note that C_1^s is positive while the two branch coefficients, C_2^s and C_3^s are negative. This is expected. The degree to which each of the output branches affects the concentration is also in proportion to the amount of flux carried by the branch. This means that a branch that only carries a small amount of flux relative to J_1 will have little effect on the branch species concentration.

Both flux control coefficients, $C_1^{J_2}$ and $C_2^{J_2}$, are positive which we would expect. The flux control coefficient, $C_3^{J_2}$ however is negative, indicating that changes in the activity of E_3 decreases the flux in the other limb, J_2. This means there is **competition** between the output branch for flux. If one branch becomes more active, then it can 'steal' flux from the other branch. The amount stolen will depend on the various kinetic properties of the branch enzymes.

To answer what determines the competition between the output branches, we must look at the control equations in more detail. In particular, we must look at how the distribution of control is affected by different flux distributions, and the kinetics of the branch enzymes. In the following analysis, J_2 will be the flux we observe as a result of perturbations to the enzymes in the branched pathway.

Most Flux Through J_3

The first situation to consider is the case when most of the flux moves along J_3 and only a small amount goes through the upper limb J_2, that is, $\alpha \to 0$ and $1 - \alpha \to 1$, see Figure 7.2(b). Let us examine how the small amount of flux through J_2 is influenced by the two branch limbs, E_2 and E_3.

As $\alpha \to 0$ and $1 - \alpha \to 1$, then:

$$C_{e_2}^{J_2} \quad \to \quad \frac{\varepsilon_1 - \varepsilon_3}{\varepsilon_1 - \varepsilon_3} = 1$$

$$C_{e_3}^{J_2} \quad \to \quad \frac{\varepsilon_2}{\varepsilon_1 - \varepsilon_3}$$

The first thing to note is that E_2 tends to acquire proportional influence over its own flux, J_2. Since J_2 only carries a very small amount of flux, any changes in E_2 will have little effect on S, hence the flux through E_2 is almost entirely governed by the activity of E_2. Because of the flux summation theorem and the fact that $C_{e_2}^{J_2} = 1$, it means that the remaining two coefficients must be equal and opposite in value. Since $C_{e_3}^{J_2}$ is negative, $C_{e_1}^{J_2}$ must be positive.

Unlike a linear pathway, the values for $C_{e_2}^{J_2}$ and $C_{e_1}^{J_2}$ are not bounded between zero and one. Depending on the values of the elasticities, it is possible for the control coefficients in a branched system to **greatly exceed** one [53, 62].

It is also possible to arrange the kinetic constants so that every step in the branch with respect to J_2 has a control coefficient of unity (one of which must be -1 in order to satisfy the summation theorem). We could therefore claim that **every step** in the pathway is a rate limiting step with respect to J_2. This clearly shows again that rate limitation is not a simple concept as traditionally thought.

> In a branched pathway it is possible to arrange the kinetic constants of the enzymes such that the feed branch has a flux control coefficient of +1, one of the output branch a coefficient of -1, and the other output branch a coefficient of +1. That is, **every step** in the pathway is equally rate limiting.

It is also possible to arrange the kinetic constants in the pathway such that the flux coefficients for E_1 and E_3 are much greater than one. This effect has been termed **ultrasensitivity** [62]. The Tellurium script 7.1 in the chapter Appendix illustrates a branched pathway with control coefficients over 8.0. Table 7.2 shows the results from the Tellurium script simulation. Note that $C_{e_1}^{J_2}$ is 8.34. That is a 1% increase in E_1 will result in an 8% increase in J_2.

The explanation for these high control coefficients is straightforward. Any changes in the two limbs that carry the high flux will have an adverse effect on the very small flux that is carried by J_2. Imagine a small stream coming off a large river. Any flooding in the large river is likely to have a huge impact on the small stream.

> In a branched pathway it is possible to arrange the kinetic constants of the enzymes such the flux control coefficients in the feed and output branch can greatly **exceed one**.

Control Coefficient	Value
$C_{e_1}^{J_2}$	8.34
$C_{e_2}^{J_2}$	0.99
$C_{e_3}^{J_2}$	-8.51

Table 7.2 Results showing high flux control coefficients in a simple branch model, see Tellurium script 7.1.

Other than with an asymmetric distribution of flux, the ability to achieve high flux sensitivity at a branch point depends on the relative values of the elasticities. For example, increasing the value ε_2 relative to ε_3 increases the sensitivity of the branch point. This could be achieved in a number of ways:

1. E_2 can show positive cooperativity with respect to the branch species. That is, any changes in E_3 become amplified through E_2.

2. v_3 is operating in a more saturated regime compared to v_2. This will make ε_3 smaller than ε_2 and amounts to ensuring that the K_m for v_2 is higher than the K_m of v_3.

3. Product inhibition on v_1 is very small.

Most Flux Through J_2

Let us now consider the other extreme, that is when most of the flux is through J_2. In other words, $\alpha \to 1$ and $1 - \alpha \to 0$, see Figure 7.2(a). Under these conditions the control coefficients become:

$$C_{e_2}^{J_2} \quad \to \quad \frac{\varepsilon_1}{\varepsilon_1 - \varepsilon_2}$$
$$C_{e_3}^{J_2} \quad \to \quad 0$$

In this situation the pathway has effectively become a simple linear chain. The influence of E_3 on J_2 is negligible. By analogy, changing the flow of water in a small stream that comes off a large river will have a negligible effect on the rate of flow in the large river.

Figure 7.2 summarizes the changes in sensitivities at a branch point.

Derivation by Implicit Differentiation

As was previously described in Chapter 3, we can also compute the control coefficients for a branched system by implicit differentiation. Start by writing out the rate of change of S at steady state for a simple branch:

$$\frac{ds}{dt} = v_1 - v_2 - v_3 = 0$$

Assuming we wish to compute the coefficients with respect to E_1, we can write the equation as:

$$0 = v_1(s(e_1), e_1) - v_2(s(e_1)) - v_3(s(e_1))$$

Differentiating with respect to E_1 gives:

$$0 = \frac{\partial v_1}{\partial s}\frac{ds}{de_1} + \frac{\partial v_1}{\partial e_1} - \frac{\partial v_2}{\partial s}\frac{ds}{de_1} - \frac{\partial v_3}{\partial s}\frac{ds}{de_1}$$

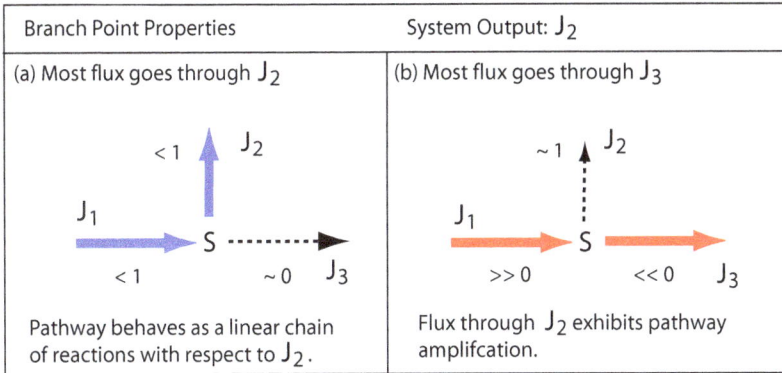

Branch Point Properties	System Output: J_2
(a) Most flux goes through J_2	(b) Most flux goes through J_3
Pathway behaves as a linear chain of reactions with respect to J_2.	Flux through J_2 exhibits pathway amplifcation.

Figure 7.2 The figure shows two flux extremes relative to the flux through branch J_2. In case (a) where most of the flux goes through J_2, the branch reverts functionally to a simple linear sequence of reactions comprised of J_1 and J_2. In case (b), where most of the flux goes through J_3, the flux through J_2 now becomes very sensitive to changes in activity at J_1 and J_3. Given the right kinetic settings, the flux control coefficients can become 'ultrasensitive' with values greater than one (less than minus one for activity changes at J_3). The values next to each reaction indicates the flux control coefficient for the flux through J_2 with respect to activity at the reaction.

Scaling, setting $\varepsilon_{e_1}^1 = 1$, and solving for $C_{e_1}^s$ yields:

$$C_{e_1}^s = \frac{1}{\varepsilon_2 \alpha + \varepsilon_3 (1 - \alpha) - \varepsilon_1}$$

where as before $\alpha = J_2/J_1$. This is the same equation as derived previously. The control coefficients for E_2 and E_3 can be derived in a similar manner.

7.2 Futile or Substrate Cycles

Closely related to branched systems are cyclic pathways. A typical cyclic pathway is shown in Figure 7.3. For cycling to occur, both forward and back reactions must operate. It is typical to find that the forward and reverse reactions are chemically distinct. Often one reaction will be driven by ATP, while the other by hydrolysis of phosphate groups. Typical examples in metabolism include the cycle between glucose and glucose-6-phosphate, and the cycling between fructose-6-phosphate and fructose 1,6-bisphosphate. Such cycles have often been called futile cycles (or perhaps more accurately substrate cycles) because of the expenditure of free energy (as ATP) without any apparent benefit. A number of suggestions have been put forth to rationalize this apparent waste of energy. These include heat production, control of flux direction, metabolite buffering, and more sensitive control of the net flux through the pathway. We will only consider the later here.

Sensitivity Control

Figure 7.3 shows a typical cyclic pathway embedded in a linear chain. Of interest is the sensitivity of the pathway rate, v_1 or v_4 to changes in v_2. The simplest assumption to make is that when we change v_2, there is no change in back rate, v_3. This could be for a number of reasons, for example v_3 is saturated by its substrate

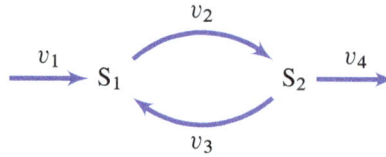

Figure 7.3 Cyclic Pathway.

S_2. One of the interesting properties of a substrate cycle is when the cycling rate is much higher compared to the net through flux. In Figure 7.4 we can see that the cycling flux is tens times higher than the through flux. Figure 7.4 illustrates two situations – a reference state in (a), and a perturbation of 5% to v_2 in (b). Assuming that all flux changes appear in output flux v_4 and that v_3 is not changed, the percentage change in v_4 (or v_1) is 100%, a twenty fold amplification. In other words, small changes in forward cycling rate can lead significant increases in the through rate.

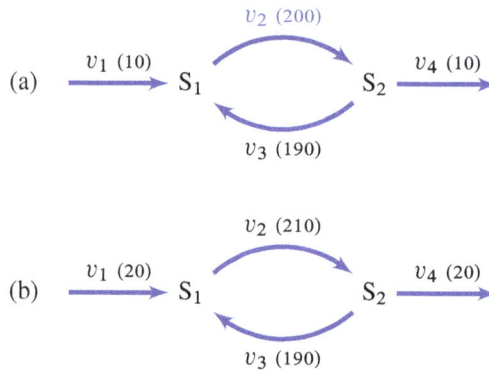

Figure 7.4 Amplification in a substrate cycle. (a) reference state, values refer to fluxes at various points, note that $v_1 = v_2 - v_3$. (b) activation of v_2 by 5% leads to a 100% change in v_1 and v_4. It assumes that v_3 is not activated by any changes in S_2.

This effect can be easily quantified as follows. First, we note that the flux constraint is:

$$v_1 = v_2 - v_3$$

We then assume that a perturbation in v_2 leads to the same change in v_1, that is:

$$\delta v_2 = \delta v_1$$

We can now compute the fractional changes in v_1 and v_2 as:

$$\frac{\delta v_1}{v_1} = \frac{\delta v_2}{v_2} \frac{v_2}{v_1}$$

The degree of amplification is then given by:

$$\frac{\delta v_1 / v_1}{\delta v_2 / v_2} = \frac{v_2}{v_1}$$

Since $v_2 = v_1 + v_3$ then:

$$\frac{\delta v_1 / v_1}{\delta v_2 / v_2} = \frac{v_1 + v_3}{v_1} = 1 + \frac{v_3}{v_1} \tag{7.3}$$

This result shows that the higher the cycling rate (v_3) compared to the through flux, the greater the amplification. This equation gives us the maximum degree of amplification possible. In practice, v_3 will not remain unchanged because S_2 rises. In addition, S_1 will fall due to higher consumption which will reduce v_2 but increase v_1 due to lower product inhibition. The resulting amplification is therefore a more complicated function than suggested by equation (7.3). However equation (7.3) gives the maximum possible amplification.

To carry out a more detailed analysis, we must turn to metabolic control analysis. We can examine the flux control coefficient for $C_2^{J_1}$ (See [32] for derivation):

$$C_2^{J_1} = \frac{\varepsilon_1^1 \varepsilon_2^4 (1 + v_3/v_1)}{D}$$

$$D = \varepsilon_1^1 \varepsilon_2^4 - \left(1 + \frac{v_3}{v_1}\right)\left(\varepsilon_1^1 \varepsilon_2^2 + \varepsilon_2^4 \varepsilon_1^2\right) + \frac{v_3}{v_1}\left(\varepsilon_1^1 \varepsilon_2^3 + \varepsilon_2^4 \varepsilon_1^3\right)$$

Simplify this equation by assuming that there is little or no product inhibition from S_2 on to v_2 and E_1 on to v_3. This means that $\varepsilon_1^3 = 0$ and $\varepsilon_2^2 = 0$. If we also multiply top and bottom by v_1, and using the relation $v_1 + v_3 = v_2$, the control equation can be simplified to:

$$C_2^{J_1} = \frac{\varepsilon_1^1 \varepsilon_2^4 v_2}{D}$$

$$D = \varepsilon_1^1 \varepsilon_2^4 v_1 - \varepsilon_2^4 \varepsilon_1^2 v_2 + \varepsilon_1^1 \varepsilon_2^3 v_3$$

Two things to note immediately from this equation. There must be product inhibition on the first step, ε_1^1, in order to get any sensitivity. If ε_1^1 is zero then so is $C_2^{J_1}$. This is because all control is now on the first step. This highlights again the danger of using rate laws in models that are product insensitive because the use of such rate laws often give misleading or trivial results of no real interest. The second relatively simple statement to make from the above equation is the importance of ε_2^3. This elasticity is the activation of the reverse arm with respect to S_2. The larger this elasticity, the smaller the degree of amplification. This is expected because any flux that flows back along the reverse cycle instead of into v_4, reduces the potential amplification factor. To analyze the equation further we can make additional simplifications.

We know that sensitivity increases when the cycling rate increases relative to the main flux, v_1 and v_4. If v_2 and v_3 are much greater than v_1, we can simplify the equation further to:

$$C_2^{J_1} = \frac{v_2}{v_3 \varepsilon_2^3 / \varepsilon_2^4 - v_2 \varepsilon_1^2 / \varepsilon_1^1}$$

If the cycling rate is so high that v_2 and v_3 are almost indistinguishable, then we can see that maximal sensitivity is achieved when:

$$\frac{\varepsilon_2^3}{\varepsilon_2^4} + \frac{\varepsilon_1^2}{\varepsilon_1^1} \ll 1$$

This tells us that substrate activation of v_4 by S_2 should be stronger than substrate activation of S_2 on v_3. Secondly, the product inhibition of S_1 on v_1 must be stronger than substrate activation of S_1 on v_2. If we think about this in a thought experiment, these results are expected.

The requirements for amplification in substrate cycles is fairly complicated and questions remain whether real pathways use this mechanism *in vivo*. At this point we leave the topic of branches and cycles. In a subsequent chapter we will consider the dynamic properties of conserved cycles.

Exercises

1. Given the simple branch in Figure 7.1, prove the following theorems:

$$C_{e_1}^{J_2}(1 - \alpha) + C_{e_3}^{J_2} = 0$$

$$C_{e_1}^{J_3}\alpha + C_{e_2}^{J_3} = 0$$

2. Prove that the following two theorems are true for the branch point in Figure 7.1:

$$C_{e_1}^{S}(1 - \alpha) + C_{e_3}^{S} = 0$$

$$C_{e_1}^{S}\alpha + C_{e_2}^{S} = 0$$

3. Create a simulation of a simple branched system and arrange the rate constants so that one of the branches is hyper sensitive to changes in the other branch.

4. Derive the flux branch theorems for the following multibranched system:

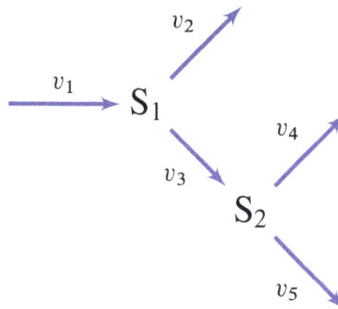

Figure 7.5 Multi-Branched Pathway.

Appendix

```
import tellurium as te
import numpy as np
r = te.loada ('''
  var S;
  ext Xo, w;
  J1: $Xo -> S; Vm1/Km1*(Xo-S/Keq)/(1+Xo/Km1+S/Km2);
  J2: S -> $w; Vm2*S^4/(Km3+S^4);
  J3: S -> $w; Vm3*S/(Km4+S);

  Xo = 9;
  S = 0.2;
  Vm1 = 1.4;
  Km1 = 0.4;
  Keq = 4.5;

  Km2 = 0.6;
  Vm2 = 0.05;
```

```
  Km3 = 0.8;
  Vm3 = 2.3;
  Km4 = 0.3;
''')

r.steadyState()
print "Flux Control Coefficients:";
print r.getCC ('J2', 'Vm1');
print r.getCC ('J2', 'Vm2');
print r.getCC ('J2', 'Vm3');
print "Elasticities:";
e1 = r.getEE ('J1', 'S');
e2 = r.getEE ('J2', 'S');
e3 = r.getEE ('J3', 'S');
print e1, e2, e3;
print "Fluxes: ", r.J1, r.J2, r.J3;
```

Listing 7.1 Simple Branched Pathway showing Flux Amplification.

8

Negative Feedback

8.1 Historical Background

Feedback is widespread in biochemical networks and physiological systems. Some form of feedback permeates almost every known biological process. On the face of it, feedback is a simple process that involves sending a portion of the output back to the input. If the portion sent back reduces the input, then the feedback is called negative feedback. Otherwise, it is called positive feedback.

> Negative feedback is where part of the output of a system is used to reduce the magnitude of the system input.

Water Clocks

The concept of feedback control goes back at least as far as the Ancient Greeks [66]. Of some concern to the ancient Greeks was the need for accurate time keeping. In about 270 BC the Greek Ktesibios invented a float regulator for a water clock. The role of the regulator was to keep the water level in a tank at a constant depth. This constant depth yielded a constant flow of water through a tube at the bottom of the tank which filled a second tank at a constant rate. The level of water in the second tank thus depended on time elapsed.

Philon of Byzantium in 250 BC [23] is known to have kept a constant level of oil in a lamp using a float regulator and in the first century AD, Heron of Alexandria experimented with float regulators for water clocks. Philon and particularly Heron (13 AD) [75] have left us with an extensive book (Pneumatica) detailing many amusing water devices that employed negative feedback.

Governors

It wasn't until the industrial revolution that feedback control, or devices for automatic control, became economically important. Probably the most famous modern device that employed negative feedback was the governor. Thomas Mead in 1787 took out a patent on a device that could regulate the speed of windmill sails. His idea was to measure the speed of the mill by the centrifugal motion of a revolving pendulum and use this to regulate the position of the sail. Very shortly afterwards in early 1788, James Watt is told of this device in a letter from his partner, Matthew Boulton. Watt recognizes the utility of the governor as a device to regulate the new steam engines that were rapidly becoming an important source of new power for the industrial revolution.

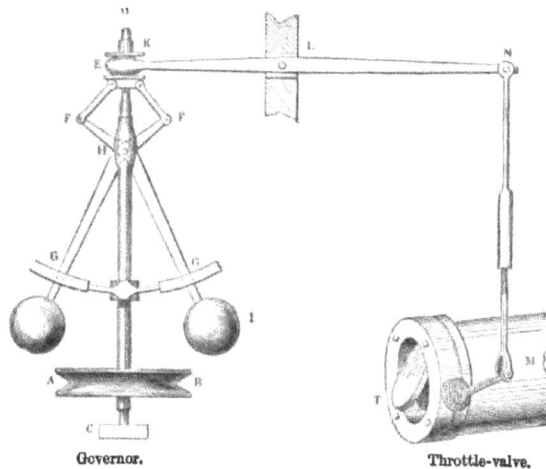

Governor.   Throttle-valve.

Figure 8.1 A typical governor from J. Farley, A Treatise on the Steam Engine: Historical, Practical, and Descriptive (London: Longman, Rees, Orme, Brown, and Green, 1827, p436)

The novel device (Figure 8.1) employed two pivoted rotating fly-balls which were flung outward by centrifugal force. As the speed of rotation increased, the flyweights swung further out and up, operating a steam flow throttling valve which slowed the engine down. Thus, a constant engine speed was achieved automatically in the face of varying loads or steam pressure. So popular was this innovation that by 1868 it is estimated that 75,000 governors[1] were in operation in England. Many similar devices were subsequently invented to control a wide range of processes, including water wheels, telescope drives, and temperature and pressure controls.

The description of the governor illustrates the operational characteristics of **negative feedback**. The output of the device, in this case the steam engine speed, is "fed back" to control the rate of steam entering the steam engine and thus influence the engine speed.

In the early days devices for automatic control were designed through trial and error and little theory existed to understand the limits and behavior of feedback control systems. One of the difficulties with feedback control is the potential for instability. As the governor became more widespread, improvements were made in manufacturing mechanical devices which reduced friction. As a result engineers began to notice a phenomena they termed hunting. This was where after a change in engine load, the governor would begin to 'hunt' in an oscillatory fashion for the new stream rate that would satisfy the load. This effect caused considerable problems with maintaining a stable engine speed and resulted in James Maxwell and independently Vyshnegradskii, undertaking the first theoretical analysis of a negative feedback system.

Until the 20th century, feedback control was generally used as a means to achieve automatic control, that is to ensure that a variable, such as a temperature or a pressure was maintained at some set value. However,

---

[1] A History of Control Engineering, 1800-1930 By Stuart Bennett, 1979

entirely new applications for feedback control emerged early in the 20th century, including artillery tracking on ships and communication. In naval warfare a major issue is being able to accurately fire a weapon from a moving ship to a moving target. In the early part of the 20th century mechanical analog computers called rangekeepers (https://en.wikipedia.org/wiki/Rangekeeper) were developed that could continuously compute a target bearing, predict the future target position and make adjustments to the weapon to account for other factors such as wind. Tracking uses negative feedback and when employed in this mode is called a **servomechanism** or servo for short.[2] Possibly the most important application of tracking came with the development of the feedback amplifier in the 1920s and 30s.

Feedback Amplifiers

Amplification is one of the most fundamental tasks one can demand of an electrical circuit. One of the challenges facing engineers in the 1920's was how to design amplifiers whose performance was robust with respect to the internal parameters of the system and which could overcome inherent nonlinearities in the implementation. This problem was especially critical to the effort to implement long distance telephone lines across the USA.

These difficulties were overcome by the introduction of the feedback amplifier, designed in 1927 by Harold S. Black [67] who was an engineer for Western Electric (the forerunner of Bell Labs). The basic idea was to introduce a negative feedback loop from the output of the amplifier to its input. At first sight, the addition of negative feedback to an amplifier might seem counterproductive. Indeed Black had to contend with just such opinions when introducing the concept. His director at Western Electric dissuaded him from following up on the idea and his patent applications were at first dismissed. In his own words, "our patent application was treated in the same manner as one for a perpetual motion machine" [4].

While Black's detractors were correct in insisting that the negative feedback would reduce the gain of the amplifier, they failed to appreciate his key insight that the reduction in gain is accompanied by increased robustness of the amplifier and improved fidelity of signal transfer.

Unlike the steam engine governor which is used to stabilize some system variable, negative feedback in amplifiers is used to accurately track and amplify an external signal. These two applications highlight the two main ways in which negative feedback can be used, namely as a **regulator** or as a **servomechanism**.

> Two modes of negative feedback:
>
> **Regulator:** Maintain a given variable at a constant level, e.g. thermostat
>
> **Servo:** Track a reference input, e.g. op amp (operational amplifier) acting as a voltage follower.

As a regulator, negative feedback is used to maintain a controlled output at some constant desired level, whereas a servomechanism will slavishly track a reference input. We can see both applications at work in the eye. On the one hand there is the need to control the level of light entering the pupil. The diameter of the pupil is controlled by two antagonistic muscles. If the external light intensity increases, the muscles respond by reducing the pupil diameter, whereas the muscles increase the pupil diameter if the light intensity falls. The pupil reflex serves as an example of negative feedback in a regulator mode. In contrast, tracking an object involves maintaining the eyeball fixed on the object. In this mode the eye functions as a servomechanism.

Both regulator and servomechanism are implemented using the same operational mechanism. Figure 8.2 shows a generic negative feedback circuit. On the left of the figure is the input, sometimes called the desired value or more often the **set point**. If the circuit is used as a servomechanism, then the output tracks the set point. As the set point changes the output follows. If the circuit is used as a regulator or homeostatic device, then the

---

[2]From the Latin servus for slave.

set point is held constant and the output is maintained at or near the set point even in the face of disturbances elsewhere in the system.

The central mechanism in the feedback circuit is the generation of the error signal, that is the difference between the desired output (set point) and the actual output. The error is fed into a controller (often something that amplifies the error), which is used to increase or decrease the activity of the process. For example, if a disturbance on the process block reduces the output, then the feedback operates by generating a positive error, which in turn increases the process activity and restores the original drop in the output.

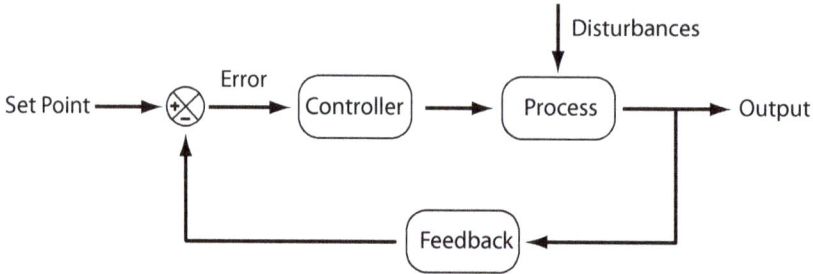

Figure 8.2 Generic structure of a negative feedback system.

8.2 Simple Quantitative Analysis

The figure of the generic negative feedback circuit (Figure 8.2) is highly stylized which makes it difficult to identify the various components in a real biological system. In addition, biological systems are invariably more complex with multiply nested feedback loops and multiple inputs and outputs. It is remarkable that even after 50 or 60 years of research, the role of many of the feedback systems in biochemical networks is still speculative.

In the remainder of this section we will consider some basic properties of negative feedback systems. The simplest way to think about feedback quantitatively is by reference to Figure 8.3.

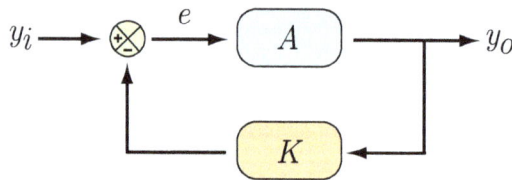

Figure 8.3 Simplest negative feedback system.

We will assume some very simple rules that govern the signal flow in this feedback system. For example, the output signal, y_o, will be given by the process A multiplied by the error, e. The feedback signal will be assumed to be proportional to y_o, that is Ky_o. Finally, the error signal, e will be given by the difference between the set point, y_i, and the feedback signal, Ky_o (Figure 8.4). These relationships are summarized below.

$$y_o = Ae, \qquad e = y_i = Ky_o$$

From these simple relations we can eliminate e to obtain:

$$y_o = \frac{Ay_i}{1 + AK} \qquad \text{or more simply} \quad y_o = Gy_i \tag{8.1}$$

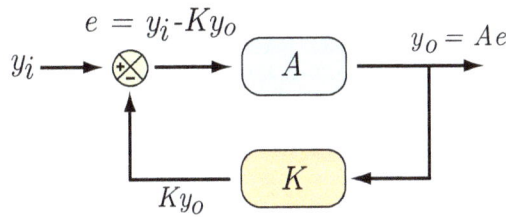

Figure 8.4 Signals at various points in the negative feedback system.

where $G = A/(A + AK)$ is called the gain of the feedback loop, often called the **closed loop gain**. *Gain* is a term that is commonly used in engineering control theory and refers to the scalar change between an input and output. Thus a gain of two means that a unit change in the input will result in twice the change in the output.

> The **gain** is a measure of the change that occurs between a signal output and its input. A gain of two means that the output will change two times in magnitude compared to a change in the input.

In addition to the closed loop gain, engineers also define two other gain factors, the **open loop gain** and the **loop gain**. The open loop gain is the gain from process, A, alone. It is the gain one would achieve if the feedback loop were absent.

> $$A = \text{open loop gain}$$
> $$AK = \text{loop gain} \tag{8.2}$$

The most important gain factor is the loop gain which is the gain from the feedback and process A combined, AK. The loop gain is an important quantity when discussing the stability and performance of feedback circuits. Figure 8.5 illustrates the different types of gain in a feedback circuit.

We can use equation (8.1) to discover some of the basic properties of a negative feedback circuit. The first thing to note is that as the loop gain, AK, increases, the system behavior becomes more dependent on the feedback loop and less dependent on the rest of the system:

$$\text{when} \quad AK \gg 1 \quad \text{then} \quad G \simeq \frac{A}{AK} = \frac{1}{K}$$

This apparently innocent effect has significant repercussions on other aspects of the system. To begin with, as the system becomes less dependent on A, so does variation in the properties of A.

> Feedback makes the performance of the system independent of variation in A.

Such variation might include noise or variation as a result of the manufacturing process or in the case of biological systems, genetic variation. To be more precise, we can compute the sensitivity of the gain G with respect to variation in A.

$$\frac{\partial G}{\partial A} = \frac{\partial}{\partial A} \frac{A}{1 + AK} = \frac{1}{(1 + AK)^2}$$

That is, the sensitivity drops as the loop gain, AK increases. If we consider the relative sensitivity we find:

$$\frac{\partial G}{\partial A} \frac{A}{G} = \frac{1}{1 + AK}$$

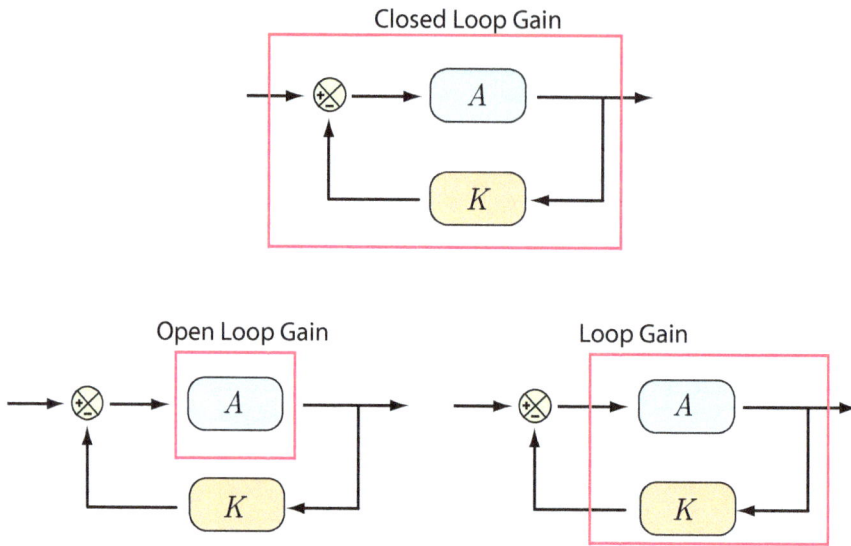

Figure 8.5 Different types of gain in a negative feedback system.

In addition to robustness to parameter variation, feedback also confers resistance to disturbances in the output. Suppose that a nonzero disturbance d changes the output. The system behavior is then described by:

$$y = Ae - d \qquad e = u - Ky$$

Eliminating e, we find:

$$y = \frac{Au - d}{1 + AK}$$

The sensitivity of the output to the disturbance is given by:

$$\frac{\partial y}{\partial d} = -\frac{1}{1 + AK}$$

The sensitivity decreases as the loop gain AK is increased. In practical terms, this means that the imposition of a load on the output, for example a current drain in an electronic circuit, protein sequestration on a signaling network, or increased demand for an amino acid, will have less of an effect on the circuit as the feedback strength increases. In electronics, this property allows engineers to divide a circuit into functional modules.

There is another, less known effect, but equally important, a linearization property. Feedback can improve the fidelity of the response. That is, for a given change in the input, a system with feedback is more likely to faithfully reproduce the input at the output than a circuit without feedback. An ability to faithfully reproduce signals is critical in electronics communications and in fact, it was this specific need that inspired the development of negative feedback in the early electronics industry.

Consider the case where the amplifier A is nonlinear. For example, a cascade pathway exhibiting a sigmoid response (Chapter 12). The behavior of the system G (now also nonlinear) is described by:

$$G(y_i) = y_o = A(e) \qquad e = y_i - Ky_o = y_i - KG(y_i)$$

Differentiating we find:

$$G'(y_i) = A'(y_i)\frac{de}{dy_i} \qquad \frac{de}{dy_i} = 1 - KG'(y_i)$$

Eliminating $\frac{de}{dy_i}$, we find:

$$G'(y_i) = \frac{A'(y_i)}{1 + A'(y_i)K}$$

We find then, that if $A'(y_i)K$ is large ($A'(y_i)K \gg 1$), then

$$G'(y_i) \approx \frac{1}{K},$$

G is approximately linear. In this case, the feedback compensates for the nonlinearities $A(\cdot)$ and the system response is not distorted. Another feature of this analysis is that the slope of $G(\cdot)$ is less than that of $A(\cdot)$, i.e. the response is "stretched out". For instance, if $A(\cdot)$ is saturated by inputs above and below a certain "active range", then $G(\cdot)$ will exhibit the same saturation, but with a broader active range[3].

A natural objection to the implementation of feedback as described above is that the system sensitivity is not actually reduced, but rather is shifted so that the response is more sensitive to the feedback K and less sensitive to the amplifier A. However, in each of the cases described above, we see that it is the nature of the loop gain AK (and not just the feedback K) which determines the extent to which the feedback affects the nature of the system. This suggests an obvious strategy. By designing a system which has a stable feedback gain, but a large "sloppy" amplifier, one ensures that the loop gain is large and the behavior of the system is satisfactory. Engineers employ precisely this strategy in the design of electrical feedback amplifiers, regularly making use of amplifiers with gains several orders of magnitude larger than the feedback gain (and the gain of the resulting system). A typical operational amplifier such as the common 741 will have an open loop gain of approximately 200,000. Given this level of gain, a 741 without any feedback is almost useless because with only the smallest change in voltage at the input, the output voltage will experience a huge voltage swing. High gain amplification is only really useful when used in conjunction with feedback.[4]

Summary of useful properties from negative feedback:

1. Amplification of signal.

2. Robustness to internal component variation.

3. High fidelity of signal transfer.

4. Low output impedance meaning the load does not affect the performance of the circuit.

8.3 Negative Feedback in Biochemical Systems

It was Umbarger (Umbarger, 1956) and Yates and Pardee (Yates & Pardee, 1956) who discovered feedback inhibition in the isoleucine biosynthesis pathway and the inhibition of aspartate transcarbamylase in *E. coli*. It wasn't long afterwards that some researchers began to investigate such feedback systems mathematically. Probably the most extensive mathematical analysis of biochemical feedback was conducted by Savageau (Savageau, 1972; Savageau, 1974; Savageau, 1976), Burns and Kacser (Burns, 1971; Kacser & Burns, 1973), and Othmer and Tyson (Othmer, 1976; Tyson & Othmer, 1978) in the 1970s, and Dibrov et. al. in the early 1980s (Dibrov et al., 1982). More recently, Cinquin and Demongeot have published an interesting review on the roles of feedback in biological systems (Cinquin & Demongeot, 2002). The topic continues to draw attention [?] but now often using empirical studies based on computer simulation.

---

[3]I wish to acknowledge Brian Ingalls for useful discussions on this topic.

[4]Even a 10 μV change at the input will, in theory result, in 200 volt change at the output. However op amps are generally powered by $\pm$ 15 volts, meaning that a small voltage change will result in a full swing from -15 to +15 volts resulting in the op amp saturating. Op amps are only really useful when feedback is added.

In the last section we considered a simple analysis of negative feedback and its behavioral effects. The treatment was however very generic and the question we wish to address here is how can we apply the same kind of analysis to biochemical feedback systems? This question is harder to answer that it seems. To begin with, biochemical systems are governed by nonlinear rate laws not the simple linear rules we used in the previous analysis. Secondly how do we map the generic diagram (Figure 8.2) onto a biochemical feedback circuit (Figure 8.6)? On the surface they look similar, but the initial impression can be misleading.

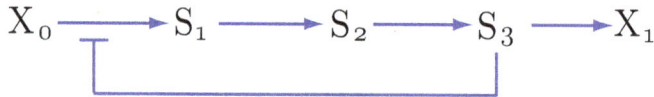

Figure 8.6 Simple four-step pathway with negative feedback.

In order to be clear, we need to identify the input (set point), output, the feedback loop and the process block in the biochemical network Figure 8.6.

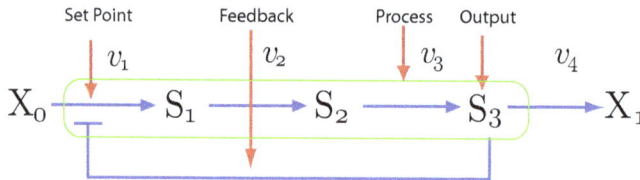

Figure 8.7 Simple four-step pathway with negative feedback.

In naturally evolved systems it is sometimes difficult to identify the various parts in a negative feedback circuit.

In biochemical pathways, control is often achieved using allosteric enzymes which have distinct binding sites for the controlling species, separate from the main active site. The most difficult element to identify is the set point. In biochemical networks the set point is embedded in the regulated enzyme, possibly related to the half saturation constant of the allosteric regulator.

When comparing the block diagram (Figure 8.2) to the biological pathway in Figure 8.7 it may not be apparent how the two representations can be matched. It is however possible to pair each component in the block diagram to an equivalent component in the biological pathway. Thus the output, y_o, in the block diagram corresponds to the concentration of S_3. The negative feedback component K corresponds to the interaction of S_3 with the allosteric enzyme in the first step. The set point, y_i, is more problematic but it is most likely embedded in the kinetic characteristics of the allosteric enzyme. Finally, the controller A is represented by the steps v_1, v_2, and v_3. The load on the system is represented by the last step, v_4, and other disturbances can be assigned to v_1, v_2, v_3, and the input concentration, X_o.

The flux through the pathway can also be considered an output because the reaction rate through the last step is a function, S_3.

Graphical Understanding of Feedback

It is possible to appreciate some of the effects of negative feedback in a biochemical network using a graphical approach. Consider the much simplified negative feedback network shown in Figure 8.9. This network only has two steps, v_1 and v_2. Figure 8.8 shows two plots, one with strong and the other with weak feedback. The strength of the feedback is modeled using a simple Hill-like equation, for stronger feedback the Hill equation has a higher Hill coefficient. The plots show the reaction rates v_1 and v_2 as a function of the intermediate

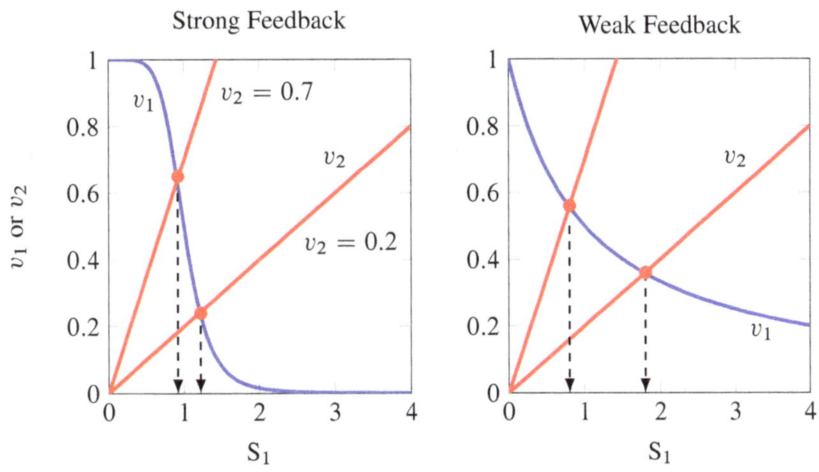

Figure 8.8 Plot of v_1 and v_2 versus the concentration of S_1 for a simple two-step pathway with negative feedback. v_2 is governed by the rate law $k_2 s_1$. Two perturbations in k_2 and its effect on v_2 are shown. In the left panel where the feedback is strong, changes in k_2 have hardly any effect on S_1. On the right panel, the same change in k_2 results in a much larger change in S_1. This illustrates the homeostatic property of negative feedback. Left Panel: $v_1 = 1/(1+s_1^4)$, Right Panel: $v_1 = 1/(1+s_1)$

species, S_1, where S_1 can negatively feedback onto the first step. We assume that the second step follows first-order kinetics so that v_2 is a straight line with rate law $k_2 s_1$. The feedback response curve shows a decline from high to low as S_1 increases. For strong feedback the decline is much steeper (left plot). If the rate constant for the second step is changed, this changes the slope of v_2. This is equivalent to a perturbation in the system. In the case of weak feedback, changes in k_2 result in significant changes to S_1 because the feedback response is shallow. In contrast, when we have strong feedback (left panel), where the slope is very steep, any changes in k_2 results in only small changes in S_1. In this way we can see how strong negative feedback can resist changes in k_2.

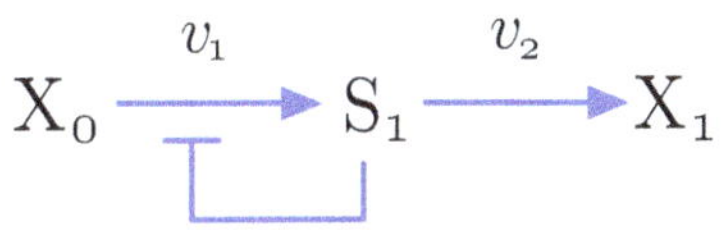

Figure 8.9 Simple two-step pathway with negative feedback: $v_1 = 1/(1+s_1^n)$, $v_2 = k_2 s_1$

Control Analysis

To apply control analysis, let us consider a three-step pathway in Figure 8.10

Just as we did earlier, we can derive the flux and concentration control coefficients in terms of the elasticities. For convenience, we will write out the theorems in matrix form, see equation (4.18)). Note the presence of the

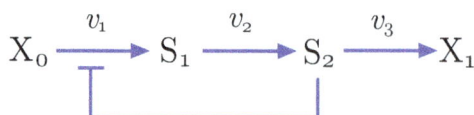

Figure 8.10 Negative feedback with three reaction steps.

feedback term, ε_2^1 in the matrix in red and that its value is negative.

$$
\begin{bmatrix} C_1^J & C_2^J & C_3^J \\ C_1^{s_1} & C_2^{s_1} & C_3^{s_1} \\ C_1^{s_2} & C_2^{s_2} & C_3^{s_2} \end{bmatrix}
\begin{bmatrix} 1 & -\varepsilon_1^1 & -\varepsilon_2^1 \\ 1 & -\varepsilon_1^2 & -\varepsilon_2^2 \\ 1 & 0 & -\varepsilon_2^3 \end{bmatrix}
=
\begin{bmatrix} 1 & 0 & 0 \\ 0 & 1 & 0 \\ 0 & 0 & 1 \end{bmatrix}
$$

Rearranging the matrix equation yields:

$$
\begin{bmatrix} C_1^J & C_2^J & C_3^J \\ C_1^{s_1} & C_2^{s_1} & C_3^{s_1} \\ C_1^{s_2} & C_2^{s_2} & C_3^{s_2} \end{bmatrix}
=
\begin{bmatrix} 1 & 0 & 0 \\ 0 & 1 & 0 \\ 0 & 0 & 1 \end{bmatrix}
\begin{bmatrix} 1 & -\varepsilon_1^1 & -\varepsilon_2^1 \\ 1 & -\varepsilon_1^2 & -\varepsilon_2^2 \\ 1 & 0 & -\varepsilon_2^3 \end{bmatrix}^{-1}
$$

Inverting the elasticity matrix yields the following equations for the flux control coefficients with and without feedback to illustrate the difference in results.

With Feedback:

$$C_{e1}^J = \frac{\varepsilon_1^2 \varepsilon_2^3}{\varepsilon_1^2 \varepsilon_2^3 - \varepsilon_1^1 \varepsilon_2^3 + \varepsilon_1^1 \varepsilon_2^2 - \varepsilon_1^2 \varepsilon_2^1}$$

$$C_{e2}^J = \frac{-\varepsilon_1^1 \varepsilon_2^3}{\varepsilon_1^2 \varepsilon_2^3 - \varepsilon_1^1 \varepsilon_2^3 + \varepsilon_1^1 \varepsilon_2^2 - \varepsilon_1^2 \varepsilon_2^1}$$

$$C_{e3}^J = \frac{\varepsilon_1^1 \varepsilon_2^2 - \varepsilon_1^2 \varepsilon_2^1}{\varepsilon_1^2 \varepsilon_2^3 - \varepsilon_1^1 \varepsilon_2^3 + \varepsilon_1^1 \varepsilon_2^2 - \varepsilon_1^2 \varepsilon_2^1}$$

Without Feedback:

$$C_{e1}^J = \frac{\varepsilon_1^2 \varepsilon_2^3}{\varepsilon_1^2 \varepsilon_2^3 - \varepsilon_1^1 \varepsilon_2^3 + \varepsilon_1^1 \varepsilon_2^2}$$

$$C_{e2}^J = \frac{-\varepsilon_1^1 \varepsilon_2^3}{\varepsilon_1^2 \varepsilon_2^3 - \varepsilon_1^1 \varepsilon_2^3 + \varepsilon_1^1 \varepsilon_2^2}$$

$$C_{e3}^J = \frac{\varepsilon_1^1 \varepsilon_2^2}{\varepsilon_1^2 \varepsilon_2^3 - \varepsilon_1^1 \varepsilon_2^3 + \varepsilon_1^1 \varepsilon_2^2}$$

Table 8.1 shows the corresponding concentration control coefficients in the presence of negative feedback. The first thing to note is that the addition of negative feedback adds a new term to the denominator, $-\varepsilon_1^2 \varepsilon_2^1$. This term is positive and therefore increases the value of the denominator.

Consider first the concentration control coefficients. To simplify matters assume that the product elasticities, ε_1^1 and ε_2^2, are zero. This greatly simplifies the equations.

The first observation to make in Table 8.1 is that $C_{e2}^{s_2}$ and C_{e2}^J are zero. That is, disturbances **inside** the feedback loop have no effect on the signal species, S_2 or the flux.

The second observation is that both $C_{e1}^{s_2}$ and $C_{e3}^{s_2}$ can be reduced to very small values if the feedback strength is significant, that is $\varepsilon_2^1 \ll 0$. This result illustrates the buffering capacity of the negative feedback loop, locking S_2 into a very narrow range, see Figure 8.8.

The response of the pathway to X_o is given by the product $C_{e1}^{s_2} \varepsilon_{x_o}^1$. If $\varepsilon_{x_o}^1$ lies between 0.5 and 1.0 for a typical Michaelian enzyme, the reactant species, X_o, will have little influence over the level of signal species, S_2. It is interesting to speculate that most enzymes regulated by feedback inhibition also show cooperativity with respect to the source species. This means that $\varepsilon_{x_o}^1$ can be of the order of 4.0 or more, suggesting that the cooperativity attempts to restore some control by the source species. In other words, it suggests some degree of control by supply over the output S_2. This leads to a complementary question of how demand can affect the pathway.

Enzyme Step	Species or Flux		
	S_1	S_2	J
E_1	$\dfrac{\varepsilon_2^3}{\varepsilon_1^2(\varepsilon_2^3 - \varepsilon_2^1)}$	$\dfrac{1}{\varepsilon_2^3 - \varepsilon_2^1}$	$\dfrac{\varepsilon_2^3}{\varepsilon_2^3 - \varepsilon_2^1}$
E_2	$\dfrac{-1}{\varepsilon_1^2}$	0	0
E_3	$-\dfrac{\varepsilon_2^1}{\varepsilon_1^2(\varepsilon_2^3 - \varepsilon_2^1)}$	$-\dfrac{1}{\varepsilon_2^3 - \varepsilon_2^1}$	$-\dfrac{\varepsilon_2^1}{\varepsilon_2^3 - \varepsilon_2^1}$

Table 8.1 Control coefficients for simple negative feedback pathway. Assumes that product inhibition on v_2 and v_3 is absent ($\varepsilon_1^1 = 0, \varepsilon_2^2 = 0$). The feedback elasticity is shown in red.

If we make the feedback elasticity larger, $\varepsilon_2^1 \ll 0$, flux control coefficient for the final step, $C_{e_3}^J$ tends to one. That is, all control moves out of the feedback loop. In terms of a steam engine analogy, it is equivalent to being able to change the work load on the steam engine without loss of power. If $C_{e_3}^J$ tends to one, then by the summation theorem, both $C_{e_1}^J$ and $C_{e_2}^J$ tend to zero. From this perspective we are forced to conclude that the regulated step **has no control over the flux**.

Steps regulated via negative feedback will tend to have small flux control coefficients.

Control over S_3 and J are complementary, namely as the feedback strength increases:

$$C_{e_3}^J \to 1$$

$$C_{e_3}^{s_2} \to 0$$

What is troubling for many is that the regulated step, $C_{e_1}^J$, has little in the way of flux control, that is, it is **not** rate-limiting. This would seem to raise a paradox. On the one hand, the regulated step must be important, Yet this importance is not reflected in the degree of influence the step has on the flux.

Traditionally, allosteric enzymes have been considered flux controllers. The analysis here suggests the opposite. Allosteric enzymes, when part of a negative feedback loop, are very poor controllers of flux. Instead, the feedback allows distal steps to be good controllers of a pathway flux. For demand driven systems this is a logical arrangement.

Most textbooks and online sites such as Wikipedia refer to phosphofructokinase as the rate-limiting or pacemaker step of glycolysis. There are many reasons why this is considered so. Phosphofructokinase is one of the earliest steps of glycolysis, *in vivo* it is a nonequilibrium reaction and most convincing of all, it is regulated by many effectors. Why are the effectors there other than to control glycolytic flux? Many allosteric enzymes such as phosphofructokinase that have many regulators, are considered flux controllers by the same reasoning.

From the previous sections we saw that regulated enzymes in unbranched pathways with feedback are in fact poor flux controllers. When determined experimentally the flux control coefficient for phosphofructokinase is found to be invariable small [41, 94, 22, 13, 106, 84, 71, 111, 64, 70] and therefore phosphofructokinase is not rate-limiting in many situations. This matches the theoretical expectation even though intuitively it seems suspect.

We therefore have a paradox (sometimes called "The PFK Paradox" [103]). Intuition suggests that phosphofructokinase should be controlling glycolytic flux particularly given the multitude of effectors that regulate it.

On the other hand, experimental evidence and theory suggests the opposite. The question is how to reconcile these two opposite views?

Intuitively, the regulated step is important, but how do we measure that importance? There are at least two answers to this. The first is that the feedback elasticity will be strongly negative. For example, if the regulated step were determined by a modified Hill-like equation such as:

$$v = \frac{V_{\max} \, x_o}{s^n + x_o + K_m}$$

where S is the feedback signal, then the elasticity of the reaction rate with respect to the signal is given by:

$$\varepsilon_s^v = \frac{n x_o K_m}{K_m + (s/K_m)^n}$$

We see that at low signal, S, the elasticity is proportional to n, the Hill coefficient. The reaction is therefore very sensitive to changes in the signal molecule.

The second way to answer the question is to consider the pathway with and without the negative feedback loop. We can compare for example the flux control coefficient on the first step with and without negative feedback. When we remove the negative feedback, the pathway will change to a new state where the concentrations of S_1 and S_2 are higher. To make the comparison fair, we will adjust the level of enzyme in the first step to restore the levels of S_1 and S_2 to the values they had before the feedback was removed. When we do this, we will also automatically restore the pathway flux to its original value. In practice this means that all the elasticities except for the feedback elasticity are exactly the same in both systems. We will call the flux control coefficient for the case without regulation, $^u C_e^J$ and with feedback control, $^n C_e^J$. The superscript u is used to indicate that this is the control coefficient for the unregulated pathway.

We can now ask the question what are the relative values for the flux control coefficients in step one? We can derive the ratio of the flux control of the regulated step in the configuration without feedback, $^u C_{e_1}^J$ to the configuration with feedback, $^n C_{e_1}^J$. This can be shown to be:

$$\frac{^u C_{e_1}^J}{^n C_{e_1}^J} = 1 - \frac{\varepsilon_1^2 \varepsilon_2^3 \varepsilon_2^1}{\varepsilon_1^1 \varepsilon_2^2 \varepsilon_3^4 - \varepsilon_1^1 \varepsilon_2^2 \varepsilon_3^3 - \varepsilon_1^1 \varepsilon_2^3 \varepsilon_3^4 + \varepsilon_1^2 \varepsilon_2^3 \varepsilon_3^4}$$

The numerator term, $\varepsilon_1^2 \varepsilon_2^3 \varepsilon_2^1$ is negative and the denominator positive. The ratio is therefore negative. However we are subtracting this negative value from one. This means that overall, the expression $^u C_{e_1}^J / {^n C_{e_1}^J}$ must be > 1. In other words, the flux control exerted by the configuration without feedback will be **greater** than the control exerted by the configuration with feedback.

To make the analysis simpler, let's assume that the product inhibition elasticities, ε_1^1 and ε_2^2, are both zero. We now take the ratio of the flux control coefficient on the first step, $^u C_1^J$ without feedback to the same coefficient with feedback:

$$\frac{^u C_1^J}{^n C_1^J} = \frac{\varepsilon_1^2 \varepsilon_2^3 - \varepsilon_2^1 \varepsilon_1^2}{\varepsilon_1^2 \varepsilon_2^3} = 1 - \frac{\varepsilon_2^1 \varepsilon_1^2}{\varepsilon_1^2 \varepsilon_2^3} = 1 - \frac{\varepsilon_2^1}{\varepsilon_2^3}$$

Given that ε_2^1 is negative, the term, $-\varepsilon_2^1/\varepsilon_2^3$ is positive. That is:

$$^u C_1^J >^n C_1^J$$

This tells us that for a negative feedback loop to be effective, the flux control coefficient in the unregulated pathway must be higher than the flux control coefficient with feedback. There is however another way to measure the importance of a regulated step, and that is to look at the loop gain, something we'll examine in the next section.

8.4 The PFK Paradox

In the last section a simple analysis was made of a feedback system using approaches developed from MCA and BST. How does this analysis and the broader literature on MCA and BST relate to the existing body of engineering control theory? This has been covered in detail by Ingalls [52] and Rao [79]. Here we will give a flavor of the connection particularly in relation to negative feedback. Not all effects of negative feedback will be discussed however, and the reader is referred to some recent articles for additional details [24, 55].

As we saw earlier in the chapter, the classic diagram often used to depict negative feedback in control theory is shown in Figure 8.11. In this diagram, K is the fraction of output fed back to the summing junction. The summing junction computes the difference between the input, and the signal fed back from the output. The output is a function of the error, often a simple proportional relationship (hence called proportional control). To put it in more concrete terms, the A block might be a heater in a room, and the output is the room's temperature. The input is the desired temperature. If the room is hotter than the desired temperature, the error signal will be negative so that the heater is turned down. The opposite happens if the room is cooler than the desired temperature.

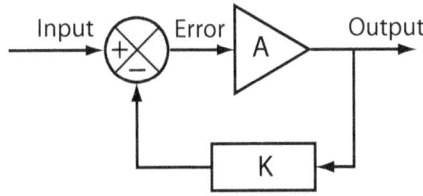

Figure 8.11 Generic negative feedback system as depicted often by engineers.

As previously shown, we can derive the relationship between the input and output in Figure 8.11 by noting the following relationships. The error signal can be written as the difference between the input and the return loop: error = input − output × K. The output can be written in terms of the gain A and the error: output = A × error. If we combine both equations and solve for the output in terms of the input, we get:

$$\text{output} = \frac{\text{input} \times A}{1 + A\,K} \tag{8.3}$$

In control theory this is called the closed-loop transfer function, see also (8.1). In practice it is often expressed in the frequency domain but this need not concern us here. What is more interesting is that the control coefficient equations are directly related to the closed-loop transfer function [52]. The only difference is that whereas the full transfer function is defined over a range of input frequencies, the control coefficients are defined only at a single frequency of zero, the so-called DC response. For a two-step pathway with feedback, the concentration control coefficient (4.17) is given by: $1/(\varepsilon_s^2 - \varepsilon_s^1)$. We therefore state the following equivalence [52, 51]:

$$C_{e_1}^s \frac{\delta e_1}{e_1} = \frac{\delta s}{s} = \frac{1}{\varepsilon_s^2 - \varepsilon_s^1} \frac{\delta e_1}{e_1} = \frac{\text{input} \times A}{1 + A\,K}$$

The input in the closed-loop transfer function matches the change in enzyme ($\delta e_1 / e_1$), but could be any input into v_1. In the case of e_1 we assume that the elasticity of v_1 with respect to e_1, $\varepsilon_{e_1}^1$, is one. For other inputs this might not be the case and the corresponding elasticity would need to be explicitly given. For example, if the input parameter is a signal P, then the term $\delta e_1 / e_1$ would be replaced with $(\delta p / p)\, \varepsilon_p^1$. The output is $\delta s / s$ which is the component we are trying to regulate.

Rearranging terms by dividing top and bottom by ε_s^2 we obtain:

$$C_{e_1}^s \frac{\delta e_1}{e_1} = \frac{\delta s}{s} = \frac{\delta e_1}{e_1} \frac{1/\varepsilon_s^2}{1 - \varepsilon_s^1/\varepsilon_s^2} = \text{input} \frac{A}{1 + A\,K}$$

From this we can determine that:

$$\text{input} = \frac{\delta e_1}{e_1}, \quad \text{output} = \frac{\delta s}{s} \quad A = \frac{1}{\varepsilon_s^2} \quad \text{and} \quad K = -\varepsilon_s^1$$

This generalizes to any length feedback loop. The feedback term, K, will always be equal to the negative of the feedback elasticity. The expression for A, $1/\varepsilon_s^2$ is the control coefficient $C_{e_1}^s$, when there is **no feedback**. $C_{e_1}^s$ is the control coefficient for the unregulated system (set ε_s^1 to zero in $C_{e_1}^s$) and is the gain of the system without feedback assuming everything else is equal. Figure 8.12 overlays the generalized feedback diagram with the equivalent elasticity terms. The error computed from the set point and feedback is the difference between the two terms $\delta e_1/e_1$ and $\varepsilon_s^1 C_{e_1}^s$ (recall that ε_s^1 is negative).

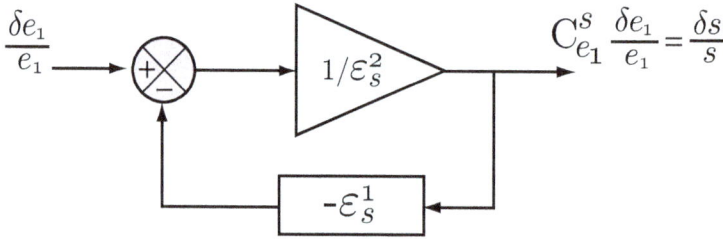

Figure 8.12 Mapping the standard engineering feedback control diagram to equivalent biochemical components.

As already introduced, the product AK is the loop gain (8.1) which has a special status in control theory. It is one of the factors that determines the stability of a system with negative feedback and general performance of the system. In the reaction pathway, $\varepsilon_s^1/\varepsilon_s^2$ is the loop gain. For the unregulated pathway, ε_s^1 will be absent and under these conditions, the control coefficient equals $1/\varepsilon_s^2$. We can therefore rewrite the loop gain as:

$$\text{Loop Gain} = AK = -\varepsilon_s^1 \, {}^u C_{e_1}^s$$

where ${}^u C_{e_1}^s$ is computed for the *unregulated pathway* and equal to $1/\varepsilon_s^2$. The minus sign comes from the observation that K is equal to the negative of the feedback elasticity. This analysis can be extended to any length pathway. As an example, consider the four-step pathway where S_3 is the return signal (Figure 8.14), the loop gain will equal:

$$\text{Loop Gain} = -\varepsilon_3^1 \, {}^u C_{e_1}^{s_3} \tag{8.4}$$

where ${}^u C_{e_1}^{s_3}$ is the control coefficient for the *unregulated* but equivalent system, and ε_3^1 is the elasticity of the feedback loop.

Control Coefficient	Numerator
$C_{e_1}^{s_3}$	$\varepsilon_1^2 \varepsilon_2^3$
$C_{e_4}^{s_3}$	$\varepsilon_1^1 \varepsilon_2^3 - \varepsilon_1^1 \varepsilon_2^2 - \varepsilon_1^2 \varepsilon_2^3$
$C_{e_1}^{J}$	$\varepsilon_1^2 \varepsilon_2^3 \varepsilon_3^4$
$C_{e_4}^{J}$	$-\varepsilon_1^1 \varepsilon_2^2 \varepsilon_3^3 - \varepsilon_1^2 \varepsilon_2^3 \varepsilon_3^1$

Table 8.2 Control coefficients and corresponding numerators of control equations for the four-step pathway in Figure 8.14. The feedback elasticity is highlighted as ε_3^1.

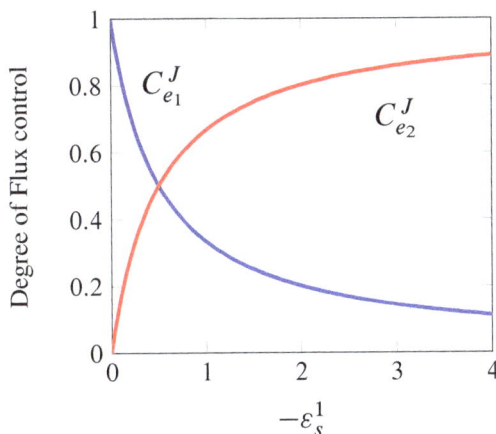

Figure 8.13 Switch-over of flux control from the first step to the second step as a function of the feedback strength.

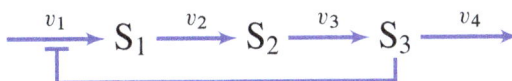

Figure 8.14 Four-step unbranched pathway with negative feedback.

The Phosphofructokinase Paradox

We first restate that phosphofructokinase in not rate-limiting when operating *in situ*. This has been shown experimentally many times [41, 94, 22, 13, 106, 84, 71, 111, 64, 70] as well as being consistent with theory. This cannot be easily disputed. And yet the literature, textbooks and online resources still claim that phosphofructokinase is rate-limiting [68]. To reconcile this difference we must introduce a different measure that describes the importance of enzymes such as PFK and their ability to regulate.

One clear possibility is to look at the loop gain [2]. As described before, the loop gain is the overall gain around the feedback loop. It was shown previously (8.3) that the loop gain is the product of the forward gain, A, and the feedback gain, K, that is AK. In terms of elasticities, the loop gain for the four-step pathway, Figure 8.14, is the product of the feedback elasticity, ε_3^1, and the concentration control coefficient, $^uC_{e_1}^{s_3}$, for the *unregulated pathway*. In this case this equals the product of the forward elasticities, $\varepsilon_1^2\varepsilon_2^3$ divided by the denominator, D, of the *unregulated pathway*, equation (8.4):

$$\text{Loop Gain} = -\varepsilon_1^2\varepsilon_2^3\varepsilon_3^1/{}^uD \tag{8.5}$$

where D is the denominator for the pathway with negative feedback.

$$^uD = \varepsilon_1^1\varepsilon_2^2\varepsilon_3^4 - \varepsilon_1^1\varepsilon_2^2\varepsilon_3^3 - \varepsilon_1^1\varepsilon_2^3\varepsilon_3^4 + \varepsilon_1^2\varepsilon_2^3\varepsilon_3^4 \tag{8.6}$$

where ^uD is the denominator for the four-step pathway without negative feedback (it simply lacks the last term).

We can illustrate the use of this measure with an example. Using the four-step pathway with negative feedback, the elasticities are set to the following values. All substrate elasticities ($\varepsilon_1^2, \varepsilon_2^3, \varepsilon_3^4$) are set to 0.5. This corresponds to substrate levels set to the K_m of each enzyme. The product inhibition elasticities ($\varepsilon_1^1, \varepsilon_2^2, \varepsilon_3^3$) are assumed to be small but not negligible and are set to -0.1. Lastly, the feedback inhibition elasticity, ε_3^1 is set to -4.0. Given these values, we can compute the flux control coefficient in the presence of negative feedback using the expressions in Table 8.2:

$$C_{e_1}^J = 0.11$$

This tells us that from the perspective of the flux control coefficient, the reaction is not rate-limiting. However we can compute the loop gain using equation (8.5), and this yields:

$$\text{Loop Gain} = 6.4$$

Note that the loop gain is significantly higher than the flux control coefficient. It means that changes to the regulator, S_3, will have a significant effect on throttling the pathway. We can compare these calculations to the same pathway but where the negative feedback strength is weak. If we set ε_3^1 to a small value such as -0.1, we obtain the following values for the flux control and loop gain:

$$^uC_{e_1}^J = 0.8$$

$$\text{Loop Gain} = 0.16$$

The flux control has increased about eight fold and is now strongly rate-limiting with respect to e_1. What is more interesting is that the loop gain has been reduced forty fold, indicating little or no regulation from the feedback loop. This leads tot he following conclusion:

> The loop gain gives a measure of how effective the regulation is. The higher the loop gain, the more effective the regulation.

It is worth noting that there are two contributions to the loop gain in equation (8.5), the action of the signal on the regulated step, ε_3^1, and the transmission of that signal to cause a change (Figure 8.15), $-\varepsilon_1^2\varepsilon_2^3/^uD$. The effectiveness of the overall regulation is therefore not just a function of the regulated step but of the entire loop. If the transmission elasticities, in this case ε_1^2 and ε_2^3, are small, then the loop gain could be significantly reduced. An examination of the elasticity of the regulated step is therefore insufficient to ascertain the effectiveness of the regulation, and it is quite possible that with weak signal transmission, even in the presence of a strongly regulated step, effective regulation could be minimal.

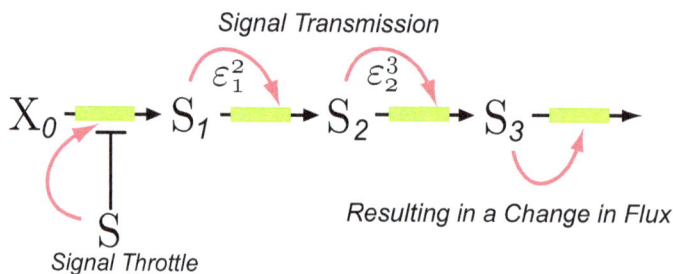

Figure 8.15 Transmission from signal to flux change.

This highlights again the idea that the behavior of a pathway cannot be judged from a single step. Just as control is a systematic property so that the influence a particular step has cannot be judged from the step alone, so it is the same with regulation. The effectiveness of a regulator cannot be judged from the regulated step alone.

> Regulation and control are systemic properties of a pathway.

8.5   Robustness and Supply/Demand

One way to look at metabolic systems is to divide them into two separate but connected blocks, the supply block and the demand block [47, 83, 49]. Negative feedback makes this division very straightforward. Consider a factory that make cars. Should the rate of car production be controlled by demand or the supply of cars? Economically the most efficient strategy is to let demand decide how many cars to make, this ensures that excess cars do not build up and thereby waste resources. In certain metabolic situations the same reasoning can be used. For example, the production of amino acids would best be determined by the demand from protein synthesis. This means that if protein synthesis slows, amino acid production should also slow. A metabolic system that supplies amino acids should be able to supply amino acids effectively at both high and low demand. One way to do this is to maintain the amino acid level at a relatively constant level independent of the demand block. Let us assume for the moment that there is no feedback regulation in the supply/demand pathway (Figure 8.16). If demand rises, this will result in the intermediate metabolite, P, falling. As P falls, the ability to supply the increased demand becomes more and more difficult. If demand falls, the flux through the pathway will fall. This will cause the intermediate metabolite, P to rise. Since the equilibrium constant across the supply block is likely to be large, the concentration of P could raise to toxic high levels as the supply block approaches equilibrium at low fluxes.

Figure 8.16 A system divided into supply and demand blocks.

The solution to avoid both problems at high and low demand is to use negative feedback (Figure 8.17). With negative feedback, high demand will result in a decrease in the intermediate metabolite, P, which in turn will release repression in the supply block to restore some loss in P. Alternatively, at low demand, the increase in P will suppress its own production preventing excessive production of P.

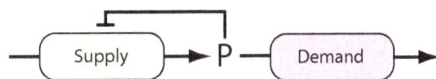

Figure 8.17 A system divided into supply and demand blocks with negative feedback.

8.6   Instability

Although negative feedback offers considerable advantages to a system, too much feedback and delays can cause instability in the form of sustained oscillations. We will discuss stability of negative feedback systems in Chapter 9.

Further Reading

1. Hofmeyr, JH and Rohwer, JM (2011) Supply-demand analysis a framework for exploring the regulatory design of metabolism. Methods Enzymology, Vol 500, 533-554

2. Sauro HM (2017) Control and regulation of pathways via negative feedback. J R Soc Interface. 2017 Feb; 14 (127)

Exercises

1. A regulated step via a negative feedback loop has a flux control coefficient of 0.9. Would you consider this system to be a well regulated pathway? Explain your answer.

2. The rate of coal production in a coal mine is driven by the supply of coal it finds in the mine. What could go wrong with such a system when compared to one that is driven by demand?

9

Stability

In the simplest terms, a stable steady state is one where if a *small* perturbation is made to a species concentration, then over time, that perturbation relaxes back to the original steady state. Figure 9.1 show the effect of two perturbations on a system. In both cases the system relaxes back to the original steady state.

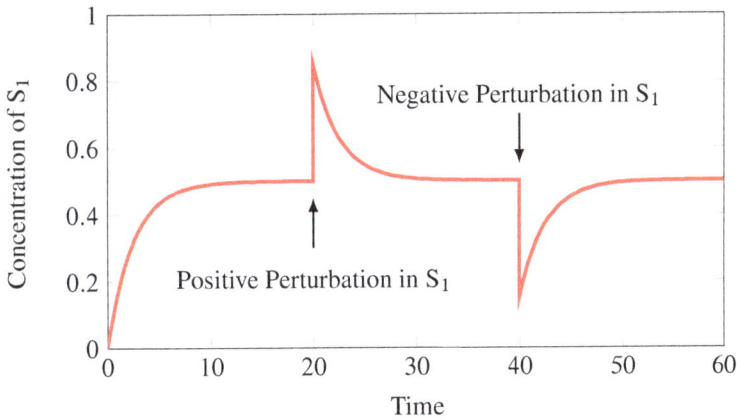

Figure 9.1 Multiple Perturbations. The steady state concentration of the species S_1 is 0.5 and a perturbation is made to S_1 by adding an additional 0.35 units of S_1 at time $= 20$ and removing 0.35 units at time $= 40$. In both cases the system relaxes back.

To understand the concept of stability more fully, consider the following simple system:

$$X_o \xrightarrow{v_1} S_1 \xrightarrow{v_2} X_1$$

where X_o and X_1 are fixed species, v_1 is given by the rate law $k_1 x_o$ and v_2 by $k_2 s_1$. The differential equation for the single floating species, S_1, is given by:

$$\frac{ds_1}{dt} = k_1 x_o - k_2 s_1 \tag{9.1}$$

By setting the differential equation to zero, the steady state solution can be determined to be:

$$s_1 = k_1 x_o / k_2 \tag{9.2}$$

An important question to ask is whether the steady state is stable or not, that is, whether a small perturbation will decay and return to the steady state. To test this we can do the following: When the system is at steady state, make a small perturbation to the steady state concentration of S_1, by an amount δs_1 and ask how δs_1 subsequently changes as a result of this perturbation. That is, what is $d(\delta s_1)/dt$? To determine this, the new rate of change equation is rewritten as follows:

$$\frac{d(s_1 + \delta s_1)}{dt} = k_1 x_o - k_2 (s_1 + \delta s_1)$$

If we insert the steady state solution for S_1, equation (9.2), into the above equation we are left with:

$$\frac{d\delta s_1}{dt} = -k_2 \delta s_1 \tag{9.3}$$

Note that the rate of change of the *disturbance itself*, δs_1, is negative. The system reduces the disturbance so that the system returns back to the original steady state. Systems with this kind of behavior are called **stable**. If the rate of change in S_1 had been positive instead of negative, the perturbation would have continued to diverge away from the original steady state and the system would then be considered **unstable**.

Let's look at this graphically by plotting the rate of change, ds_1/dt, as a function of S_1, shown in Figure 9.2. The steady state occurs when the net rate of change is zero, marked by the straight line arrow. If the concentration falls below this value, the net rate goes positive, thereby increasing the level of S_1. If the concentration rises above the steady state level, the graph shows the net rate of change going negative, so that S_1 decreases. The system is therefore stable.

This kind of stability is also called the **internal stability** because it describes the system's stability to perturbations in the internal state. We can informally define a stable system as:

Internal Stability: A biochemical pathway is internally stable if at steady state, small perturbations to the floating species relax back to the steady state.

Caveat: If the perturbed species is part of a conserved cycle (See Chapter 11), then the total mass in the cycle must remain constant during the perturbation. This may require perturbing one species in a positive direction and another in a negative direction.

Divide both sides of equation (9.3) by δs_1 and taking the limit, $\delta s_1 \to 0$, we find that:

$$\frac{\partial (ds_1/dt)}{\partial s_1} = -k_2$$

The left-hand side is the normalized rate of change of the perturbation, in this case normalized with respect to the steady state concentration of S_1. The sign of the left-hand side will still tell us whether the perturbation is expanding or contacting. The stability of this simple system can therefore be determined by inspecting the sign of $\partial (ds_1/dt)/\partial s_1$. In this case $\partial (ds_1/dt)/\partial s_1 = -k_2$ is negative, meaning the system is *stable*. It is worth noting that the larger the rate constant, k_2, the quicker the system relaxes back to steady state.

For systems with more than one species, a system's stability can be determined by looking at all the terms $\partial (ds_i/dt)/\partial s_j$ which are given collectively by the matrix expression:

$$\frac{d(d\mathbf{s}/dt)}{d\mathbf{s}} = \boldsymbol{J} \tag{9.4}$$

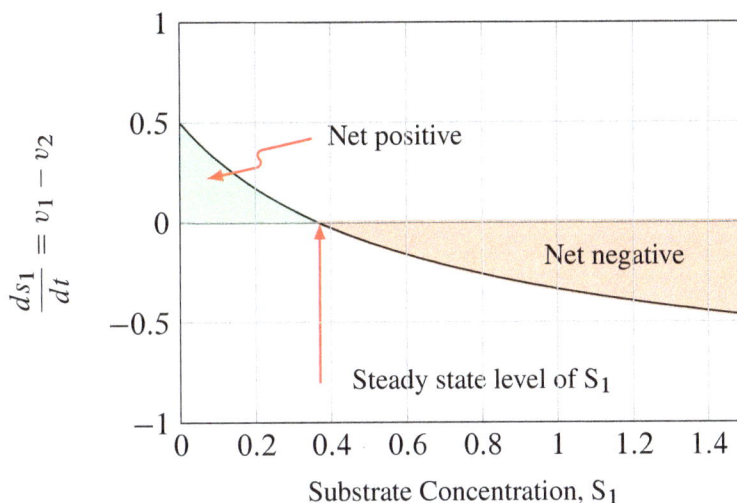

Figure 9.2 Rate of change as a function of S_1. The arrow indicates the steady state for S_1. When S_1 is below the steady state value, the net change is positive meaning that S_1 will increase. When S_1 is above the steady state value, the net change is negative meaning that S_1 will decrease. The system is therefore stable.

where J is called the **Jacobian matrix** containing elements of the form $\partial(ds_i/dt)/\partial s_j$. Equation (9.3) can also be generalized to:

$$\frac{d(\delta\mathbf{s})}{dt} = J\delta\mathbf{s} \tag{9.5}$$

where J is given by:

$$\begin{bmatrix} \frac{\partial(ds_1/dt)}{\partial s_1} & \cdots & \frac{\partial(ds_1/dt)}{\partial s_m} \\ \vdots & \ddots & \vdots \\ \frac{\partial(ds_m/dt)}{\partial s_1} & \cdots & \frac{\partial(ds_m/dt)}{\partial s_m} \end{bmatrix}$$

Equation (9.5) is an example of an *unforced* linear differential equation and has the general form:

$$\frac{d\mathbf{x}}{dt} = A\mathbf{x}$$

Solutions to unforced linear differential equations are well known and take the form:

$$x_j(t) = c_1 K_1 e^{\lambda_1 t} + c_2 K_2 e^{\lambda_2 t} + \cdots c_n K_n e^{\lambda_n t}$$

The solution involves a sum of exponentials, $e^{\lambda_i t}$, constants c_i, and vectors, K_i. The exponents of the exponentials are given by the eigenvalues of the matrix, A, and K_i, the corresponding eigenvectors. The c_i terms are related to the initial conditions assigned to the problem. It is possible for the eigenvalues to be complex, but in general if the real parts of the eigenvalues are negative, the exponents will decay. If they are positive, the exponents will grow. We can therefore determine the stability properties of a given model by computing the eigenvalues of the Jacobian matrix and looking for any positive eigenvalues. Note that the elements of the Jacobian matrix will often be a function of the species levels; it is therefore important that the Jacobian be evaluated at the steady state of interest.

Example 9.1

Assume that following system:

$$S_1 \rightarrow S_2 \rightarrow$$

is governed by the set of differential equations:

$$\frac{ds_1}{dt} = -2s_1$$

$$\frac{ds_2}{dt} = 2s_1 - 4s_2$$

If we assume that at time zero, the concentrations of S_1 and S_2 equal 1 and 2 respectively, then the solution to this system can be derived using Mathematica or by using standard algebraic methods for solving linear homogeneous systems. The solution can be found to be:

$$\begin{pmatrix} s_1 \\ s_2 \end{pmatrix} = c_1 \begin{pmatrix} 1 \\ 1 \end{pmatrix} e^{-2t} + c_2 \begin{pmatrix} 0 \\ 1 \end{pmatrix} e^{-4t}$$

$$s_1 = c_1 e^{-2t}$$

$$s_2 = c_1 e^{-2t} + c_2 e^{-4t}$$

Since the exponents are all negative (-2, -2 and -4), the system is stable to perturbations in S_1 and S_2.

We can formally define the internal stability of a biochemical system as follows:

The steady state for the biochemical system:

$$\frac{d\mathbf{s}}{dt} = \mathbf{N}v(\mathbf{s}, \mathbf{p}) \tag{9.6}$$

is stable if all the eigenvalues of the system's Jacobian matrix have negative real parts. The system is unstable if at least one of the eigenvalues has a positive real part.

There are many software packages that compute the eigenvalues of a matrix, and there are a small number of packages that can compute the Jacobian directly from a biochemical model. For example, the script below is taken from Tellurium. It defines a simple model, initializes the model values, computes the steady state, and then prints out the eigenvalues of the Jacobian matrix (Listing 9.1). For a simple one variable model, the Jacobian matrix only has a single entry and the eigenvalue corresponds to that entry. The output from running the script is given below, showing that the eigenvalue is -0.3. Since we have a negative eigenvalue, the pathway must be stable to perturbations in S_1.

```
import tellurium as te

r = te.loada ('''
    $Xo -> S1;  k1*Xo;
```

```
    S1 -> $X1; k2*S1;

    // Set up the model initial conditions
    Xo = 1;    X1 = 0;
    k1 = 0.2;  k2 = 0.3;
''')

# Evaluation of the steady state
r.getSteadyStateValues()

# print the eigenvalues of the Jacobian matrix
print r.getEigenvalues()

# Output follows:
[[-0.3  0. ]]
```

Listing 9.1 Computing eigenvalues to determine stability.

Example 9.2

The following system:

$$\rightarrow S_1 \rightarrow S_2 \rightarrow$$

is governed by the set of differential equations:

$$\frac{ds_1}{dt} = 3 - 2s_1$$

$$\frac{ds_2}{dt} = 2s_1 - 4s_2$$

The Jacobian matrix is computed by differentiating the equations with respect to the steady state values of S_1 and S_2:

$$J = \begin{bmatrix} -2 & 0 \\ 2 & -4 \end{bmatrix}$$

The eigenvalues for this matrix are: -2 and -4, respectively. Since both eigenvalues are negative, the system is stable to small perturbations in S_1 and S_2.

Example 9.3

Consider the system:

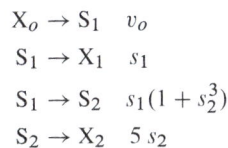

$$\begin{aligned}
X_o &\rightarrow S_1 & v_o \\
S_1 &\rightarrow X_1 & s_1 \\
S_1 &\rightarrow S_2 & s_1(1 + s_2^3) \\
S_2 &\rightarrow X_2 & 5\,s_2
\end{aligned}$$

where X_o, X_1, and X_2 are fixed species. At steady state $s_1 = 2.295$ and $s_2 = 1.14$ with parameter values $v_o = 8$. Note that S_2 is raised to the power of three in the third equation. Determine whether this steady state is stable or not. The differential equations for the system are given by:

$$\frac{ds_1}{dt} = v_o - s_1 - s_1(1 + s_2^3)$$

$$\frac{ds_2}{dt} = s_1(1 + s_2^3) - 5s_2$$

The Jacobian matrix is computed by differentiating the equations with respect to the steady state values of S_1 and S_2 and substituting S_1 and S_2 with their steady state values:

$$J = \begin{bmatrix} -2 - s_2^3 & -3s_1s_2^2 \\ 1 + s^3 & -5 + 3s_1s_2^2 \end{bmatrix} = \begin{bmatrix} -3.4815 & -8.948 \\ 2.482 & 3.948 \end{bmatrix}$$

The eigenvalues for this matrix are: $0.2333 + 2.9i$ and $0.2332 - 2.9i$, respectively.

Since the real parts of the eigenvalues are positive, the system is unstable to small perturbations in S_1 and S_2. The eigenvalues are also complex, meaning they have imaginary terms ($2.9i$). This indicates that the system displays some kind of periodic behavior.

The pattern of eigenvalues tells us a lot about stability, but also about the kind of transients that occur after a perturbation. The following sections will investigate this subject further.

9.1 Jacobian for Biochemical Systems

For a given set of differential equations, we can compute the Jacobian by differentiating the equations with respect to the model variables. However for biochemical networks, the Jacobian can be written in a special way that highlights the importance of the network structure and kinetics of the biochemical reaction steps. To do this, recall that the unscaled elasticity (see (2.3)) is given by:

$$\varepsilon_s^v = \frac{\partial v}{\partial s}$$

where v is a reaction rate and S an effector of the reaction. For example, if $v = k_1 s$, the unscaled elasticity, $\varepsilon_s^v = k_1$. The matrix of unscaled elasticities can be defined as:

$$\frac{\partial \mathbf{v}}{\partial \mathbf{s}} = \begin{bmatrix} \varepsilon_{s_1}^{v_1} & \varepsilon_{s_2}^{v_1} & \cdots & \varepsilon_{s_m}^{v_1} \\ \varepsilon_{s_1}^{v_2} & \varepsilon_{s_2}^{v_2} & \cdots & \varepsilon_{s_m}^{v_2} \\ \vdots & \vdots & & \vdots \\ \varepsilon_{s_1}^{v_n} & \varepsilon_{s_2}^{v_n} & \cdots & \varepsilon_{s_m}^{v_n} \end{bmatrix}$$

where n is the number of reactions and m the number of species. An elasticity matrix has n rows representing n reactions, and m columns represent m species, representing all possible combinations of reactions and species. Many entries in the elasticity matrix will often be zero because not all effectors influence every reaction. For example, consider the pathway:

$$X_o \xrightarrow{v_1} S_1 \xrightarrow{v_2} S_2 \xrightarrow{v_3} X_1$$

where X_o and X_1 are fixed species. The pathway has three reactions, which we will designate v_1, v_2, and v_3 and two floating species, S_1 and S_2. The unscaled elasticity matrix will therefore be a three by two matrix. If we assume reversibility or product inhibition in all three reactions, the entries in the matrix will be:

$$\frac{\partial \mathbf{v}}{\partial \mathbf{s}} = \begin{bmatrix} \varepsilon_{s_1}^{v_1} & 0 \\ \varepsilon_{s_1}^{v_2} & \varepsilon_{s_2}^{v_2} \\ 0 & \varepsilon_{s_2}^{v_3} \end{bmatrix}$$

Note that the entries $\varepsilon_{s_2}^{v_1}$ and $\varepsilon_{s_1}^{v_3}$ are zero because S_2 has no direct effect on v_1, and S_1 has no direct effect on v_3. Some of the unscaled elasticities will also be negative. For example, $\varepsilon_{s_2}^{v_2}$ will be negative because increases in S_2 will slow down the v_2 reaction rate due to product inhibition.

Recall that an element of the Jacobian is defined as:

$$\frac{\partial (ds_j/dt)}{\partial s_j}$$

That is, the differential equation is differentiated with respect to a species. However, we also know that the vector of rates of change is given by the system equation:

$$\frac{d\mathbf{s}}{dt} = \mathbf{N} v(\mathbf{s}, p)$$

Differentiating this with respect to $\mathbf{s}$ yields[1]:

$$\frac{\partial}{\partial \mathbf{s}} \left(\frac{d\mathbf{s}}{dt} \right) = \mathbf{N} \frac{\partial \mathbf{v}}{\partial \mathbf{s}}$$

The left-hand side is the Jacobian, hence the Jacobian is the product of the stoichiometry matrix and the unscaled elasticity matrix:

$$\mathbf{J} = \mathbf{N} \frac{\partial \mathbf{v}}{\partial \mathbf{s}} \tag{9.7}$$

This is a very important result. Given that stability is determined from the Jacobian, this result indicates that stability is a function of network topology **and** the kinetics of the individual reactions. The result indicates that it is not always possible to discern the functional dynamics of a motif just from the topological structure. The dynamics also depends on the kinetics of the constituent parts.

The dynamics of a network is a function of the network topology **and** the kinetics of its constituent parts.

Example 9.4

Consider following branched system:

---

[1] This is modified if there are conserved cycles in the pathway.

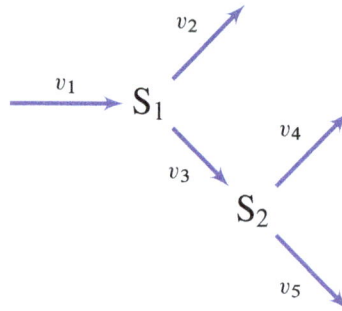

The stoichiometry matrix $\mathbf{N}$, and the matrix of unscaled elasticities, $\partial\mathbf{v}/\partial\mathbf{s}$, are given by:

$$
\mathbf{N} = \begin{array}{c} \\ \\ \end{array}\begin{array}{ccccc} v_1 & v_2 & v_3 & v_4 & v_5 \\ \end{array} \\
\mathbf{N} = \left[\begin{array}{ccccc} 1 & -1 & -1 & 0 & 0 \\ 0 & 0 & 1 & -1 & -1 \end{array} \right] \begin{array}{c} S_1 \\ S_2 \end{array}
\qquad
\frac{\partial\mathbf{v}}{\partial\mathbf{s}} = \left[\begin{array}{cc} \varepsilon_{s_1}^{v_1} & 0 \\ \varepsilon_{s_1}^{v_2} & 0 \\ \varepsilon_{s_1}^{v_3} & \varepsilon_{s_2}^{v_3} \\ 0 & \varepsilon_{s_2}^{v_4} \\ 0 & \varepsilon_{s_2}^{v_5} \end{array} \right]
$$

The Jacobian matrix can be determined from the product, $\mathbf{N}\partial\mathbf{v}/\partial\mathbf{s}$ which yields:

$$
\mathbf{N}\frac{\partial\mathbf{v}}{\partial\mathbf{s}} = \left[\begin{array}{ccccc} 1 & -1 & -1 & 0 & 0 \\ 0 & 0 & 1 & -1 & -1 \end{array} \right] \left[\begin{array}{cc} \varepsilon_{s_1}^{v_1} & 0 \\ \varepsilon_{s_1}^{v_2} & 0 \\ \varepsilon_{s_1}^{v_3} & \varepsilon_{s_2}^{v_3} \\ 0 & \varepsilon_{s_2}^{v_4} \\ 0 & \varepsilon_{s_2}^{v_5} \end{array} \right] = \left[\begin{array}{cc} \varepsilon_{s_1}^{v_1} - \varepsilon_{s_1}^{v_2} - \varepsilon_{s_1}^{v_3} & -\varepsilon_{s_2}^{v_3} \\ \varepsilon_{s_1}^{v_3} & \varepsilon_{s_2}^{v_3} - \varepsilon_{s_2}^{v_4} - \varepsilon_{s_2}^{v_5} \end{array} \right]
$$

9.2 External Stability

There is one other type of stability that is useful with respect to biochemical systems, namely **external stability**. This refers to the idea that if a system is externally stable, then a finite change to an input of the system should elicit a finite change to the internal state of the system. In control theory this is called BIBO, or Bounded Input Bounded Output stability. It is very important to bear in mind that the finite change in the internal state refers to a linearized system.[2] When the system is nonlinear, the output may be bounded by physical constraints and nonlinearities.

A system that is internally unstable will also be unstable to changes in the systems inputs. In biochemical systems such inputs could be the boundary species that feed a pathway, a drug intervention, or the total mass of

---

[2]See Systems Biology: Introduction to Pathway Modeling, Sauro (2014) for details on linearization.

a conserved cycle. External stability can be determined using the same criteria used for internal stability, that is the real parts of eigenvalues of the Jacobian matrix should all be negative.

9.3 Phase Portraits

The word **phase space** refers to a space where all possible states are shown. For example, in a biochemical pathway with two species, S_1 and S_2, the phase space consists of all possible trajectories of S_1 and S_2 in time. For a two dimensional system with species S_1 and S_2, the phase space can be conveniently displayed with S_1 on one axis and S_2 on the other. A line on a two dimensional plane will represent how S_1 and S_2 move with respect to each other in time. Figure 9.3 shows a time course plot for a simple three-step pathway with two species and the corresponding trajectory in the phase plot. In a real phase plot we would have many trajectories shown rather than just one, each trajectory generated from a different set of initial conditions. Figure 9.4 shows the same phase plot but this time with forty-four trajectories. Note they all converge on a single point that represents the system's steady state.

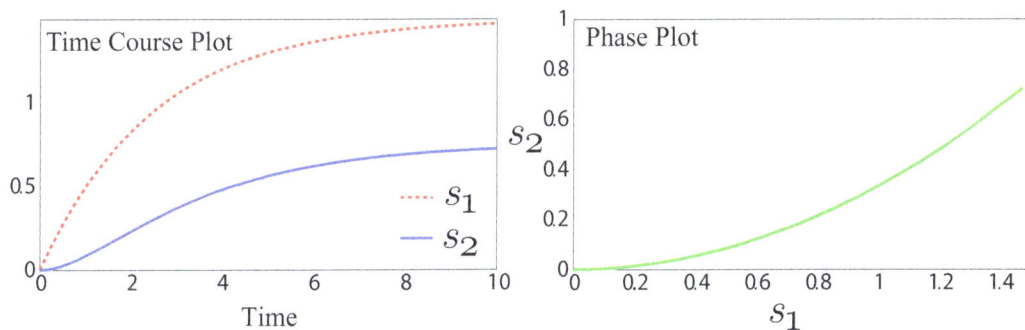

Figure 9.3 Time course simulation plot and corresponding phase plot.

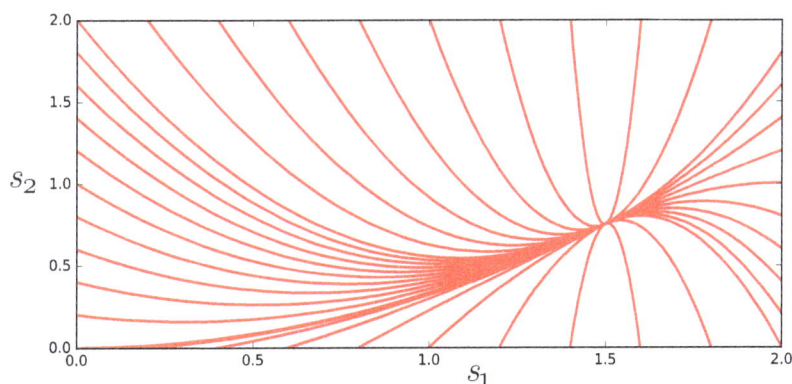

Figure 9.4 Multiple trajectories plotted on the phase plot. Each trajectory was started at a different initial condition for S_1 and S_2. See Tellurium Listing: 9.5.

A visual representation of the phase space is often called a **phase portrait** or **phase plane**.

Let us first consider the most general two dimensional linear set of differential equations:

$$\frac{ds_1}{dt} = a_{11}s_1 + a_{12}s_2$$

$$\frac{ds_2}{dt} = a_{21}s_1 + a_{22}s_2$$

A two dimensional linear system of differential equations has solutions of the form:

$$s_1 = c_1 k_1 e^{\lambda_1 t} + c_2 k_2 e^{\lambda_2 t}$$

$$s_2 = c_3 k_3 e^{\lambda_3 t} + c_4 k_4 e^{\lambda_4 t}$$

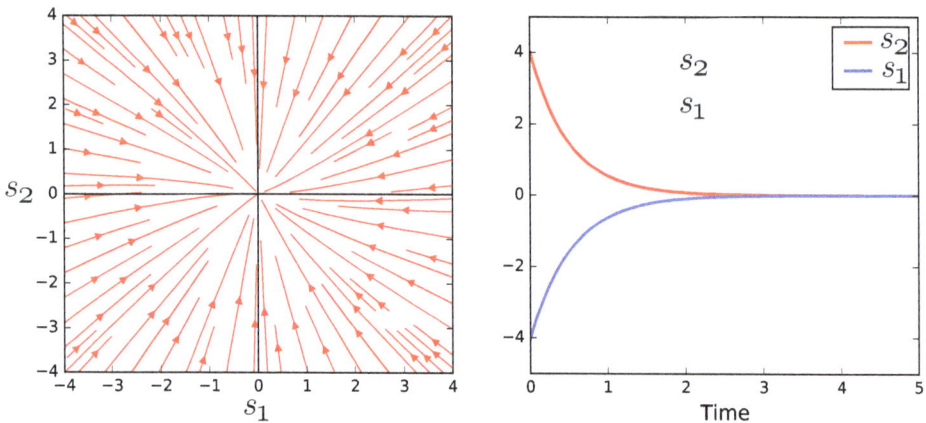

Figure 9.5 Trajectories for a two variable system. On the left is the phase plot and on the right, a single transient as a function of time. This system illustrates a stable node corresponding to **Negative Eigenvalues** in the Jacobian. Jacobian: $a_{11} = -2, a_{12} = 0, a_{21} = -0.15, a_{22} = -2$. Corresponding eigenvalues: $\lambda_1 = -2, \lambda_2 = -2$. The symmetry in the trajectories is due to eigenvalues of the same magnitude. All phase portraits were made using the Tellurium script: 9.6. This can be modified to accommodate the different plots by changing the coefficients in the differential equations.

That is, a sum of exponential terms. The c_i and k_i terms are constants related to the initial conditions and eigenvectors, respectively, but the λ_i terms or eigenvalues determine the qualitative pattern that a given behavior might exhibit. Note that the eigenvalues can be complex or real numbers. In applied mathematics, e raised to a complex number immediately suggests some kind of periodic behavior. Let us now consider different possibilities for the eigenvalues.

• **Both Eigenvalues have the same sign, different magnitude but are real**. If both eigenvalues are negative, the equations describe a system known as a **stable node**. All trajectories move towards the steady state point. If the eigenvalues have the same magnitude and the c_i terms have the same magnitude, the trajectories move to the steady state in a symmetric manner as shown in Figure 9.5.

If the two eigenvalues are both positive, the trajectories move out from the steady state reflecting the fact that the system is unstable. Such a point is called an **unstable node**. If the two eigenvalues have different magnitudes but are still positive, the trajectories twist as shown in Figure 9.6.

• **Real Eigenvalues but of opposite sign**. If the two eigenvalues are real but of opposite sign, we see behavior called a **saddle-node** shown in Figure 9.7. This is where the trajectories move towards the steady state in one

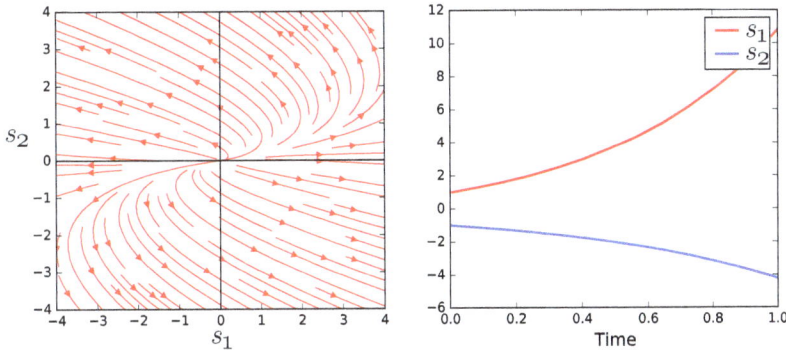

Figure 9.6 Trajectories for a two variable system. On the left is the phase plot and on the right a single transient as a function of time. This system illustrates an unstable node, also called an improper node corresponding to **Positive Eigenvalues**. Jacobian: $a_{11} = 1.2, a_{12} = -2, a_{21} = -0.05, a_{22} = 1.35$. Corresponding eigenvalues: $\lambda_1 = 1.6, \lambda_2 = 0.95$.

direction, called the stable manifold, and form a stable ridge. In all other directions trajectories move away, resulting in an unstable manifold. Since trajectories can only move towards the steady state if they are exactly on the stable ridge; saddle nodes are considered unstable.

Description	Eigenvalues	Behavior
Both Positive	$r_1 > r_2 > 0$	Unstable
Both Negative	$r_1 < r_2 < 0$	Stable
Positive and Negative	$r_1 < 0 < r_1$	Saddle point
Complex Conjugate	$r_1 > r_1 > 0$	Unstable spiral
Complex Conjugate	$r_1 < r_1 < 0$	Stable spiral
Pure Imaginary	$r_1 = r_1 = 0$	Center

Table 9.1 Summary of Node Behaviors.

• **Complex Eigenvalues.** Sometimes the eigenvalues can be complex, that is of the form $a + ib$ where i is the imaginary number. It may seem strange that the solution to a differential equation that describes a physical system can admit complex eigenvalues. To understand what this means, we must recall Euler's formula:

$$e^{i\theta} = cos(\theta) + i \sin(\theta)$$

which can be extended to:

$$e^{(a+bi)t} = e^{at} \cos(bt) + i e^{bt} \sin(bt) \tag{9.8}$$

When the solutions are expressed in sums of sine and cosine terms, the imaginary parts cancel out, leaving just trigonometric terms with real parts (the proof is provided at the end of the chapter). This means that systems with complex eigenvalues show periodic behavior.

Figures 9.8, 9.9, and 9.10 show typical trajectories when the system admits complex eigenvalues. If the real parts are positive, the spiral trajectories move outwards away from the steady state. Such systems are unstable. In a pure linear system, the trajectories will expand out forever. They will only stop and converge to a stable oscillation if the system has nonlinear elements which limits the expansion. In these cases we observe limit cycle behavior.

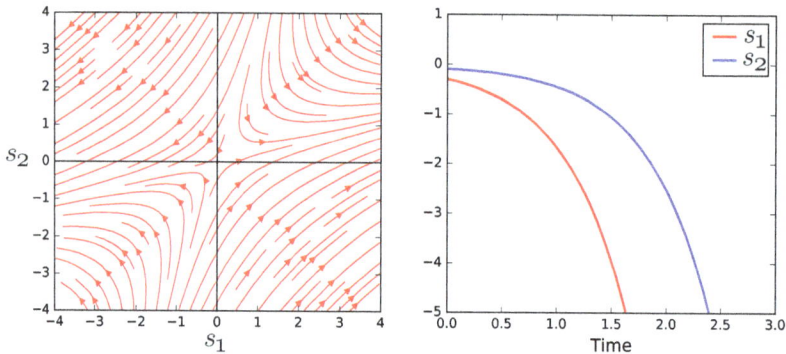

Figure 9.7 Trajectories for a two variable system. On the left is the phase plot and on the right a single transient as a function of time. This system illustrates a saddle node corresponding to **One Positive and One Negative Eigenvalue**. Jacobian: $a_{11} = 2, a_{12} = -1, a_{21} = 1, a_{22} = -2$. Corresponding eigenvalues: $\lambda_1 = -1.73, \lambda_2 = 1.73$.

If the real parts of the eigenvalues are negative, the spiral trajectory moves into the steady state and is therefore considered stable.

• **Imaginary Eigenvalues with Zero Real Parts.** It is possible for the pair of eigenvalues to have no real component but retain an imaginary part. In this situation the behavior is called a center. This is where the trajectory orbits the steady state. The oscillation is an unusual one in the sense that it implies zero dampening in the system, in other words, zero energy loss.

Conjugate Pair

A complex conjugate pair is a complex number of the form: $a \pm bi$. The eigenvalues for a two variable linear system with matrix A can be computed directly using the relation:

$$\lambda = \frac{\text{tr}(A) \pm \sqrt{\text{tr}^2(A) - 4\det(A)}}{2}$$

where $\text{tr}(A) = a + d$, and $\det(A) = ad - bc$. If the term in the square root is negative, the eigenvalues will always come out as a conjugate pair owing to the $\pm$ term. If $\text{tr}^2(A) - 4\det(A) < 0$, then the solution will be the conjugate pair:

$$\lambda = \frac{\text{tr}(A)}{2} \pm \frac{\sqrt{\text{tr}^2(A) - 4\det(A)}}{2}$$

Therefore a complex eigenvalue will always be accompanied by its conjugate partner.

Such a situation is rare in biology and even in non-living systems, such behavior tends to be idealized. For example, a pendulum in a vacuum with no friction at the fulcrum. The other unusual aspect of a center is that the oscillation depends on the starting condition. Again, we can relate this to a pendulum where the swing depends on how much force we initially apply to the pendulum. Biological oscillators are invariably energy dependent and the frequency is independent of the initial conditions. Biological oscillators therefore tend not to be center types.

Focus now on a specific system such as the simple three step pathway:

$$\xrightarrow{v_o} S_1 \xrightarrow{k_1 \, s_1} S_2 \xrightarrow{k_2 \, s_2}$$

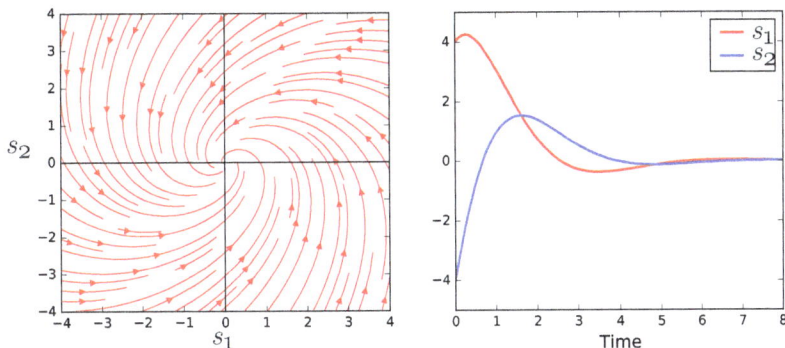

Figure 9.8 Trajectories for a two variable system. On the left is the phase plot, and on the right, a single transient as a function of time. This system illustrates a stable spiral node corresponding to **Negative Complex Eigenvalues**. Jacobian: $a_{11} = -0.5, a_{12} = -1, a_{21} = 1, a_{22} = -1$. Corresponding eigenvalues: $\lambda_1 = -0.75 + 0.97i, \lambda_2 = -0.75 - 0.97i$.

with two linear differential equations:

$$\frac{ds_1}{dt} = v_o - k_1 s_1$$

$$\frac{ds_2}{dt} = k_1 s_1 - k_2 s_2$$

The Jacobian for this system is given by:

$$J = \begin{bmatrix} -k_1 & 0 \\ k_1 & -k_2 \end{bmatrix}$$

The eigenvalues of the Jacobian are $-k_1$ and $-k_2$. This corresponds to a stable node. A simple three step pathway with linear mass-action kinetics cannot exhibit any behavior other than a stable node.

9.4 Bifurcation Plots

In its simplest form, a bifurcation plot is just a plot of the steady state value of a system variable, such as a concentration or flux versus a parameter of the system. For example, we know that the steady state solution for the simple system (9.1):

$$\frac{ds_1}{dt} = k_1 x_o - k_2 s_1 \tag{9.9}$$

is given by:

$$s_1 = k_1 x_o / k_2 \tag{9.10}$$

We can plot the steady state value of S_1 as a function of k_2 as shown in Figure 9.12. This isn't a particularly interesting bifurcation plot however, and misses one of the most important characteristics.

Equation (9.10) shows that the simple system (9.9) only has one steady state for a given set of parameters. That is, if we set values to X_o, k_1, and k_2, we find there is only *one value* of S_1 that satisfies these parameter

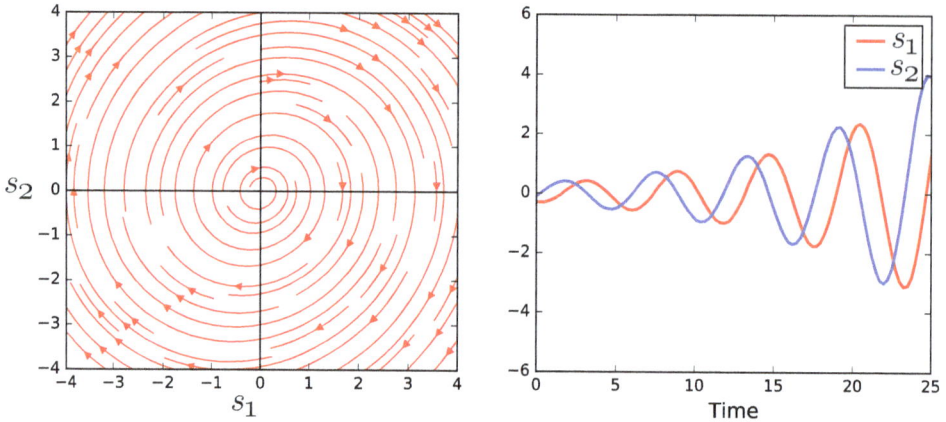

Figure 9.9 Phase portrait for a two variable system. Unstable spiral node. **Positive Complex Eigenvalues**. Jacobian: $a_{11} = 0, a_{12} = 1.0, a_{21} = -1.2, a_{22} = 0.2$. Corresponding eigenvalues: $\lambda_1 = 0.1 + 1.09i, \lambda_2 = 0.1 - 1.09i$.

settings. This is what Figure 9.12 demonstrates. What is more interesting is when a system admits multiple steady state values for a given set of parameter values. To illustrate this behavior, let us look at a common system that can admit three possible steady states. It is in these cases that bifurcation plots become particularly useful and more interesting.

Bistable Systems

Bifurcation plots can be useful for identifying changes in qualitative behavior, particularly for systems that have multiple steady states. Consider the system shown in Figure 9.13 where the species S stimulates its own production forming a positive feedback loop. As species S accumulates, the rate of formation of S increases.

At first glance this would seem to be a very unstable situation. One might imagine that the concentration of S would continue to increase without limit. However, physical constraints ultimately limits the upper value for the concentration of S. To investigate the properties of these networks, we will construct a simple model. This model uses the following kinetic laws for the synthesis and degradation steps:

$$v_1 = b + k_1 \frac{s_1^4}{k_2 + s_1^4}$$

$$v_2 = k_3 s_1$$

v_1 is a Hill like equation with a Hill coefficient of four and a fixed basal rate of b. v_2 is a simple irreversible mass-action rate law. The differential equation for the model is:

$$\frac{ds_1}{dt} = v_1 - v_2$$

We now plot both rate laws as a function of the species S_1. When we do this we obtain Figure 9.14. The intersection points marked with filled circles indicate the steady state solutions because at these points, $v_1 = v_2$. If we vary the slope of v_2 by changing k_3, the intersection points will change (Figure 9.15). At a high k_3 value, only one intersection point remains (Panel C), the low intersection point. If the value of k_3 is low, only

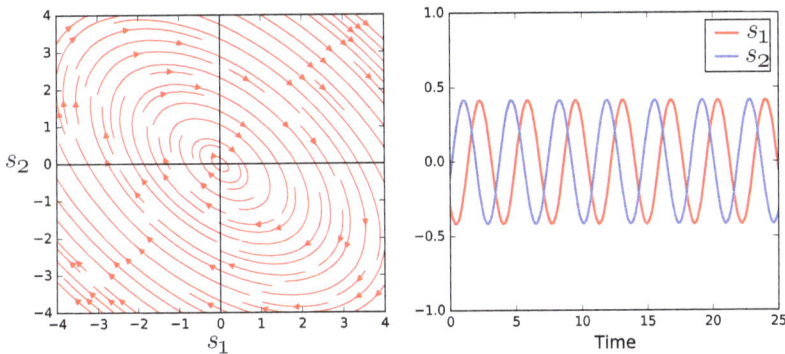

Figure 9.10 Phase portrait for a two variable system. Center node. **Complex Eigenvalues, Zero Real Part**. Jacobian: $a_{11} = 1, a_{12} = 2.0, a_{21} = -2, a_{22} = -1$. Corresponding eigenvalues: $\lambda_1 = 0 + 1.76i, \lambda_2 = 0 - 1.76i$. See Tellurium script:9.6

the high intersection point remains (Panel A). However, with the right set of parameter values, we can make a system possess three steady state values (Panel B).

We can determine the three different steady state stabilities by doing a simple graphical analysis on Figure 9.14. Figure 9.16 shows the same plot but with perturbations.

Starting with the first steady state in the lower left corner of Figure 9.16, consider a perturbation made in s, δs. This means that both v_1 and v_2 increase, however $v_2 > v_1$ meaning that after the perturbation, the rate of change in s is *negative*. Since it is negative, this restores the perturbation back to the steady state. The same logic applies to the upper steady state. This tells us that the lower and upper steady states are both stable.

What about the middle steady state? Consider again a perturbation, δs. This time $v_1 > v_2$ which means that the rate of change of s is *positive*. Since it is positive, the perturbation, instead of falling back, continues to grow until s reaches the upper steady state. We conclude that the middle steady state is therefore unstable. This system possess three steady states, one unstable and two stable. Such a system is known as a **bistable system** because it can rest in one of two stable states but not the third.

Another way to observe the different steady states is to run a time course simulation at many different starting points. Figure 9.17 shows the plots generated using the script in Listing 9.2. The plots show two steady states, a high state at around 40, and a low state at around 3. Notice that there is no third state observed. As we have already discussed, the middle steady state is unstable, and all trajectories diverge from this point. It is therefore not possible, when doing a time course simulation, to observe an unstable steady state since there is no way to reach it.[3]

```
import tellurium as te
import numpy

r = te.loada ('''
    J1: $Xo -> x; 0.1 + k1*x^4/(k2+x^4);
    x -> $w; k3*x;

    k1 = 0.9;
    k2 = 0.3;
```

---

[3]The author has been reliably informed that running time backwards in a time course simulation will cause the simulation to converge on the unstable steady state. The author has not tried this himself, however.

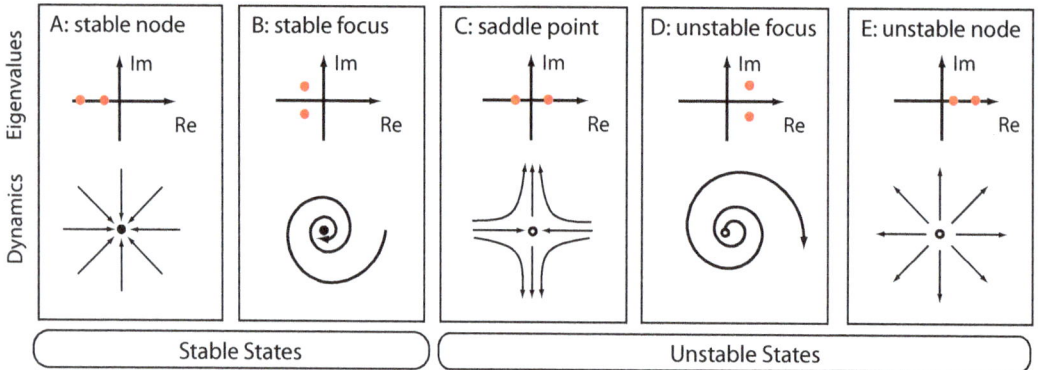

Figure 9.11 Summary of behaviors including dynamics and associated eigenvalues for a two dimensional linear system. Adapted from "Computational Models of Metabolism: Stability and Regulation in Metabolic Networks", Adv in Chem Phys, Vol 142, Steuer and Junker.

```
    k3 = 0.7;
    x = 0.05;
''')

m = r.simulate(0, 15, 100)
for i in range(1, 10):
    r.x = i*0.2
    mm = r.simulate(0, 15, 100, ["x"])
    m = numpy.hstack((m, mm))
te.plotArray(m)
```

Listing 9.2 Tellurium script used to generate Figure 9.17.

We can get an estimate for the values of all three steady states from Figure 9.14. Reading from the graph we find s values at $0.145, 0.683$, and 1.309. It is also possible to use the steady state solver from Tellurium to locate the steady states. Listing 9.3 shows a simple script to compute them. By setting an appropriate initial condition, we can use Tellurium to pinpoint all three steady states. For example, if we use an initial value of S_1 at 0.43, the steady state solver will locate the third steady state at 0.683. Steady state solvers such as the one included with Tellurium can be used to find unstable states, providing the initial starting point is close enough.

```
import tellurium as te

r = te.loada ('''
    $Xo -> x; 0.1 + k1*x^4/(k2+x^4);
    x -> $w; k3*x;

    // Initialization here
    k1 = 0.9; k2 = 0.3;
    k3 = 0.7;
''')

# Compute steady state
```

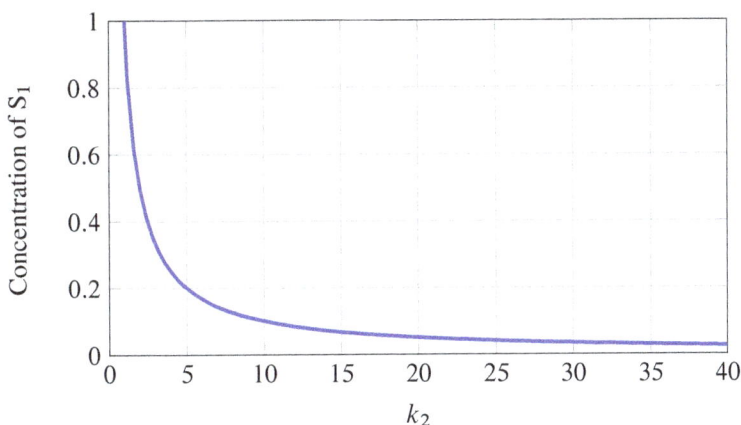

Figure 9.12 Steady state concentration of S_1 as a function of k_2 for the system, $ds_1/dt = k_1 x_o - k_2 s_1$.

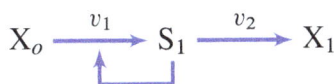

Figure 9.13 System with positive feedback.

```
print r.getSteadyStateValues()
```

Listing 9.3 Basic bistable model.

Stability of Positive Feedback

What determines the stability of a positive feedback system? Let us consider the same network with positive feedback as before (Figure 9.13). The differential equation for this system is:

$$\frac{ds}{dt} = v_1(s) - v_2(s)$$

where we have explicitly shown that each reaction rate is a function of s. To determine whether the system is stable to small perturbations, we can differentiate the equation with respect to s to form the Jacobian. Notice there is only one element in the Jacobian because we only have one variable, S_1:

$$\frac{ds/dt}{ds} = \frac{\partial v_1}{\partial s} - \frac{\partial v_2}{\partial s}$$

The terms on the right are unscaled elasticities (2.3). If the expression is positive, the system is unstable because it means that ds/dt is increasing if we increase s. We can scale both sides to yield:

$$J_s = \varepsilon_s^1 - \varepsilon_s^2$$

where the right-hand term now includes the scaled elasticities (2.2). The criteria for instability is that $\varepsilon_s^1 - \varepsilon_s^2 < 0$, or $\varepsilon_s^1 > \varepsilon_s^2$. Therefore, if the positive feedback is stronger than the effect of s on the degradation step v_2, the system will be unstable.

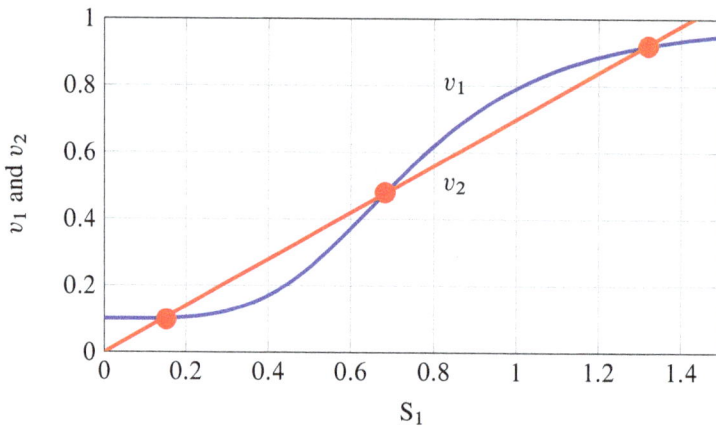

Figure 9.14 Reaction velocities, v_1 and v_2, as a function of s for the system in Figure 9.13. The intersection points marked by full circles indicate possible steady states. $k_1 = 0.9; k_2 = 0.3; k_3 = 0.7; b = 0.1$.

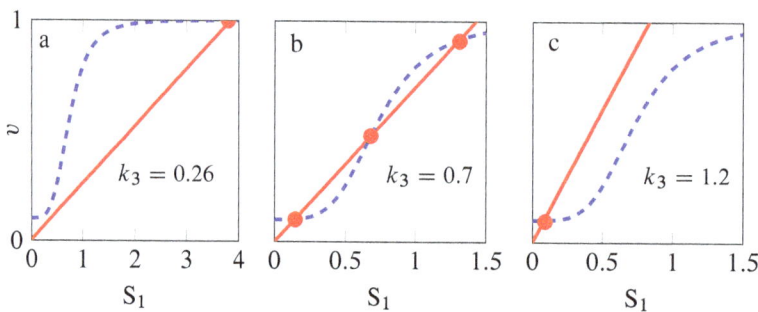

Figure 9.15 v_1 and v_2 plotted against s concentration. Intersection points on the curves mark the steady state points. Panel a) One intersection point at a high steady state; b) Three steady states; c) One low steady state.

> If the positive feedback elasticity is larger than the elasticity of the degradation step then the system can admit unstable states.

Recall that the elasticities are a measure of the kinetic order of the reaction. Thus an elasticity of one means the reaction is first-order. A saturable irreversible Michaelis-Menten reaction will have a variable kinetic order between one and zero (near saturation). A Hill equation can, depending on the Hill coefficient, can have kinetic orders greater than one (Table 9.2). Knowing this information, there are at least two ways to make sure that the elasticity for the feedback elasticity, v_1, is greater than the elasticity for the degradation step, v_2:

1. v_1 is modeled using a Hill equation with a Hill coefficient > 1 and v_2 is first-order or less.

2. A Hill coefficient = 1 on v_1, with Michaelis-Menten saturable kinetics on v_2 to ensure less than first-order kinetics on v_2.

By substituting the three possible steady state values for s into the equation for ds/dt, we can compute the

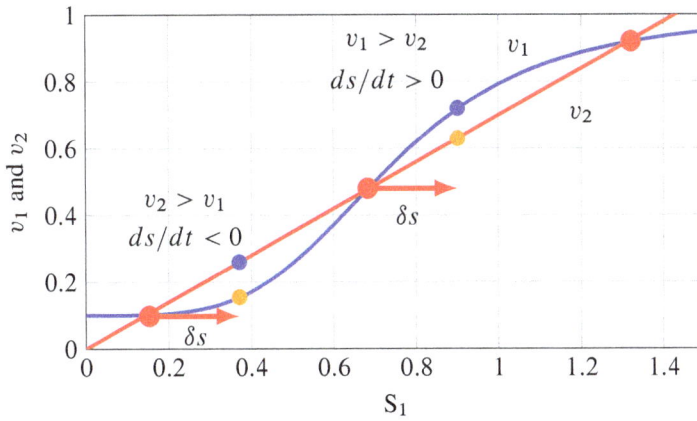

Figure 9.16 A graphical understanding of the stability of steady states. See text for details. Computed using the SBW rate law plotter. $k_1 = 0.9; k_2 = 0.3; k_3 = 0.7; b = 0.1$.

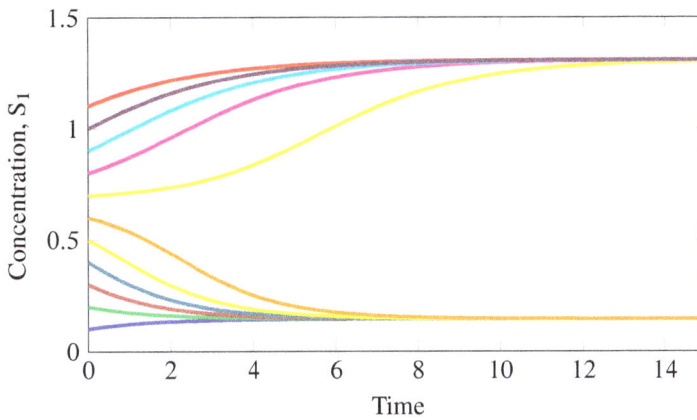

Figure 9.17 Time course data generated from Tellurium model 9.2. Each line represents a different initial concentration for S_1. Some trajectories transition to the low state while others to the upper state.

Kinetic Order	Elasticity
First-Order	1.0
Zero-Order	0.0
Sigmoidal	> 1.0

Table 9.2

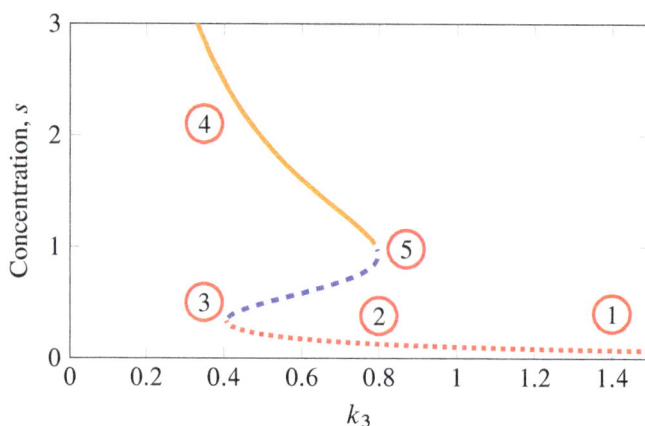

Figure 9.18 Plotting intersection points from Figure 9.14 as a function of k_3. Dotted line marks the lower intersection point, dashed line the middle intersection points, and solid line the upper intersection point. The Tellurium script for generating this graph can be found in listing 9.7.

value for the Jacobian element in each case (Table 9.3).

Steady State s	Jacobian: $(ds/dt)/ds$	Elasticity, ε_s^1
0.145	-0.664	0.052
0.683	0.585	1.835
1.309	-0.47	0.33

Table 9.3 Table of steady state values of s and corresponding values for the Jacobian element. Negative Jacobian values indicate a stable steady state, positive elements indicate an unstable steady state. The table shows one stable and two unstable steady states.

The unstable steady state at $s = 0.683$ has an elasticity for v_1 of 1.835. Note this value is greater than the elasticity of the first-order degradation reaction, v_2, which equals one. Therefore this state is unstable.

Bifurcation Plot

Let's now return to the question of plotting a bifurcation graph for the bistable system in Figure 9.13. Figure 9.15 shows both reaction rates, v_1 and v_2 plotted as a function of the intermediate species, S. In this figure we see three intersection points, marking the three possible steady states. By varying the degradation constant k_3, we can change the behavior of the system so that it exhibits a single high steady state, three separate steady states, or a single low steady state (See Figure 9.14).

If we track the intersection points as we vary the value of the rate constant k_3, we obtain the bifurcation plot shown in Figure 9.18.

Figure 9.18 shows that at some value of the parameter k_3, the system has three possible steady states, outside this range only a single steady state persists. Bifurcation diagrams are extremely useful for uncovering and displaying such information. Drawing bifurcation diagrams is not easy, however, and there are some software tools that can help. Figure 9.18 for example was generated using the SBW Auto C# tool[4]. Another useful tool for drawing bifurcation diagrams is Oscill8[5]. Both tools can read SBML. Figure 9.18 was generated first by entering the model into Tellurium (Shown in Listing 9.3) to generate the SBML. The model was then passed to Auto C# to produce the bifurcation diagram.

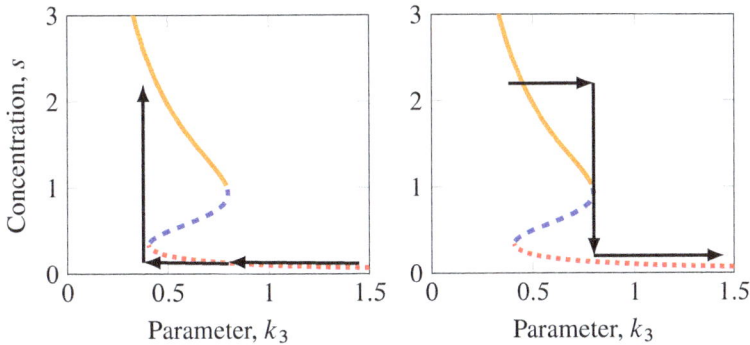

Figure 9.19 Depending on whether we increase or decrease k_3, the steady state path we traverse will be different. This is a characteristic of hysteresis.

The bifurcation plot in Figure 9.18 shows how the steady state changes as a function of a parameter, in this case k_3. Of interest is the following observation. If we start k_3 at a high value of 1.4 (Marker 1), we see that there is only one low steady state. As k_3 is reduced in value, we pass the point at approximately $k_3 = 0.8$ (Marker 2) where three steady states emerge. We continue lowering k_3, and see that the concentration of s rises very slowly until about 0.4 (Marker 3). At this point the system jumps to a single steady state, but now at a high level (Marker 4). The interesting observation is that if we now increase the value of k_3, we do not traverse the same path. As we increase k_3 beyond 0.4, we do not drop back to the low state, but continue along the high state until we reach $k_3 = 0.8$ (Marker 5), at which point we jump down to the low state (Marker 2). The direction in which we traverse the parameter k_3 affects the type of behavior we observe. This special phenomena is called **hysteresis**, Figure 9.19.

> Hysteresis is where the behavior of a system depends on its past history.

Irreversible Bistability

It is possible to design an irreversible bistable switch. Figure 9.21 shows the bifurcation plot for such a system. This is modified from the 'Mutual activation' model in the review by Tyson, Chen and Noval [110], Figure 1e. In this example increasing the signal results in the system switching to the high state at around 2.0. If we reduce the signal from a high level, we traverse the top arc. If we assume the signal can never be negative, we will remain at the high steady state even if the signal is reduced to zero. The bifurcation plot in the negative quadrant of the graph is physically inaccessible. This means it is not possible to transition to the low steady state by decreasing the signal. As a result, the bistable system is **irreversible**, that is, once it is switched on, it will always remain on.

```
import tellurium as te
```

---

[4]http://jdesigner.sourceforge.net/Site/Auto_C.html
[5] http://oscill8.sourceforge.net/

```
r = te.loada ('''
    $X -> R1;   k1*EP + k2*Signal;
    R1 -> $w;   k3*R1;
    EP -> E;    Vm1*EP/(Km + EP);
    E -> EP;    ((Vm2+R1)*E)/(Km + E);

    Vm1 = 12; Vm2 = 6;
    Km = 0.6;
    k1 = 1.6; k2 = 4;
    E = 5; EP = 15;
    k3 = 3; Signal = 0.1;
''')

result = r.simulate(0, 20, 500)
r.plot()
```

Listing 9.4 Script for Figure 9.21.

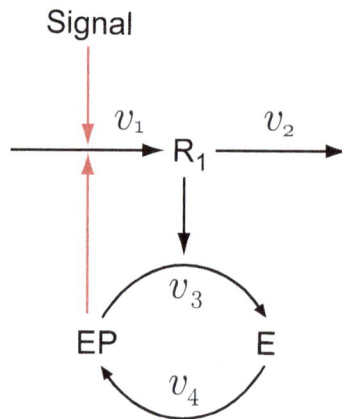

Figure 9.20 System with Positive Feedback using a covalent modification cycle, E, EP.

Further Reading

1. Edelstein-Keshet L (2005) Mathematical Model sin Biology. SIAM Classical In Applied Mathematics. ISBN-10: 0-89871-554-7.

2. Fall CP, Marland ES, Wagner JM, Tyson JJ (2000) Computational Cell Biology. Springer: Interdisciplinary Applied Mathematics. ISBN 0-387-95369-8.

3. Steuer R and Junker BH (2009). Computational models of metabolism: stability and regulation in metabolic networks. Advances in chemical physics, 142, 105.

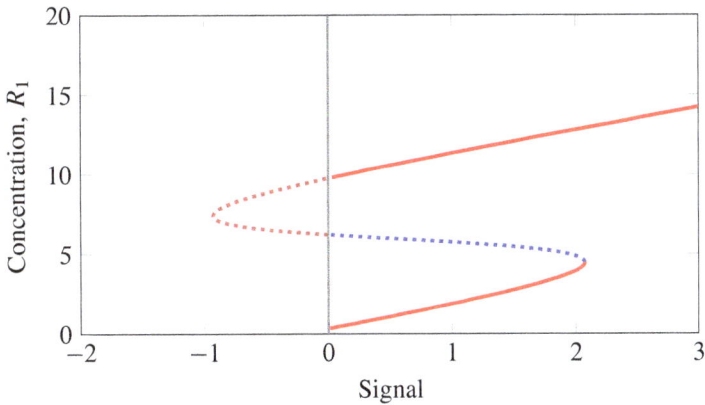

Figure 9.21 Bifurcation diagram for species R_1 with respect to the signal. Signal from the model shown in Tellurium script 9.4. The continuous line represents stable steady state points, the dotted line the unstable steady states. Plotted using Oscil8 http://oscill8.sourceforge.net/.

Exercises

1. Determine the Jacobian matrix for the following two systems:

   a) $\dfrac{dx}{dt} = x^2 - y^2 \quad \dfrac{dy}{dt} = x(1-y)$

   b) $\dfrac{dx}{dt} = y - xy \quad \dfrac{dy}{dt} = xy$

2. Compute the steady state solutions to the two systems in the previous question.

3. Determine the stability of the solutions from the previous question.

4. Determine the Jacobian in terms of the elasticities and stoichiometry matrix for the following systems:

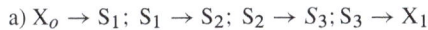

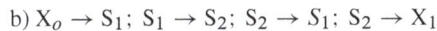

   a) $X_o \rightarrow S_1; \ S_1 \rightarrow S_2; \ S_2 \rightarrow S_3; S_3 \rightarrow X_1$

   b) $X_o \rightarrow S_1; \ S_1 \rightarrow S_2; \ S_2 \rightarrow S_1; \ S_2 \rightarrow X_1$

   c) $S_1 \rightarrow S_2; \ S_2 \rightarrow S_3$

   Assume all reactions are product insensitive, X_i species are fixed, and in c) S_3 regulates the first step.

5. Show that the following system is stable to perturbations in S_1 and S_2 by computing the eigenvalues at steady state (See Listing 9.1):

   $$X_o \rightarrow S_1 \rightarrow S_2 \rightarrow X_1$$

The three rate laws are given by:

$$v_1 = \frac{V_{m1}\, x_o}{Km_1 + x_o + s_1/K_i}$$

$$v_2 = \frac{V_{m2}\, s_2}{Km_2 + s_1 + s_2/K_j}$$

$$v_2 = \frac{V_{m3}\, s_3}{Km_3 + s_2}$$

Assign the following values to the parameters: $x_o = 1; x_1 = 0; V_{m1} = 1.5; V_{m2} = 2.3; V_{m3} = 1.9; Km_1 = 0.5;, Km_2 = 0.6; Km_3 = 0.45; K_i = 0.1; K_j = 0.2.$

6. Show that the following system is unstable or stable depending on the value of v_2.

  a) $\rightarrow S_1;\ v_o$

  b) $S_1 \rightarrow;\ v_2 = k_2 s_1$

  c) $S_1 \rightarrow;\ v_3 = k_3 \dfrac{s_1}{1 + (s_1/K_I)^n}$

7. Show that the following system is unstable. What kind of unstable dynamics does it have?

```
import tellurium as te

rr = te.loada ('''
    J0: $X0 -> S1; VM1*(X0-S1/Keq1)/(1+X0+S1+pow(S4,h));
    J1: S1 -> S2; (10*S1-2*S2)/(1+S1+S2);
    J2: S2 -> S3; (10*S2-2*S3)/(1+S2+S3);
    J3: S3 -> S4; (10*S3-2*S4)/(1+S3+S4);
    J4: S4 -> $X1; Vm4*S4/(KS4+S4);

    X0 = 10;        X1 = 0;
    S1 = 0.973182; S2 = 1.15274;
    S3 = 1.22721;   S4 = 1.5635;
    VM1 = 10; Keq1 = 10;
    h = 10;    Vm4 = 2.5;
    KS4 = 0.5;
''')
```

8. Show that the following system is unstable. What kind of unstable dynamics does it have?

```
import tellurium as te

r = te.loada ('''
    // var R, X, E, EP;
    // ext src, S, Waste;

    J0:  $src -> X;    k1*S;
    J1:  X -> R;       (kop + ko*EP)*X;
    J2:  R -> $waste;  k2*R;
```

```
    J3:   E -> EP;        Vmax_1*R*E/(Km_1 + E);
    J4:   EP -> E;        Vmax_2*EP/(Km_2 + EP);

    src = 0;        kop = 0.01;
    ko =  0.4;      k1 = 1;
    k2 = 1;         R = 1;
    EP = 1;         S = 0.2;
    Km_1 = 0.05;   Km_2 = 0.05;
    Vmax_2 = 0.3;  Vmax_1 = 1;
    KS4 = 0.5;
''')

result = r.simulate(0, 300, 1000)
r.plot()
```

9.5 Appendix

Prove that the presence of imaginary numbers in the solution to a set of differential equations means that the solution is periodic (Equation (9.8)). Consider the system:

$$x(t) = c_1 z_1 e^{(\lambda + i\mu)t} + c_2 z_2 e^{(\lambda - i\mu)t}$$

where z_1 and z_2 are corresponding conjugate eigenvectors. Using Euler's formula, $e^{i\mu} = \cos(\mu) + i \sin(\mu)$ and that $e^{(\lambda + i\mu)t} = e^{\lambda t} e^{i\mu t}$ we obtain:

$$x(t) = c_1 z_1 e^{\lambda t} (\cos(\mu t) + i \sin(\mu t))$$
$$+ c_2 z_2 e^{\lambda t} (\cos(\mu t) - i \sin(\mu t))$$

Writing the conjugate eigenvectors as $z_1 = a + bi$ and $z_2 = a - bi$, we get:

$$x(t) = c_1 (a + bi) e^{\lambda t} (\cos(\mu t) + i \sin(\mu t))$$
$$+ (a - bi) e^{\lambda t} (\cos(\mu t) - i \sin(\mu t))$$

Multiply out and separate the real and imaginary parts:

$$x(t) = e^{\lambda t} [c_1 (a \cos(\mu t) - b \sin(\mu t) + i(a \sin(\mu t) + b \cos(\mu t)))$$
$$+ c_2 (a \cos(\mu t) - b \sin(\mu t) - i(a \sin(\mu t) + b \cos(\mu t)))]$$

The complex terms cancel leaving only the real parts. If we set $c_1 + c_2 = k_1$ and $(c_1 - c_2)i = k_2$ then:

$$x(t) = e^{\lambda t} [k_1 (a \cos(\mu t) - b \sin(\mu t))$$
$$k_2 (a \sin(\mu t) + b \cos(\mu t))]$$

The solution is real when the constants c_1 and c_2 are real. This will only be the case when the eigenvalues are a conjugate pair, $(a \pm ib)$, which is the case we are considering. Therefore, systems that admit a complex pair of conjugate eigenvalues result in periodic real solutions.

```
# Plot a phase portrait for a simple two species pathway
import tellurium as te
```

```python
import  matplotlib.pyplot as plt

r = te.loada ('''
    $Xo -> S1;   k1*Xo;
    S1 -> S2;    k2*S1;
    S2 -> $X1;   k3*S2;
    k1 = 0.6; Xo = 1;
    k2 = 0.4; k3 = 0.8;
''')

plt.figure(figsize=(9,4))
S1Start = 0
S2Start = 0
for i in range(1, 11):
    r.S1 = S1Start
    r.S2 = S2Start
    m = r.simulate(0, 10, 120, ["S1", "S2"])
    p = te.plotArray(m, show=False)
    plt.setp (p, color='r')
    S1Start = S1Start + 0.2
S1Start = 2
S2Start = 0
for i in range(1, 11):
    r.S1 = S1Start
    r.S2 = S2Start
    m = r.simulate(0, 10, 120, ["S1", "S2"])
    p = te.plotArray(m, show=False)
    plt.setp (p, color='r')
    S2Start = S2Start + 0.2
S2Start = 0
S1Start = 0
for i in range(1, 11):
    r.S1 = S1Start
    r.S2 = S2Start
    m = r.simulate(0, 10, 120, ["S1", "S2"])
    p = te.plotArray(m, show=False)
    plt.setp (p, color='r')
    S2Start = S2Start + 0.2
S1Start = 0
S2Start = 2
for i in range(1, 11):
    r.S1 = S1Start
    r.S2 = S2Start
    m = r.simulate(0, 10, 120, ["S1", "S2"])
    p = te.plotArray(m, show=False)
    plt.setp (p, color='r')
    S1Start = S1Start + 0.2
plt.xlabel ('S1', fontsize=16)
plt.ylabel ('S2', fontsize=16)
plt.savefig ("plot.pdf")
plt.show()
```

Listing 9.5 Script for Figure 9.4.

```
import numpy as np
import matplotlib.pyplot as plt
import tellurium as te

Y, X = np.mgrid[-4:4:100j, -4:4:100j]

r = te.loada('''
    x' = 1*x + 2*y;
    y' = -2*x - 1*y;

    x = -0.3; y = -0.1;
''')

m = r.simulate (0, 25, 200)

U =  1*X + 2*Y
V = -2*X - 1*Y

plt.subplots(1,2, figsize=(10,4))
plt.subplot(121)
plt.xlabel('x', fontsize='16')
plt.ylabel('y', fontsize='16')
plt.streamplot(X, Y, U, V, density=[1, 1])

plt.ylim((-4,4))
plt.xlim((-4,4))

plt.axhline(0, color='black')
plt.axvline(0, color='black')

plt.subplot(122)

plt.ylim((-1,1))
plt.xlim((0,25))
plt.xlabel('Time', fontsize='13')


plt.plot (m[:,0], m[:,1], color='r', linewidth=2, label='x')
plt.plot (m[:,0], m[:,2], color='b', linewidth=2, label='y')
plt.legend()

plt.savefig ('c:\\tmp\\phase.pdf')

plt.show()
```

Listing 9.6 Script for Figure 9.10.

```
import tellurium as te
import roadrunner
from rrplugins import *

r = te.loada ('''
    $Xo -> s; b + k1*s^4/(k2 + s^4);
    s -> $w; k3*s;

    // Initialization here
    k1 = 0.9; k2 = 0.3;
    k3 = 0.7; b = 0.1;
''')

r.conservedMoietyAnalysis = True

auto = Plugin("tel_auto2000")

auto.setProperty("SBML", r.getSBML())
auto.setProperty("NMX", 5000)
auto.setProperty("ScanDirection", "Negative")
auto.setProperty("PrincipalContinuationParameter", 'k3')
auto.setProperty("PCPLowerBound", 0.34)
auto.setProperty("PCPUpperBound", 1.4)

auto.execute()

pts     = auto.BifurcationPoints
lbls    = auto.BifurcationLabels
biData  = auto.BifurcationData
biData.plotBifurcationDiagram(pts, lbls)
```

Listing 9.7 Script for Figure 9.18.

10

Stability of Negative Feedback Systems

10.1   Introduction

Section 10.3 includes some advanced mathematical techniques and may be omitted in an introductory class.

In a negative feedback system most disturbances are damped due to the action of the feedback. However, what if the feedback mechanism takes too long to respond, so that by the time the negative feedback acts, the disturbance has already abated? In this situation the feedback would attempt to restore a disturbance that is no longer present, resulting in an incorrect action. Now imagine that the feedback system acts in the opposite direction to the disturbance because of the delay. This would cause the disturbance to grow rather than decline. If there is sufficient gain in the feedback loop to amplify or at least maintain the original disturbance, the result would be an unstable situation. If the loop gain amplifies, then the disturbance will grow until it reaches the physical limits of the system or nonlinearities intervene. At this point the loop gain is likely to fall and the disturbance likewise will fall. The net result is a continuous growth and decline in the original disturbance, that is, a sustained oscillation.

The key elements for sustaining an oscillation is a sufficient loop gain (at least 1.0) and a delay in the system response of exactly -360°. Delays are measured by inputting a periodic signal such as a sine wave, and measuring the shift in phase that results. A -360° phase shift means that the sine wave has been delayed by a full cycle. In a system with negative feedback, the delay is caused by two components. The first shift is -180°, which comes from the inverting negative feedback. That is, if a rising disturbance reaches the feedback signal, the reaction rate at the signal target must fall and vice versa. The second delay emerges as the disturbance propagates through the reaction steps of the system. Each reaction step can contribute up to -90° of delay; two reaction steps can contribute up to -180°, and three reaction steps up to -270° and so on. However, the degree of delay caused by the reaction steps is a function of the frequency components in the disturbance. Random disturbances in the system will occur at all frequencies, therefore disturbances in a system that has sufficient loop gain will quickly locate the point where the delay caused by the reaction steps is exactly -180°. Adding this to the negative feedback delay, we arrive at a total of -360°. This results in a rapid destabilization of the system as disturbances are reinforced instead of damped.

Another way to look at this is that if a $-360°$ delay occurs, it means that the negative feedback is effectively behaving as a positive feedback, and positive feedback is normally destabilizing.

10.2   Stability using MCA

In this section we will look at the stability of negative feedback systems using metabolic control analysis. This is a fairly advanced topic which can be omitted if necessary although the analysis of the three-step pathway is reasonably straightforward. Consider first a simple three step pathway with negative feedback. This system (Figure 10.1) has just two steps within the feedback loop. S_2 is the feedback signal that inhibits v_1.

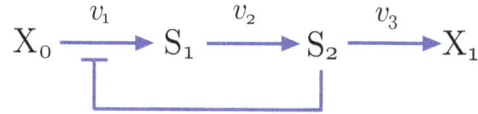

Figure 10.1 Negative feedback with three reaction steps.

We can compute the Jacobian using the expression introduced in the last chapter (9.7):

$$N \frac{\partial v}{\partial s}$$

where N is the stoichiometry matrix, and $\partial v / \partial s$ is the matrix of unscaled elasticities. The stoichiometry matrix N for the model in Figure 10.1 is:

$$N = \begin{bmatrix} 1 & -1 & 0 \\ 0 & 1 & -1 \end{bmatrix}$$

As before, the symbol $\mathcal{E}_j^i$ is used to denote the unscaled elasticity for reaction i with respect to species j, so that the elasticity matrix for the feedback model is given by:

$$\frac{\partial v}{\partial s} = \begin{bmatrix} 0 & \mathcal{E}_{s_2}^{v_1} \\ \mathcal{E}_{s_1}^{v_2} & 0 \\ 0 & \mathcal{E}_{s_2}^{v_3} \end{bmatrix}$$

where $\mathcal{E}_{s_2}^{v_1}$ is the feedback elasticity. Replace each unscaled elasticity, $\mathcal{E}$, with the equivalent term involving the scaled elasticity, ε. This can be done by substituting each unscaled elasticity with the term:[1]

$$\mathcal{E}_j^i = \varepsilon_j^i \frac{v_i}{s_j}$$

If the reactions v_2 and v_3 are assumed to be first-order and irreversible, then the corresponding elasticities, ε_1^2, and ε_2^3, can be set to one. This means that the $\partial v / \partial s$ matrix can be written as:

$$\frac{\partial v}{\partial s} = \begin{bmatrix} 0 & \varepsilon_{s_2}^{v_1} F_1 \\ F_2 & 0 \\ 0 & F_3 \end{bmatrix}$$

---

[1]Recall $\varepsilon = \mathcal{E} s / v$

where F_i are the scaling factors, v_i/s_j. Given the stoichiometry and unscaled elasticity matrix, the Jacobian matrix can be written as:

$$\mathbf{N}\frac{\partial \mathbf{v}}{\partial \mathbf{s}} = \begin{bmatrix} -F_2 & \varepsilon_{s_2}^{v_1} F_1 \\ F_2 & -F_3 \end{bmatrix}$$

To determine the stability of this system, the eigenvalues of the Jacobian must be evaluated. The standard approach to evaluating the eigenvalues is to compute the determinant of the expression:

$$\lambda \boldsymbol{I} - \boldsymbol{A} \tag{10.1}$$

where $\boldsymbol{A}$ is the Jacobian and λ the eigenvalue.

For a square matrix, $\boldsymbol{A}$, if it is possible to find a vector $\mathbf{v}$ such than when we multiple the vector by $\boldsymbol{A}$ we get a scaled version of $\mathbf{v}$, then we call the vector $\mathbf{v}$ the eigenvector of $\boldsymbol{A}$ and the scaling value, λ, the eigenvalue of $\boldsymbol{A}$. For a matrix of dimension n, there will be at most n eigenvalues and n eigenvectors. We can write out the eigenvalue/eiegenvector equation as:

$$\boldsymbol{A}\mathbf{v} = \lambda \mathbf{v}$$

This equation can be rearranged as follows:

$$\boldsymbol{A}\boldsymbol{v} = \lambda \boldsymbol{I} \boldsymbol{v}$$
$$\boldsymbol{A}\boldsymbol{v} - \lambda \boldsymbol{I} \boldsymbol{v} = 0$$
$$(\boldsymbol{A} - \lambda \boldsymbol{I})\boldsymbol{v} = 0$$

From linear algebra we know that there will be non-zero solutions to $(\boldsymbol{A} - \lambda \boldsymbol{I})\boldsymbol{v} = 0$ if $\det(\boldsymbol{A} - \lambda \boldsymbol{I}) = 0$. We can use this observation to compute the eigenvalues and eigenvectors of a matrix by evaluating the determinant.

Using (10.1) it follows that:

$$\begin{bmatrix} \lambda & 0 \\ 0 & \lambda \end{bmatrix} - \begin{bmatrix} -F_2 & \varepsilon_{s_2}^{v_1} F_1 \\ F_2 & -F_3 \end{bmatrix} = \begin{bmatrix} \lambda + F_2 & -\varepsilon_{s_2}^{v_1} F_1 \\ -F_2 & \lambda + F_3 \end{bmatrix} \tag{10.2}$$

The determinant for a 2 by 2 matrix is given by: $a_{11}a_{22} - a_{21}a_{12}$, so that from (10.2):

$$\lambda^2 + \lambda(F_2 + F_3) + F_2 F_3 - F_1 F_2 \varepsilon_{s_2}^{v_1} = 0$$

This is a quadratic equation so there will be at most two eigenvalues. The criterion for negative roots (and hence stability) in a quadratic equation is that all the coefficients have the same sign. We should note that all F_i factors are positive, therefore it should be clear that the first two coefficients (attached to λ^2 and λ) are positive. The right most coefficient is $-F_1 F_2 \varepsilon_{s_2}^{v_1}$ where the elasticity $\varepsilon_{s_2}^{v_1}$ is negative because it represents the negative feedback. The expression, $-F_1 F_2 \varepsilon_{s_2}^{v_1}$, is therefore positive overall. Since all the coefficients are positive the two solutions, λ_1 and λ_2, must be both negative, therefore the system must be stable. There is no possibility of this system **ever** showing instability. Although it was assumed that the reaction v_2 and v_3 were first-order, relaxing this condition makes no difference to the signs of the terms in the quadratic. Stability is therefore also ensured under these less stringent conditions.

A linear pathway of three steps with a negative feedback loop from the second species to the first reaction will always be stable so long as the elasticities do no change sign.

Four Step Pathway

Let us look at a slightly modified system shown in Figure 10.2 where we have added an additional step in the feedback loop so that the pathway is four steps long.

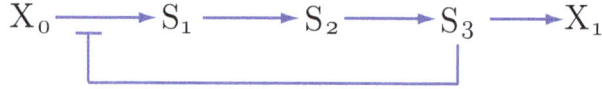

Figure 10.2 Negative feedback in a four species pathway.

The stoichiometry matrix and elasticity matrix are given by:

$$\mathbf{N} = \begin{bmatrix} 1 & -1 & 0 & 0 \\ 0 & 1 & -1 & 0 \\ 0 & 0 & 1 & -1 \end{bmatrix}$$

$$\frac{\partial \mathbf{v}}{\partial \mathbf{x}} = \begin{bmatrix} 0 & 0 & \varepsilon_{s_3}^{v_1} \\ \varepsilon_{s_1}^{v_2} & 0 & 0 \\ 0 & \varepsilon_{s_2}^{v_3} & 0 \\ 0 & 0 & \varepsilon_{s_3}^{v_4} \end{bmatrix}$$

As before replace the unscaled elasticities, $\mathcal{E}$, using the expression:

$$\mathcal{E}_j^i = \varepsilon_j^i \frac{v_i}{s_j}$$

where the ratio v_i / s_j is in turn replaced with the factor F_i. As a point of reference the replacements will be:

$$\mathcal{E}_3^1 = \varepsilon_3^1 \frac{v_1}{s_3} = \varepsilon_3^1 F_1 \qquad \mathcal{E}_1^2 = \varepsilon_1^2 \frac{v_2}{s_1} = \varepsilon_1^2 F_2$$

$$\mathcal{E}_2^3 = \varepsilon_2^3 \frac{v_3}{s_2} = \varepsilon_2^3 F_3 \qquad \mathcal{E}_3^4 = \varepsilon_3^4 \frac{v_4}{s_3} = \varepsilon_3^4 F_4$$

To make matters simpler, assume first-order kinetics for ε_1^2 and ε_3^4, so that these elasticities equal one. The product of the stoichiometry and elasticity matrix with therefore yield:

$$\mathbf{N} \frac{\partial \mathbf{v}}{\partial \mathbf{s}} = \begin{bmatrix} -F_2 & 0 & \varepsilon_3^1 F_1 \\ F_2 & -F_3 & 0 \\ 0 & F_3 & -F_4 \end{bmatrix}$$

To compute the eigenvalues we evaluate the determinant of $\lambda \mathbf{I} - \mathbf{A}$, where $\mathbf{A}$ is the Jacobian matrix:

$$\lambda \mathbf{I} - \mathbf{A} = \begin{bmatrix} \lambda + F_2 & 0 & \varepsilon_3^1 F_1 \\ F_2 & \lambda + F_3 & 0 \\ 0 & F_3 & \lambda + F_4 \end{bmatrix}$$

The determinant of the above matrix yields the following cubic equation:

$$\lambda^3 + v^2(F_2 + F_3 + F_4) + s(F_2 F_4 + F_2 F_3 + F_3 F_4)$$
$$+ F_1 F_2 F_3 - F_1 F_2 F_3 \, \varepsilon_3^1 = 0$$

Since the pathway is linear, at steady state we can assert that the reaction rates, $v_1, v_2, v_3,$ and v_4 are equal to each other. This allows us to make a useful simplification. Recall that $F_1 = v_1/s_3$ and $F_4 = v_4/s_3$, meaning $F_1 = F_4$. This changes the characteristic equation to:

$$\lambda^3 + \lambda^2(F_2 + F_3 + F_4) + s(F_2 F_4 + F_2 F_3 + F_3 F_4)$$
$$+ F_2 F_3 F_4(1 - \varepsilon_3^1) = 0$$

where F_1 has been replaced with F_4. For a general cubic equation such as:

$$a_0 x^3 + a_1 x^2 + a_2 x + a_3 = 0$$

the condition for all negative roots is:

$$a_0 > 0, \; a_1 > 0, \; a_3 > 0 \quad \text{and} \quad a_1 a_2 > a_0 a_3$$

Negative roots of course indicate stability. We note that the first three conditions are satisfied because the first two coefficients are positive (all F_i terms are positive), and the expression $(1 - \varepsilon_{s3}^{v_1})$ is net positive (note $\varepsilon_{s3}^{v_1} < 0$). With these conditions satisfied, the requirement for stability ($a_1 a_2 > a_0 a_3$) becomes:

$$(F_2 + F_3 + F_4)(F_2 F_4 + F_2 F_3 + F_3 F_4) > F_2 F_3 F_4(1 - \varepsilon_3^1)$$

Dividing both sides by $F_2 F_3 F_4$ (this is positive, we therefore do not flip the inequality) yields a new inequality:

$$\frac{F_2}{F_3} + \frac{F_2}{F_4} + \frac{F_3}{F_4} + \frac{F_3}{F_2} + \frac{F_4}{F_3} + \frac{F_4}{F_2} + 3 > 1 - \varepsilon_3^1$$

Subtracting one from both sides yields:

$$\frac{F_2}{F_3} + \frac{F_2}{F_4} + \frac{F_3}{F_4} + \frac{F_3}{F_2} + \frac{F_4}{F_3} + \frac{F_4}{F_2} + 2 > -\varepsilon_3^1 \tag{10.3}$$

This is the condition for stability. So long as the left-hand side is greater than the negative of the feedback elasticity, ε_3^1, the system will be stable. The question now is what values can the left-hand side have? One way to pin down the inequality further is to try all combinations of F_i to discover what range of values the left-hand side can yield. Figure 10.3 shows combinations of F_i terms. The graph clearly shows that no combination falls below 8. Eight is therefore a lower bound for the left-hand term. This means that the feedback elasticity must be greater than -8 in order to achieve stability. Values less that -8 will cause instability.

Another way to look at equation (10.3) is to recall the relation:

$$\frac{1}{a} + a \geqslant 2$$

Looking more closely at (10.3), we see that each pair of terms is in the form $(1/a) + a$. Therefore each pair of terms must be $\geqslant 2$. The lowest value each term can have is 2, therefore the minimum value for the sum of three of the paired terms must be 6. With the addition of the remaining 2 in the equation, it means that the minimum value for the left-hand side of equation (10.3) is 8. This matches the result from the numerical study in Figure 10.3. Swapping the negative signs in the inequality to make interpretation easier, we can summarize that if:

$$\varepsilon_3^1 > -8 \tag{10.4}$$

then the system is stable.

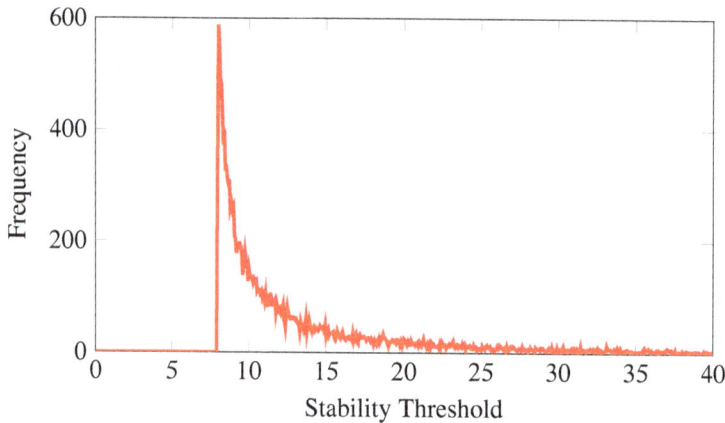

Figure 10.3 Stability Threshold for equation (10.3). Only feedback elasticities greater than eight result in instability.

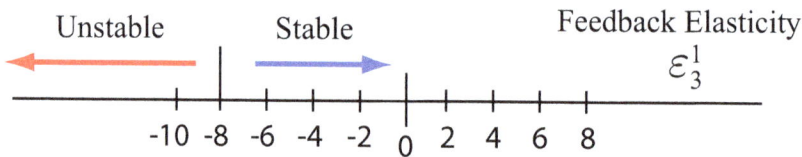

Figure 10.4 Threshold for Stability.

Figure 10.4 illustrates the threshold in diagrammatic form. Recall that the more negative ε_3^1 is, the stronger the feedback. For example if $\varepsilon_3^1 = -2.0$, then the system will be stable since -2 is greater than -8. If the strength of the negative feedback is less than -8, the system will be unstable. For example if $\varepsilon_3^1 = -10$, then the system will be unstable because $-10 < -8$. This means that instability in this system requires a fairly strong degree of feedback. Note that many allosteric enzymes tend to have elasticities for effectors in the range of -1 to -4. A similar analysis can be done on pathways of other lengths, which will be considered in the next section.

From this analysis we can also make a few more observations. We have assumed that the elasticities for the effect of S_1 on v_2 and S_2 on v_3 was first-order. What if these values are closer to 0.5 where the substrate levels are near the K_m of the enzymes? The elasticities can be reintroduced so that the threshold for stability becomes:

$$\varepsilon_3^1 \varepsilon_1^2 \varepsilon_2^3 > -8$$

If the two elasticities have values around 0.5, then the threshold for instability will be greater and it becomes more difficult to reach an unstable situation. To illustrate this, if ε_1^2 and ε_2^3 have values of 0.5 each, the threshold equation becomes:

$$\frac{1}{4}\varepsilon_3^1 > -8$$

That is for stability:

$$\varepsilon_{s_3}^1 > -32$$

The threshold for stability is now much higher. It requires significant feedback strength to make such a system unstable. Also note that if either step two or three are close to equilibrium, the associated elasticities go to infinity, dashing any hopes of instability. Another way to look at this is that if one of the reactions is close to equilibrium, the pathway is effectively shortened by one step, which takes us back to a three-step and stable pathway.

10.3 Effect of Pathway Length

Consider the negative pathway shown in Figure 10.5, which has n species and $n + 1$ steps:

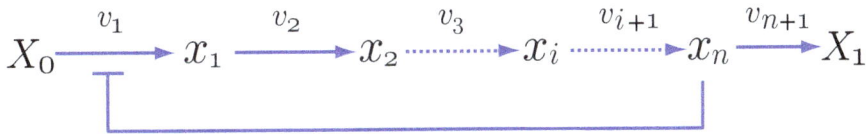

Figure 10.5 Negative feedback system with n species.

The stoichiometry and elasticity matrices have the general form:

$$\mathbf{N} = \begin{bmatrix} 1 & -1 & 0 & 0 & \ldots \\ 0 & 1 & -1 & 0 & \ldots \\ 0 & 0 & 1 & -1 & \ldots \\ \vdots & \vdots & \vdots & \vdots & \ddots \end{bmatrix} \qquad \frac{\partial \mathbf{v}}{\partial \mathbf{x}} = \begin{bmatrix} \dfrac{\partial v_1}{\partial s_1} & 0 & 0 & 0 & \ldots & \dfrac{\partial v_1}{\partial s_n} \\ \dfrac{\partial v_2}{\partial s_1} & \dfrac{\partial v_2}{\partial s_2} & 0 & 0 & 0 & \ldots \\ 0 & \dfrac{\partial v_3}{\partial s_2} & \dfrac{\partial v_3}{\partial s_3} & 0 & 0 & \ldots \\ \vdots & \vdots & \vdots & \vdots & \vdots & \ddots \\ 0 & 0 & 0 & 0 & 0 & \dfrac{\partial v_{n+1}}{\partial s_n} \end{bmatrix}$$

Replacing $\partial v / \partial s$ with $\mathcal{E}$ and assuming all reactions are product insensitive, we obtain:

$$\frac{\partial \mathbf{v}}{\partial \mathbf{x}} = \begin{bmatrix} 0 & 0 & 0 & 0 & 0 & \mathcal{E}_n^1 \\ \mathcal{E}_1^2 & 0 & 0 & 0 & 0 & \ldots \\ 0 & \mathcal{E}_2^3 & 0 & 0 & 0 & \ldots \\ \vdots & \vdots & \vdots & \vdots & \vdots & \ddots \\ 0 & 0 & 0 & 0 & 0 & \mathcal{E}_n^{n+1} \end{bmatrix}$$

The Jacobian, $\mathbf{N} \partial \mathbf{v} / \partial \mathbf{x}$ is given by:

$$\mathbf{N}\frac{\partial \mathbf{v}}{\partial \mathbf{x}} = \begin{bmatrix} -\mathcal{E}_1^2 & 0 & 0 & 0 & \cdots & \mathcal{E}_n^1 \\ \mathcal{E}_1^2 & -\mathcal{E}_2^3 & 0 & 0 & \cdots & 0 \\ 0 & \mathcal{E}_2^3 & -\mathcal{E}_3^4 & 0 & \cdots & 0 \\ \vdots & \vdots & \vdots & \vdots & \vdots & \ddots \\ 0 & 0 & 0 & 0 & \cdots & -\mathcal{E}_n^{n+1} \end{bmatrix}$$

We now replace terms $\mathcal{E}_j^i$ with $\varepsilon_j^i F_j$ and as before compute $\lambda I - \mathbf{N}\partial \mathbf{v}/\partial \mathbf{x}$ from which we can derive the characteristic equation. Note that in the dynamical systems community the matrix is computed using $(A - \lambda I)$ so that the signs will be reversed, however the result is identical:

$$\lambda I - \mathbf{N}\frac{\partial \mathbf{v}}{\partial \mathbf{x}} = \begin{bmatrix} \lambda + \varepsilon_1^2 F_2 & 0 & 0 & 0 & \cdots & -\varepsilon_n^1 F_1 \\ -\varepsilon_1^2 F_2 & \lambda + \varepsilon_2^3 F_3 & 0 & 0 & \cdots & 0 \\ 0 & -\varepsilon_2^3 F_3 & \lambda + \varepsilon_3^4 F_4 & 0 & \cdots & 0 \\ 0 & 0 & -\varepsilon_3^4 F_4 & \lambda + \varepsilon_4^5 F_5 & \cdots & 0 \\ \vdots & \vdots & \vdots & \vdots & \vdots & \ddots \\ 0 & 0 & 0 & 0 & 0 & \lambda + \varepsilon_n^{n+1} F_{n+1} \end{bmatrix}$$

To make matters simpler, let us assume that all reactions other than the feedback step are first-order with *equal* rate constants. In such a situation the substrate elasticities, $\varepsilon_j^i = 1$ and all steady state species concentrations, s_j, *are the same*. This means that $\varepsilon_1^2 = \varepsilon_2^3 = \varepsilon_3^4 = \varepsilon_j^i = 1$ and $F_1 = F_2 = F_3 = F_n = F$. The matrix can therefore be rewritten as:

$$\begin{bmatrix} \lambda + F & 0 & 0 & 0 & \cdots & 0 & -\varepsilon_n^1 F \\ -F & \lambda + F & 0 & 0 & \cdots & 0 & 0 \\ 0 & -F & \lambda + F & 0 & \cdots & 0 & 0 \cdot \\ 0 & 0 & -F & \lambda + F & \cdots & 0 & \lambda + F \end{bmatrix}$$

As before, the characteristic equation is obtained from the determinant which in this case has a simple form:

$$(\lambda + F)^n - \varepsilon_n^1 F^n = 0$$

Following Savageau [93], we now solve for λ using de Moivre's theorem. We first express the above equation by rearranging and taking the nth root on both sides.

$$(\lambda + F)^n = \varepsilon_n^1 F^n$$

$$\sqrt[n]{(\lambda + F)^n} = \sqrt[n]{\varepsilon_n^1 F^n}$$

$$\lambda + F = \left(\varepsilon_n^1\right)^{1/n} F$$

$$\lambda = \left(\varepsilon_n^1\right)^{1/n} F - F$$

$$\lambda = \left[(\varepsilon_n^1)^{1/n} - 1\right] F$$

Because ε_n^1 is negative, the roots of $(\varepsilon_n^1)^{1/n}$ are complex. To simplify the analysis, we can add a negative sign to ε_n^1 and insert $(-1)^{1/n}$ to compensate:

$$\lambda = \left[(-1)^{1/n} \left(-\varepsilon_n^1 \right)^{1/n} - 1 \right] F \tag{10.5}$$

This reduces the problem to finding the roots of $(-1)^{1/n}$. The n roots of $(-1)^{1/n}$ are well known and can be obtained using de Moivre's theorem where the mth root is given by:

$$\lambda_m = \left[\left(-\varepsilon_n^1 \right)^{1/n} \left(\cos\left(\frac{2m+1}{n}\pi \right) + i \sin\left(\frac{2m+1}{n}\pi \right) \right) - 1 \right] F \quad \text{for } m = 0, 1, 2, \ldots n - 1$$

For the system to be stable, the *real part* of the λ_m must be negative. Since F is positive, for stability it must be true that the real part of the expression must be negative:

$$\left(-\varepsilon_n^1 \right)^{1/n} \cos\left(\frac{2m+1}{n}\pi \right) - 1 < 0$$

The transition to *instability* will occur when the largest λ_m becomes positive. This will happen at $m = 0$ since the cosine term is maximum at this point. Setting $m = 0$, stability is assured when:

$$\left(-\varepsilon_n^1 \right)^{1/n} \cos\left(\frac{\pi}{n} \right) - 1 < 0$$

For stability, the first term, not including the -1 on the left-hand side, must be *less* than +1 to make the overall expression on the left be less than 0. That is, for the system to be stable it must be true that:

$$\left(-\varepsilon_n^1 \right)^{1/n} \cos\left(\frac{\pi}{n} \right) < 1 \tag{10.6}$$

Recall that since n is the number of species, then $n + 1$ is the number of steps in the pathway. Let us consider the cases where $n = 1$ (two steps, one species), and $n = 2$ (three steps, two species) which we know to be stable from the previous sections. When $n = 1$, the cosine term on the left-hand side of equation (10.6) is -1. Since the elasticity term will always be positive, the left-hand side will always be negative and thus never exceed 1. That is, the system will *always* be stable.

When $n = 2$ (three steps), the left-hand side is 0 ($\cos(\pi/2) = 0$) and again the left-hand side will *always* be less than one. That is, the system will *always* be stable.

When $n = 3$ (four steps), the cosine term is 0.5. Therefore for the expression to exceed one, the term containing the elasticity must equal at least 2. This happens when the elasticity has a value of -8 because $8^{1/3} = 2$. That is, the system becomes unstable when the elasticity is less than -8.

Another way to look at expression (10.6), and one commonly found in the literature, is to rearrange the inequality, noting the sign change rules, to obtain a slightly different expression:

$$-\varepsilon_n^1 < \frac{1}{\cos^n(\pi/n)} = \sec^n(\pi/n) \tag{10.7}$$

We therefore arrive at the central result [38, 92, 109, 107]:

Condition for stability where n is the number of species:

$$-\varepsilon_n^1 < \sec^n\left(\frac{\pi}{n} \right)$$

When $n = 3$ (four steps), $\sec^3(\pi/3)$ equals 8, that is:

$$\varepsilon_3^1 < -8$$

Number of Steps in Pathway	Instability Threshold $-\varepsilon_{\text{feedback}}$
2	stable
3	stable
4	8
5	4.0
6	2.9
7	2.4
8	2.1
9	1.9
$\vdots$	$\vdots$
∞	1.0

Table 10.1 Relationship between pathway length and the degree of feedback inhibition on the threshold for stability (See Figure 10.5). $-\varepsilon_{\text{feedback}}$ is the elasticity of the feedback inhibition. The analysis assumes first-order kinetics for all reactions other than the feedback reaction. Additionally, it also assumes that the rate constants for the first-order reactions all have the same value. This ensures that the species levels are the same which greatly simplifies the algebra.

for stability. This confirms the results we obtained previously in equation (10.4). That is, as long as the elasticity is larger than -8, the network is stable.

To give another example, if the number of steps in the pathway is eight ($n = 7$), then $\sec^7(\pi/7) = 2.08$. Thus the feedback elasticity must be less then -2.08 for the system to be unstable. Note this threshold is much less stringent. As the length of the pathway increases, it becomes easier for the system to be unstable.

Sometimes this condition is expressed as [42]:

$$n_H > \frac{A}{\cos^n\left(\frac{\pi}{n}\right)}$$

When expressed this way, n_H is the term in the inhibition rate law, $v_o/(1 + kS^{n_H})$, and is therefore is always positive (this explains the change in the inequality sign). Equation (10.7) is however more general and does not depend on a specific mechanism for the negative feedback.

Table 10.1 summarizes some of the threshold points for a negative feedback system with varying pathway lengths.

Much of the work described in this chapter was taken, with modification, from Savageau's book, Biochemical Systems Analysis, 1976

Further Reading

1. Edelstein-Keshet L (2005) Mathematical Models in Biology. SIAM Classical In Applied Mathematics. ISBN-10: 0-89871-554-7

2. Fall CP, Marland ES, Wagner JM, Tyson JJ (2000) Computational Cell Biology. Springer: Interdisciplinary Applied Mathematics. ISBN 0-387-95369-8.

3. Steuer R and Junker BH (2009). Computational models of metabolism: stability and regulation in metabolic networks. Advances in chemical physics, 142, 105.

4. M Savageau (1976) Biochemical systems analysis: a study of function and design in molecular biology, Addison-Wesley. Note: The text was republished in 2010 and is available on Amazon.

Exercises

1. Run simulations of a four-step pathway to show that the threshold for stability is when $\varepsilon_3^1 < -8$. Assume first-order irreversible kinetics for each reaction except for the regulated step, v_1. The rate law for the first step can be set to $v_o/(1 + k s_3^h)$ where h is the Hill coefficient. Set all rate constants for the reactions steps to the same value. Investigate the stability of the systems as a function of changes to the Hill coefficient and hence ε_3^1.

2. Investigate the stability threshold for longer pathways using simulation. For example investigate a pathway with six steps and confirm the theoretical prediction made in Table 10.1 are correct.

11

Moiety Conservation Laws

11.1 Moiety Constraints

Many cell processes operate on different time scales. For example, metabolic processes tend to operate on a faster time scale than protein synthesis and degradation. Such time scale differences have a number of implications to model builders, software designers and model behavior. In this chapter we will examine these aspects in relation to species conservation laws. To introduce this topic, consider a simple protein phosphorylation cycle such as the one shown in Figure 11.1. This shows a protein undergoing phosphorylation (upper limb) and dephosphorylation (lower limb) via a kinase and phosphatase, respectively.

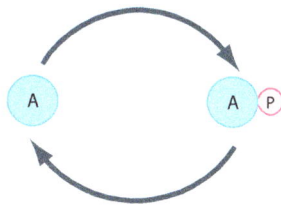

Figure 11.1 Phosphorylation and dephosphorylation cycle forming a moiety conservation cycle between unphosphorylated (left species) and phosphorylated protein (right species).

The depiction in Figure 11.1 is a simplification. The ATP used during phosphorylation is not shown, as well as the release of free phosphate during the dephosphorylation event. Furthermore, synthesis and degradation of protein is also absent. In many cases we can leave these aspects out of the picture. ATP for instance is held at a relatively constant level by strong homeostatic forces from metabolism so that within the context of the cycle, changes in ATP isn't something we need worry about. More interesting is that within the time scale of phosphorylation and dephosphorylation, we can assume that the rate of protein synthesis and degradation is negligible. This assumption leads to the emergence of a new property of the cycle called **moiety conservation** [82].

In chemistry a moiety is described as a subgroup of a larger molecule. In Figure 11.1 the moiety is a protein.

During the interconversion between the phosphorylated and unphosphorylated protein, the amount of moiety (protein) remains constant. More abstractly, we can draw a cycle in the following way (Figure 11.2), where S_1 and S_2 are the cycle species:

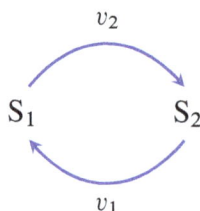

Figure 11.2 Simple conserved cycle where $s_1 + s_2 =$ constant. Note: This assumes unit volume for both S_1 and S_2 so that concentrations can be summed.

The two species, S_1 and S_2, are conserved because the total $s_1 + s_2$ remains constant over time (at least over a time scale shorter than protein synthesis and degradation). Such cycles are collectively called **moiety conserved cycles** or conserved cycles for short.

Protein signalling pathways abound with conserved cycles such as these, although many are more complex and may involve multiple phosphorylation reactions. In addition to protein networks, other pathways also possess conservation cycles. One of the earliest conservation cycles to be recognized was the adenosine triphosphate (ATP) cycle. ATP is a chain of three phosphate residues linked to a nucleoside adenosine group, Figure 11.3.

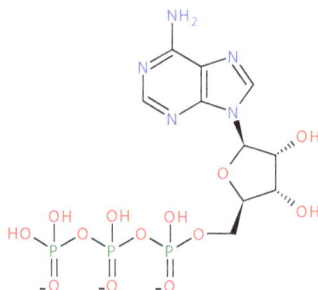

Figure 11.3 Adenosine Triphosphate: Three phosphate groups plus an adenosine subgroup.

The linkage between the phosphate groups involves phosphoric acid anhydride bonds that can be cleaved by hydrolysis one at a time, leading in turn to the formation of adenosine diphosphate (ADP) and adenosine monophosphate (AMP), respectively. The hydrolysis provides much of the free energy to drive endergonic processes in the cell. Given the insatiable need for energy, there is a continual and rapid interconversion between ATP, ADP and AMP as energy is released or captured. One thing that is constant during these interconversions is the amount of adenosine group (Figure 11.4). That is, adenosine is a conserved moiety. Over longer time scales there is also the slower process of AMP degradation and biosynthesis via the purine nucleotide pathway, but for some models, we assume that this process is negligible compared to the ATP turnover by energy metabolism.

There are many other examples of conserved moieties such enzyme/enzyme-substrate complexes, NAD/-NADH, phosphate and coenzyme A. In all these cases the basic assumption is that the interconversions of the subgroups is rapid compared to their net synthesis and degradation. We should emphasize that in reality conserved moieties do not exist since all molecular subgroups will at some point be subject to synthesis and degradation. However, over sufficiently short time scales, the sum total of these groups can be considered constant. In this chapter we will consider conserved moieties in detail. In particular, we will look at how to detect them in our models and how they influence the design of simulation software. We will wait until Chapter 12 to discuss their effect on pathway dynamics and their relationship to MCA.

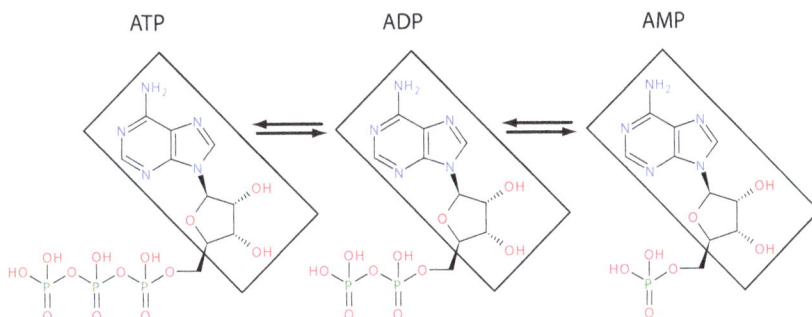

Figure 11.4 The adenosine moiety, indicated by the boxed molecular group, is conserved during the interconversion of ATP, ADP and AMP.

Moiety:	A subgroup of a larger molecule.
Conserved Moiety:	A subgroup whose interconversion through a sequence of reactions leaves it unchanged.

11.2 Moiety Conserved Cycles

Any chemical group that is preserved during a cyclic series of interconversions is called a **conserved moiety**. Examples of conserved moiety subgroups include species such as phosphate, acyl, nucleoside groups or covalently modifiable proteins. As a moiety gets redistributed through a network, the **total amount** of the moiety is constant and does not change during the time evolution of the system. For any particular subgroup, the total amount is determined solely by the initial conditions imposed on the model.

There are rare cases when a 'conservation' relationship arises out of a non-moiety cycle. This does not affect the mathematic analysis, only the physical interpretation of the relationship. For example, in Figure 11.6 the constraint $b - c = T$ applies even though there is no moiety involved.

The presence of conserved moieties is an approximation introduced into a model, however, over the time scale in which the conservation holds, their existence can have a profound effect on the dynamic behavior of the model. For example, the hyperbolic response of a simple enzyme (in the form of enzyme conservation between E and ES), or the sigmoid behavior observed in protein signalling networks, is due in significant part to moiety conservation laws (see Chapter 12).

In the subsequent discussion it will be assumed that all species reside in a unit volume. This makes it possible to express the conservation laws using concentrations. If the volumes are non-unity and of varying magnitude, the conservation laws would have to be adjusted with respect to the volume because it is amounts that are

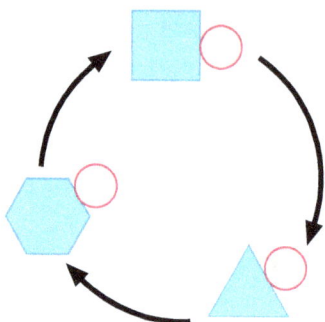

Figure 11.5 Conserved moiety in a cyclic network. The darker blue species are modified as they traverse the reaction cycle, but the lighter subgroup (smaller circle), remains unchanged. This creates a conserved cycle, where the total number of moles of moiety (smaller circle) stays constant.

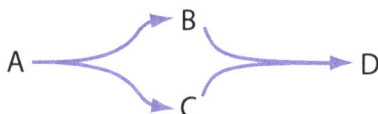

Figure 11.6 Conservation due to stoichiometric matching. In this system, $b - c =$ constant.

strictly conserved, not concentration. For an introductory discussion, non-unity volumes are an unnecessary distraction to the focus of the chapter and we will therefore assume unit volume in all cases.

Figure 11.7 illustrates the simplest possible network which displays a conserved moiety, the total mass, $s_1 + s_2$, is constant during the evolution of the network.

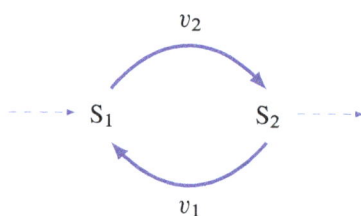

Figure 11.7 Simple conserved cycle. The dotted lines signify negligible levels of synthesis and degradation, therefore over short time scales, $s_1 + s_2 =$ constant.

The system equations for the conserved cycle in Figure 11.7 are written as:

$$\frac{ds_1}{dt} = v_1 - v_2$$

$$\frac{ds_2}{dt} = v_2 - v_1$$

(11.1)

From these equations it should be evident that the rate of appearance of S_1 must equal the rate of disappearance of S_2, that is $ds_1/dt = -ds_2/dt$. This means that whenever S_1 changes, S_2 must change in the opposite direction by **exactly** the same amount. During this time, the sum of S_1 and S_2 will therefore remain unchanged.

When solving the system by computer we only need to numerically integrate one of the species because the other species can be computed from the conservation relationship. Whichever differential equation is chosen, the species left out must be computed algebraically using the conservation law. Therefore, the system (11.1) can be reduced to one differential and one linear algebraic equation (11.2) compared to the two differential equations in the original formulation:

$$s_2 = T - s_1$$

$$\frac{ds_1}{dt} = v_1 - v_2 \tag{11.2}$$

T in the algebraic equation (11.2) refers to the total amount of S_1 and S_2. This value is computed from the initial amounts given to S_1 and S_2 at the start of a simulation.

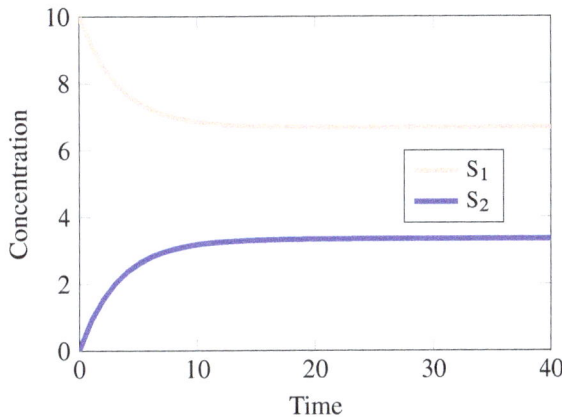

Figure 11.8 Simulation of the simple cycle shown in Figure 11.7. The total moiety remains constant at 10 concentration units. Model: S1 -> S2; k1*S1; S2 -> S1; k2*S2; S1=10; k1=0.1; k2=0.2.

11.3 Basic Theory

The question we want to address here is how to determine whether a given network contains conserved cycles, and if so what are they? The key to this question is the stoichiometry matrix, **N**. In the example shown in Figure 11.7 the stoichiometry matrix is given by:

$$\mathbf{N} = \begin{bmatrix} 1 & -1 \\ -1 & 1 \end{bmatrix}$$

Since either row can be derived from the other by multiplication by -1, the rows are called **linearly dependent rows** (See Box 11.0). The linear dependence among the rows reflects the earlier relationship $ds_1/dt = -ds_2/dt$. Any linear dependence among the rows of the stoichiometry matrix means that there are linear relationships among the rates of change, ds/dt. In turn, relationships among the rates of change mean there are mass constraints on how the individual amounts can change. In the previous example it was shown that for

Box 11.0 Linear Dependence and Independent - Recap

One of the most important ideas in linear algebra is the concept of linear dependence and independence. Take three vectors, say $[1, -1, 2]$, $[3, 0, -1]$ and $[9, -3, 4]$. If we look at these vectors carefully it should be apparent that the third vector can be generated from a combination of the first two, that is $[9, -3, 4] = 3[1, -1, 2] + 2[3, 0, -1]$. Mathematically we say that these vectors are *linearly dependent*.

In contrast, the following vectors, $[1, -1, 0]$, $[0, 1, -1]$ and $[0, 0, 1]$, are independent because there is no combination of these vectors that can generate even one of them. Mathematically we say that these vectors are *linearly independent*.

every mole of S_1 that was consumed, one mole of S_2 was produced, and for every mole of S_2 consumed, one mole of S_1 was produced. Therefore since at steady state $ds_1/dt + ds_2/dt$ must be zero, this implies that the sum $s_1 + s_2$ remains constant.

In conclusion, whenever a network exhibits conserved moieties, there will be dependencies among the rows of N. A measure of such dependencies is the rank, denoted $rank(N)$ (See Box 11.1). Any dependencies among the rows will be reflected in a rank that is less than m, the number of rows of N.

It is possible to arrange the rows of N so that the first $rank(N)$ rows are linearly independent. The metabolites which correspond to these rows are called the **independent species** (s_i). The remaining $m - rank(N)$ rows correspond to the **dependent species** (s_d).

In terms of the rank, the Number of dependent species = Total number of species - rank (N).

In the simple conserved cycle, Figure 11.7, there is one independent species, S_1 and one dependent species, S_2.

Example 11.1

Figure 11.5 illustrates a three species cycle. What is the conservation law for this pathway? The stoichiometry matrix for this system is given by:

$$N = \begin{matrix} & \begin{matrix} v_1 & v_2 & v_3 \end{matrix} & \\ \begin{bmatrix} -1 & 0 & 1 \\ 1 & -1 & 0 \\ 0 & 1 & -1 \end{bmatrix} & \begin{matrix} S_1 \\ S_2 \\ S_3 \end{matrix} \end{matrix} \qquad (11.3)$$

Inspection reveals that the sum of the three rows is zero meaning that:

$$\frac{ds_1}{dt} + \frac{ds_2}{dt} + \frac{ds_3}{dt} = 0$$

or that the total $s_1 + s_2 + s_3$ is constant. There are no other relationships between the rows other than this one.

Example 11.2

A linear pathway has the following stoichiometry matrix:

$$N = \begin{bmatrix} 1 & -1 & 0 \\ 0 & 1 & -1 \end{bmatrix}$$

Does the pathway contain any conserved cycles? No, because neither row in the matrix can be derived from the other by a linear combination. The rows are linearly independent, therefore the pathway has no conserved cycles.

To illustrate this idea on a more complicated example, consider the pathway shown in Figure 11.9. This pathway includes four species, S_1, S_2, E and ES. The right-hand side of the figure shows the moiety composition. When drawn this way, it is not difficult to discern the different moieties. We can see there is a square moiety conserved by reactions v_2 and v_3, and also a circle moiety that is conserved by reactions, v_1, v_2 and v_3.

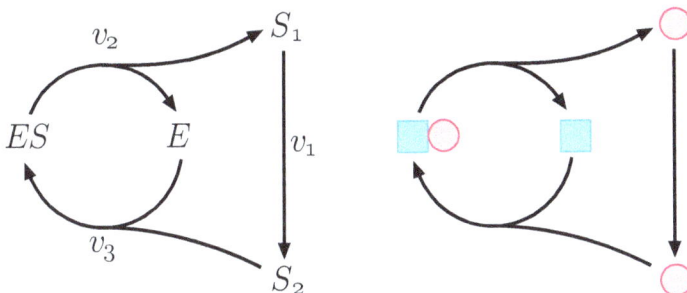

Figure 11.9 Linked Conserved Cycles. The network rendered on the right shows the moiety composition of the participating species.

The mass-balance equations of this model can be written as:

$$\frac{de}{dt} = v_2 - v_3 \qquad\qquad \frac{des}{dt} = v_3 - v_2$$

$$\frac{ds_1}{dt} = v_2 - v_1 \qquad\qquad \frac{ds_2}{dt} = v_1 - v_3$$

A visual inspection of the mass-balance equations reveals the following two relationships:

$$\frac{de}{dt} + \frac{des}{dt} = 0$$

$$\frac{des}{dt} + \frac{ds_1}{dt} + \frac{ds_2}{dt} = 0$$

(11.4)

These relationships tell us that there are two conservation laws, $e + es$ and $es + s_1 + s_2$. This means that given the amount of ES, the amount of E can be computed. In addition, given the amount of ES and S_1, the amount of S_2 can be computed. Therefore ES and S_1 can be designated the independent species, and E and S_2 the dependent species. What this means in practical terms is that in a modeling program only two differential

Box 11.1 The Rank of a Matrix - Recap

Closely related to linear independence (Box 11.0) is the concept of Rank. Consider the three vectors described in Box 11.0, $[1, -1, 2]$, $[3, 0, 1]$ and $[9, -3, 4]$ and stack them one atop each other to form a matrix:

$$\begin{bmatrix} 1 & -1 & 2 \\ 3 & 0 & 1 \\ 9 & -3 & 4 \end{bmatrix}$$

then the Rank is the number of linear independent vectors that make up the matrix. In this case the Rank is 2, because there are only two linear independent row vectors in the matrix.

equations need be solved instead of four. The reduced model equations will look like:

$$e = T_1 - es$$

$$s_2 = T_2 - s_1 - es$$

$$\frac{des}{dt} = v_3 - v_2$$

$$\frac{ds_1}{dt} = v_2 - v_1$$

where T_1 is the total amount of E type moiety, and T_2 is the total amount of S type moiety.

The stoichiometry matrix for the model in Figure 11.9 is given by:

$$\mathbf{N} = \begin{matrix} & \begin{matrix} v_1 & v_2 & v_3 \end{matrix} & \\ \begin{bmatrix} 1 & 0 & -1 \\ 0 & -1 & 1 \\ -1 & 1 & 0 \\ 0 & 1 & -1 \end{bmatrix} & & \begin{matrix} S_2 \\ ES \\ S_1 \\ E \end{matrix} \end{matrix} \qquad (11.5)$$

Examining the stoichiometry matrix reveals conservation laws as relationships among the matrix rows. The 4th row (E) can be formed by multiplying the 2nd row (ES) by -1, and the 3rd row (S_1) can be formed by multiplying the first row by -1 and adding it to the 4th row (ES).

These simple examples show that it is possible to derive conservation laws by looking for dependencies among the rows of the stoichiometry matrix. For simple cases this can be done by inspection, but for large pathways this approach is not practical. Instead a more systematic theory for deriving the conservation laws must be developed.

11.4 Computational Approaches

There are a number of related methods for computing the conservation laws of a given pathway, some are simple such as the one shortly to be described, while others are more sophisticated and are used to determine the conservation laws in very large stoichiometry matrices.

The easiest method to derive conservation laws is to use row reduction [76, 19, 20]. This is based on forward elimination which is the first part of Gaussian Elimination. Gaussian Elimination is a traditional way to solve simultaneous linear equations by eliminating one unknown at a time, and is a technique often taught in high school. Elimination is carried out by applying a series of simple manipulations called **elementary operations**. These operations include interchanging two equations (exchange), multiplying an equation through by a nonzero number (scaling), and adding an equation one or more times to another equation (replacement). In practice the equations are cast into a matrix form so that the elementary operations (See Box 11.2) are applied to the values in the matrix where each row of the matrix represents an equation. Thus, interchanging two equations is equivalent to swapping two rows in the matrix. The elementary operations are carried out on the matrix until a particular arrangement, called the echelon form, is established (See Box 11.3).

Elementary operations are often represented in matrix form and are then called elementary matrices (See Box 11.2). Applying a particular elementary operation then becomes equivalent to multiplying by an elementary matrix. The technique for finding conservation laws works as follows. Consider the network in Figure 11.9. The system equation for this network is:

$$
\begin{matrix} S_2 \\ ES \\ S_1 \\ E \end{matrix}
\begin{bmatrix} 1 & 0 & -1 \\ 0 & -1 & 1 \\ -1 & 1 & 0 \\ 0 & 1 & -1 \end{bmatrix}
\begin{bmatrix} v_1 \\ v_2 \\ v_3 \end{bmatrix}
=
\begin{bmatrix} ds_2/dt \\ des/dt \\ ds_1/dt \\ de/dt \end{bmatrix}
$$

We will cast the equation in the following form where an identity matrix has been added to the right-hand side.

$$
\mathbf{N}v = \mathbf{I}\frac{d\mathbf{s}}{dt}
$$

Written out fully the system equation will look like:

$$
\begin{matrix} S_2 \\ ES \\ S_1 \\ E \end{matrix}
\begin{bmatrix} 1 & 0 & -1 \\ 0 & -1 & 1 \\ -1 & 1 & 0 \\ 0 & 1 & -1 \end{bmatrix}
\begin{bmatrix} v_1 \\ v_2 \\ v_3 \end{bmatrix}
=
\begin{bmatrix} 1 & 0 & 0 & 0 \\ 0 & 1 & 0 & 0 \\ 0 & 0 & 1 & 0 \\ 0 & 0 & 0 & 1 \end{bmatrix}
\begin{bmatrix} ds_2/dt \\ des/dt \\ ds_1/dt \\ de/dt \end{bmatrix}
$$

Note that multiplying by the identity matrix doesn't change the equations.

Forward elimination is first applied to the stoichiometry matrix. To do this a series of elementary operations to the left-hand side is applied such that the stoichiometry matrix is reduced to echelon form. For consistency we apply the same set of elementary operations to the right-hand side so that the identity matrix records whatever operations we carry out. This amounts to multiplying both sides by a set of elementary matrices. We only need to reduce the matrix to its row echelon form, not to its reduced echelon form (See Box 11.3).

Reducing a matrix to echelon form raises the possibility of generating zero rows in the matrix if there are dependencies in the rows (See Box 11.3).

With this being the case, the system equation after forward elimination can be expressed in the following way:

$$
\begin{bmatrix} \mathbf{M} \\ \mathbf{0} \end{bmatrix} v = E\frac{d\mathbf{s}}{dt} \tag{11.6}
$$

where the identity matrix has been shown transformed into the matrix E which represents the product of all elementary operations that were applied to the left-hand side. The left-hand side has itself been transformed into an echelon form which is represented as a partitioned matrix of M and $\mathbf{0}$. The E matrix can also be partitioned row-wise to match the partitioning in the echelon matrix, that is:

$$
\begin{bmatrix} \mathbf{M} \\ \mathbf{0} \end{bmatrix} v = \begin{bmatrix} X \\ Y \end{bmatrix} \frac{d\mathbf{s}}{dt} \tag{11.7}
$$

Multiplying out the lower partition, we obtain:

$$Y \frac{d\mathbf{s}}{dt} = \mathbf{0} \tag{11.8}$$

This general result is equivalent to the equations shown in (11.4), that is (11.8) represents the set of conservation laws. Determining the conservation laws therefore involves reducing the stoichiometry matrix and extracting the lower portion of the modified identity matrix.

Let us now proceed with an example to illustrate this method. We will use the stoichiometry matrix from equation (11.5). For convenience, the stoichiometry and identity matrix are placed next to each other in the following sequence of elementary operations. An elementary operation carried out on the stoichiometry matrix is simultaneously applied to the identity matrix.

1. Stoichiometry matrix on the left and identity matrix on the right.

$$\begin{bmatrix} 1 & 0 & -1 \\ 0 & -1 & 1 \\ -1 & 1 & 0 \\ 0 & 1 & -1 \end{bmatrix} \quad \begin{bmatrix} 1 & 0 & 0 & 0 \\ 0 & 1 & 0 & 0 \\ 0 & 0 & 1 & 0 \\ 0 & 0 & 0 & 1 \end{bmatrix}$$

2. Add the 1st row to the third row to yield:

$$\begin{bmatrix} 1 & 0 & -1 \\ 0 & -1 & 1 \\ 0 & 1 & -1 \\ 0 & 1 & -1 \end{bmatrix} \quad \begin{bmatrix} 1 & 0 & 0 & 0 \\ 0 & 1 & 0 & 0 \\ 1 & 0 & 1 & 0 \\ 0 & 0 & 0 & 1 \end{bmatrix}$$

3. Add the 2nd row to the third and forth rows to yield:

$$\begin{bmatrix} 1 & 0 & -1 \\ 0 & -1 & 1 \\ 0 & 0 & 0 \\ 0 & 0 & 0 \end{bmatrix} \quad \begin{bmatrix} 1 & 0 & 0 & 0 \\ 0 & 1 & 0 & 0 \\ 1 & 1 & 1 & 0 \\ 0 & 1 & 0 & 1 \end{bmatrix}$$

4. Multiply the second row by -1 to yield the final echelon form:

$$\begin{bmatrix} 1 & 0 & -1 \\ 0 & 1 & -1 \\ 0 & 0 & 0 \\ 0 & 0 & 0 \end{bmatrix} \quad \begin{bmatrix} 1 & 0 & 0 & 0 \\ 0 & -1 & 0 & 0 \\ 1 & 1 & 1 & 0 \\ 0 & 1 & 0 & 1 \end{bmatrix}$$

The final operation achieves the goal of reducing the stoichiometry matrix to an echelon form (in this case it happens to be a reduced echelon form). Note that the operation has resulted in two zero rows appearing in the reduced stoichiometry matrix. These two rows correspond to the Y partition in equation (11.7). The lower two rows can be extracted from the right-hand matrix (what was once the identity matrix) to construct equation (11.8), thus:

$$\begin{bmatrix} 1 & 1 & 1 & 0 \\ 0 & 1 & 0 & 1 \end{bmatrix} \begin{bmatrix} ds_2/dt \\ des/dt \\ ds_1/dt \\ de/dt \end{bmatrix} = 0$$

Or:

$$\frac{ds_2}{dt} + \frac{des}{dt} + \frac{ds_1}{dt} = 0$$

$$\frac{des}{dt} + \frac{de}{dt} = 0$$

From the above equations the following conservation laws should be evident:

$$s_2 + es + s_1 = T_1$$
$$es + e = T_2$$

(11.9)

In summary, the algorithm for deriving the conservation laws is as follows:

1. Apply elementary operations to the stoichiometry matrix until the matrix is reduced to its row echelon form. Simultaneously apply the elementary operations to an identity matrix. The size of the identity matrix should be equal to the number of rows in the stoichiometry matrix.

2. If there are zero rows at the bottom of the reduced stoichiometry matrix then there are conservation laws in the network; otherwise there are not. The number of conservation laws will be equal to the number of zero rows.

3. Extract the rows in the transformed identity matrix that correspond to the position of the zero rows in the reduced stoichiometry matrix. The extracted rows represent the conservation laws.

There are two points worth making when applying this algorithm. The first is that any row swaps made using the row reduction in the stoichiometry matrix will not translate to swaps in the names of the species on the right-hand side of the equation. This means that when reading the conservation rows, the names on the columns are not changed by any row exchanges in the stoichiometry matrix. The second point to make is that when carrying out the elementary row operations, it is recommended to eliminate, whenever possible, terms below a leading entry by adding rather than subtracting. This will ensure that entries in the transforming identity matrix remain positive and that the resulting conservation laws will be made up of positive terms. Sometimes the ability to add will not be possible and subtractions will be necessary. This results in negative terms appearing in the conservation laws which may make them more difficult to interpret physically.

A useful strategy to avoid negative terms in the conservation equations is to order the rows of the stoichiometry matrix such that any species that is likely to appear in more than one conservation relationship is placed at the bottom of the stoichiometry matrix. In the case of the previous example we would make sure that ES becomes the bottom row of the stoichiometry matrix. This ordering ensures that the independent species (top rows) are represented by the so-called free variables, and the dependent species (bottom rows) by the so-called shared variables. This means that the shared or dependent variables (i.e. complexes) will then be a function of the free variables which is more likely to result in positive terms [89]. A more brute force method is to try all permutations of the matrix rows until a positive set of conservation laws is found. For small models (< 10 species) this approach is a viable option.

Although it is possible to manually reduce a stoichiometry matrix, it is far easier to use specialized math software such Scilab, Octave, Python, Matlab and Mathematica or even advanced modern pocket calculators. All these tools offer a `rref()` command for generating a reduced row echelon. The following examples will illustrate the use of the freely available Scilab application (www.scilab.org) to compute the conservation laws.

Example 11.3

Row reduction using Scilab/Matlab. Given the following stoichiometry matrix, use Scilab functions to row reduce and extract the conservation laws.

$$\mathbf{N} = \begin{array}{c} S_2 \\ ES \\ S_1 \\ E \end{array} \left[\begin{array}{ccc} 1 & 0 & -1 \\ 0 & -1 & 1 \\ -1 & 1 & 0 \\ 0 & 1 & -1 \end{array} \right]$$

Enter the stoichiometry matrix into the software:

```
-->n = [1 0 -1; 0 -1 1; -1 1 0; 0 1 -1];
```

Augment the matrix with the identity matrix, this will allow us to record row reduction operations in the identity matrix part of the augmented matrix.

```
-->ni = [n, eye(4,4)]
ni  =
     1.     0.   - 1.     1.     0.     0.     0.
     0.   - 1.     1.     0.     1.     0.     0.
   - 1.     1.     0.     0.     0.     1.     0.
     0.     1.   - 1.     0.     0.     0.     1.
-->
```

Row reduce the augmented matrix:

```
-->rni = rref (ni)
 rni  =
     1.     0.   - 1.     0.     0.   - 1.     1.
     0.     1.   - 1.     0.     0.     0.     1.
     0.     0.     0.     1.     0.     1.   - 1.
     0.     0.     0.     0.     1.     0.     1.
```

The left partition of the reduced matrix contains two zero rows, therefore there are two conservation laws. These laws correspond to the two bottom rows in the right partition. We extract the rows in the right partition to yield:

```
-->c = rni(3:4,4:7)
 c  =
     1.     0.     1.   - 1.
     0.     1.     0.     1.
```

The species column order is the same as the species row order in the original matrix, that is S_2, ES, S_1, and E, therefore:

$$s_2 + s_1 - e = T_1$$
$$es + e = T_2$$

Note the negative E term in the first conservation law. At first glance this does not appear to be the same set of conservation laws that were derived earlier. However, if we substitute E from the second equation into the first, we will get the same set of conservation laws: $s_1 + s_2 + es = T$, showing us that the two sets are identical. To avoid negative terms appearing in the conservation laws, we can use the rule that all complex species (that is shared species) such as ES, be moved to the bottom of the matrix (See next example).

The previous example can be also accomplished in Python using the sympy package, although its not as straight-forward [1]:

---

[1] Although Python is touted as being one of the easiest languages to learn, it certainly does make some operations unnecessarily complex. Python coders also tend to enjoy writing the most unreadable code, possibly making the entire point of Python somewhat redundant.

```
import sympy
import numpy as np

n = sympy.Matrix ([[1,0,-1],[0,-1,1],[-1,1,0],[0,1,-1]])

fs = sympy.Matrix (np.hstack ((n, sympy.eye (4))))
rni = fs.rref()[0]
print rni

c = rni[2:4,3:7]
print c
```

Example 11.4

Row reduction using Scilab/Matlab. Given the following stoichiometry matrix, use Scilab functions to row reduce and extract the conservation laws. In this example the shared species ES has been moved to the bottom of the matrix.

$$N = \begin{matrix} S_2 \\ S_1 \\ E \\ ES \end{matrix} \begin{bmatrix} 1 & 0 & -1 \\ -1 & 1 & 0 \\ 0 & 1 & -1 \\ 0 & -1 & 1 \end{bmatrix}$$

The reduced augmented matrix is:

```
-->rni = rref (ni)
 rni  =
    1.    0.   - 1.    0.   - 1.    0.   - 1.
    0.    1.   - 1.    0.     0.    0.   - 1.
    0.    0.     0.    1.     1.    0.     1.
    0.    0.     0.    0.     0.    1.     1.
```

Once again there are two zero rows, but this time the corresponding conservation laws all have positive entries, yielding the following equations:

$$s_2 + s_1 + es = T_1$$
$$es + e = T_2$$

The Scilab/Matlab code shown in Figure 11.10 will find the conservation laws for any stoichiometry matrix.

Similar calculations can be done using Python. However in Python one should import the sympy library rather than the scipy.linalg library. The sympy library is the python symbolic manipulation library. For whatever reason the creators of scipy made the explicit decision not to include support for functions such as rref or even for computing the null space. Users of these libraries are expected to write their own functions to accomplish these tasks.

The equivalent code using sympy in Python to compute the conservation laws is given in Figure 11.11. If we use only functions from the sympy library, the code is quite compact. Two helper functions, `getNumColumns` and `getNumRows` were defined to make the code easier to read for those not familiar with the shape syntax. Many of these computations can be done automatically using Tellurium and example is shown in Figure 11.12.

Row reduction of the augmented stoichiometry is probably the easiest way to derive the conservation laws. The main advantage of this method is its simplicity, but we also have the ability to direct the calculation by setting the order of rows (species) in the stoichiometry matrix. However, it has one disadvantage which is the possibility of numerical instability for large systems. In particular, for large genomic style stoichiometry models [74] that involve many hundreds or even thousands of reactions and species, the method can suffer dramatic failures due to rounding errors during row reduction. There are more robust methods that rely on QR factorization [112] and Singular Value Decomposition (SVD). The main disadvantage of these other methods is that sometimes, depending on the particular algorithm, the row order cannot be easily prescribed. Further details can be found in the last section of this chapter as well as the companion text book "Linear Algebra for Systems Biology" [88]. In any event, there are some simple tests one can do to check that the computed conservation laws are correct, one such test will be described next.

Null Space of N^T

To complete this section let us consider in more detail the algebraic nature of the Y partition in equation (11.8). The elementary matrix, E, reduced the stoichiometry matrix to a row echelon form, that is to:

$$E N = \left[\begin{array}{c} M \\ 0 \end{array} \right] \tag{11.10}$$

The E matrix corresponds to the same E matrix in equation (11.7), so that we can partition the elementary matrix, E row-wise into X and Y partitions (equation (11.7)).

$$\left[\begin{array}{c} X \\ Y \end{array} \right] N = \left[\begin{array}{c} M \\ 0 \end{array} \right]$$

From which we can immediately see that:

$$Y N = 0$$

Taking the transpose we obtain:

$$N^T Y^T = 0$$

The Y partition is therefore the null space (See Box 11.4) of the transpose of the stoichiometry matrix [2] This is a significant result for a number of reasons. It gives a very concise definition of the conservation matrix, but more importantly, it opens up the possibility of using other computational approaches.

Box 11.4 The Null Space

Given a matrix equation of the form $A x = 0$ where A is an $m \times n$ matrix and x is a column vector of n elements, the solution, that is the set of vectors x that satisfy this equation, is called the **null space** of A.

The minimum number of vectors required to fully describe the null space is called the **dimension** and is equal to the rank of the matrix rank(A) minus the number of columns, n. These vectors form what is called a **basis** for the space. Linear combinations of these vectors can generate any other vector in the null space. In order to form a basis, the vectors must also be linearly independent.

Many tools can compute the basis for the null space, for example `null (A, 'r')` will compute the basis in Matlab, while `NullSpace[A]` can be used to compute the basis in Mathematica.

---

[2] *cf.*. Chapter 4, Section Computing the Null Space in Introduction to Linear Algebra for Systems Biology (2015), Sauro

The other point of interest is that this result can be used to test whether a set of conservation laws were correctly derived or not. To do this we multiply the transpose of N by the transpose of the conservation matrix Y and make sure the product equals zero.

Many software packages such as Matlab, Scilab or Mathematica supply commands to compute the null space. This makes it easy to compute the conservation laws by computing the null space of the transpose of the stoichiometry matrix. For example, the following session shows how we can use Scilab to compute the conservation laws for the example matrix we used in previous examples.

```
-->N = [1 0 -1; -1 1 0; 0 1 -1; 0 -1 1]
 N   =
    1.    0.   - 1.
    0.   - 1.    1.
  - 1.    1.    0.
    0.    1.   - 1.
--> ns = kernel (N')
ans   =
    0.             0.6324555
    0.             0.6324555
    0.7071068   - 0.3162278
    0.7071068     0.3162278
--> // Convert the orthonormal set
--> // into a rational basis using rref
-->rref (ns')'
ans   =
    1.    0.
    1.    0.
    0.    1.
    1.    1.
```

The null space command in Scilab is `kernel`, in Matlab it is `null`, and in Mathematical it is `NullSpace`. In Python the null space command can be found in the sympy library by using the `nullspace` command. Like many null space commands implemented in mathematical software, the kernel command in Scilab has the drawback of generating an orthonormal set[3]. In order to generate a rational basis we must row reduce the kernel, this results in a more interpretable set of conservation laws. In Matlab it is possible to use the modified null space command, `null (N, 'r')`, which will automatically generate a rational basis (neither Octave or Scilab support this format). Interestingly, Mathematica's (v7.0) null space function does generate a rational basis, however, the algorithm that Mathematica uses is unknown, raising its own issues. More recent version of Mathematica (v11) appear to generate orthonormal sets. The Python implementation, based on Sympy, does generate a normalized rational basis which is useful, for example:

```
sympy.Matrix.nullspace (sympy.Matrix.transpose (n))
[Matrix([[1],[1],[1],[0]]), Matrix([[0],[1],[0],[1]])]
```

Given that we can now compute the conservation laws for arbitrary networks, one question to consider is whether conservation laws have any behavioral consequences. The answer to this question will be considered in Chapter 12.

---

[3]An orthonormal set is one where the vectors are orthogonal to each other and of magnitude one.

11.5   Advanced Theory – Optional

In this section we will look at further aspects of conservation law analysis using a more formal approach. In a later section we will also consider more advanced numerical methods for computing conservation laws.

Let us begin by assuming that the rows of the stoichiometry matrix have been arranged so that the top rows, m_o, include the independent rows, and the bottom, $m - m_o$ rows the dependent rows. If we designate the top rows with the symbol N_R and the bottom rows by N_0, we can write the stoichiometry matrix as:

$$\mathbf{N} = \left[\begin{array}{c} N_R \\ N_0 \end{array} \right]$$

where the submatrix N_R is full rank, and each row of the submatrix N_0 can be derived by a linear combination of the rows of N_R. We can also reorder the columns of the stoichiometry matrix for which there will also be m_o independent columns (column and rows ranks are equal). We will denote the partition of $\mathbf{N}$ that contains the last m_o columns, the N_C matrix. Finally, we will designate the partition of $\mathbf{N}$ that includes only the independent rows and columns the N_{RC} matrix. The N_{RC} matrix will be a $m_o \times m_o$ square invertible matrix. N_{RC} must be invertible because all rows and columns are independent. The graphical depiction of this partitioning is given in Figure 11.13.

If there are no conserved cycles in the network, then the rank $(\mathbf{N}) = m$ (i.e. full rank) and $\mathbf{N}$ equals N_R. Following Reder [81], Ehlde [25] and Hofmeyr [46], we make the following construction. Since the rows of N_0 are linear combinations of the rows of N_R, we can define a **link-zero matrix**, L_0 which satisfies:

$$N_0 = L_0 N_R. \tag{11.11}$$

L_0 will have dimensions $(m - m_o) \times m_o$. We can combine L_0 with the identity matrix – of dimension $rank(\mathbf{N})$ – to form the $m \times m_o$ link matrix, L, thus:

$$L = \left[\begin{array}{c} I \\ L_0 \end{array} \right]$$

When $\mathbf{N}$ has full rank, L equals the identity matrix. Using equation (11.11) and the link matrix we can write:

$$\mathbf{N} = \left[\begin{array}{c} N_R \\ N_0 \end{array} \right] = \left[\begin{array}{c} I \\ L_0 \end{array} \right] N_R = L N_R$$

For networks without conserved moieties the L matrix reduces to the identity matrix, I. If we delete the dependent columns of $\mathbf{N}$ and N_R we obtain:

$$N_C = L_0 N_{RC} \quad \text{or} \quad L = N_C N_{RC}^{-1}$$

By partitioning the stoichiometry matrix into a dependent and independent set we also partition the system equation. The full system equation which describes the dynamics of the network is thus:

$$\left[\begin{array}{c} I \\ L_0 \end{array} \right] N_R v = \frac{d\mathbf{s}}{dt} = \left[\begin{array}{c} \frac{ds_i}{dt} \\ \frac{ds_d}{dt} \end{array} \right]$$

where the terms ds_i/dt and ds_d/dt refer to the independent and dependent rates of change, respectively. From the above equation, we see that:

$$\frac{ds_d}{dt} = L_0 \frac{ds_i}{dt}.$$

Integrating this last equation, we find:

$$s_d(t) - s_d(0) = L_0 \left[s_i(t) - s_i(0) \right]$$

This equation is true for all time t, where $s_d(0)$ and $s_i(0)$ are the amounts of the dependent and independent species respectively at time zero. Introducing the constant vector $T = s_d(0) - L_0 s_i(0)$, we can write the above equation as:

$$\begin{bmatrix} -L_0 & I \end{bmatrix} \begin{bmatrix} s_i \\ s_d \end{bmatrix} = T \tag{11.12}$$

If $s = (s_i, s_d)$, we can introduce $\Gamma = [-L_0 \ I]$, and write this concisely as:

$$\Gamma s = T$$

We will call Γ the **conservation matrix** and is equivalent to the Y matrix in equation (11.8). Each row of the conservation matrix relates to a particular conserved cycle and thus the number of rows indicates the number of conserved cycles in the network. The elements in a particular row indicate which metabolite species contribute to a particular cycle.

The relationship, $N_0 = L_0 N_R$ (11.11) can be reexpressed in the following form:

$$\begin{bmatrix} -L_0 & I \end{bmatrix} \begin{bmatrix} N_R \\ N_0 \end{bmatrix} = 0 \tag{11.13}$$

However since the conservation matrix: $\Gamma = [-L_0 \ I]$, the above relation can be rewritten as: $\Gamma N = 0$. Taking the transpose of this gives us:

$$N^T \Gamma^T = 0 \tag{11.14}$$

We have already seen this equation in a previous section (11.4) and tells us that the conservation matrix is the null space of the transpose of the stoichiometry matrix. An equivalent way to state this is that the conservation matrix is the *left null space* of the stoichiometry matrix ($\Gamma N = 0$).

The significance of equation (11.14) is that there are many software tools that allow one to compute the null space very easily. For example Matlab, Mathematica, Python, Maple, or Scilab can easily compute the null space of a matrix and thus derive the conservation laws. Some of these tools however, for example Scilab and Matlab, do not normalize the null space so that a second stage is required, but this is easily accomplished with the command `rref`. As mentioned, before Matlab has a variant on the null command, `null (A, 'r')` which generates what is called a rational basis. In Scilab one would enter, `cm = rref (kernel (N')')`.

The final transpose in the equation that is applied is to reorientate the conservation matrix for better viewing. The Python based Tellurium `tellurium.analogmachne.org` offers an integrated simulation environment that includes routines to automatically compute these matrices (See Figure 11.14) as well as labeling the resulting conservation matrix. PySCeS is another tool that supports a range of methods including computing the L_0. Similarly the `libStructural` python package [?] also supports computation of a wide variety of stoichiometrically derived terms.

Returning once again to the network shown in Figure 11.9, equation (11.12) can be rearranged so that the dependent species can be computed from the independent species, that is:

$$s_d = L_0 s_i + T \tag{11.15}$$

The complete set of conservation law equations for this model is therefore, equation (11.15):

$$\begin{bmatrix} s_1 \\ e \end{bmatrix} = \begin{bmatrix} -1 & -1 \\ 0 & -1 \end{bmatrix} \begin{bmatrix} s_2 \\ es \end{bmatrix} + \begin{bmatrix} T_1 \\ T_2 \end{bmatrix}$$

$$\begin{bmatrix} \dfrac{ds_2}{dt} \\ \dfrac{des}{dt} \end{bmatrix} = \begin{bmatrix} 1 & 0 & -1 \\ 0 & -1 & 1 \end{bmatrix} \begin{bmatrix} v_1 \\ v_2 \\ v_3 \end{bmatrix} \tag{11.16}$$

To emphasize this point again, even though there appears to be four variables in this system, there are in fact only two independent variables, $\{ES, E_1\}$, and thus two differential equations and two linear constraints. When solving the system in time, only two differential equations need to be explicitly integrated.

Scaled L

In metabolic control analysis [54, 81, 30] the link matrix, L plays a central role in formulating the sensitivities. In such cases the scaled version of L, denoted, $\mathcal{L}$ is often used [46].

$\mathcal{L}$ is defined as:

$$\mathcal{L} = (D^s)^{-1} \cdot L \cdot D^{sI}$$

where D represents a diagonal matrix of either the reciprocals of species, D^s or a diagonal of the independent species, D^{sI}. For the previous example, $\mathcal{L}$ would be given by:

$$\mathcal{L} = \begin{bmatrix} 1/s_2 & 0 & 0 & 0 \\ 0 & 1/es & 0 & 0 \\ 0 & 0 & 1/s_1 & 0 \\ 0 & 0 & 0 & 1/e \end{bmatrix} \begin{bmatrix} 1 & 0 \\ 0 & 1 \\ -1 & -1 \\ 0 & -1 \end{bmatrix} \begin{bmatrix} s_2 & 0 \\ 0 & es \end{bmatrix} = \begin{bmatrix} 1 & 0 \\ 0 & 1 \\ -1 & -es/s_1 \\ 0 & -es/e \end{bmatrix}$$

11.6 Numerical Methods

In a previous section 11.4, a simple method based on forward elimination was described to derive the conservation laws. This method has a number of advantages but for large matrices, it can potentially be numerically unstable. In this section we will review alternative methods that, although not always as flexible as forward elimination, are still well suited for the analysis of large matrices.

These methods fall into two groups, those based on QR factorization and those based on Singular Value Decomposition (SVD). The method based on SVD is the simplest and will be described first.

SVD

Singular Value Decomposition, or SVD is a very useful method for decompiling a matrix into the four orthonormal fundamental subspaces. These subspaces include the row space, columns, null space and the left null space. SVD is based on the following factorization:

$$A = USV^\mathsf{T}$$

where A is a $m \times n$ matrix of real numbers, U is a $m \times m$ orthonormal matrix, V is an $n \times n$ orthonormal matrix, and S a $m \times n$ diagonal matrix with entries $\sigma_1 \geq \sigma_2 \geq \dots \sigma_p$ where p is either m or n, whichever is the smallest ($p = \min\{m, n\}$). The numbers σ_i are called the singular values and are positive. The columns of U and V form the left and right-hand singular vectors.

Of more interest here is the fact that the lower rows of V^T, which correspond to the zero singular values in S, form an orthonormal basis for the null space of A. Therefore one way to obtain the null space of a given matrix is to extract these lower rows from the V^T matrix. The number of rows in V^T that correspond to the null space vectors will equal $n - r$, where r is the rank and n the number of columns of A. If there are no zero rows in the S matrix, then the null space is empty.

Example 11.5 ━━

Obtain an estimate for the null space of the *transpose* of the following stoichiometry matrix using SVD. Since we will be working on the transpose, the null space vectors will represent the conservation laws.

$$\mathbf{N} = \begin{bmatrix} 1 & 0 & -1 \\ -1 & 1 & 0 \\ 0 & 1 & -1 \\ 0 & -1 & 1 \end{bmatrix}$$

Many math applications such as Scilab or Matlab have svd functions. Here we will use the svd function from Scilab.

```
-->[U, S, V] = svd (N')
V  =
  -0.316229  -0.707107   0.632456   0.
  -0.316229   0.707107   0.632456   0.
  -0.632456   0.        -0.316229   0.707107
   0.632456   0.         0.316229   0.707107
 S  =
   2.236068   0.         0.         0.
   0.         1.7320508  0.         0.
   0.         0.         1.587D-16  0.
 U  =
  -1.886D-16  -0.8164966  0.5773503
  -0.7071068   0.4082483  0.5773503
   0.7071068   0.4082483  0.5773503
```

We can extract the null space from V^T. The number of zero rows in the **S** matrix is two, therefore we must extract the bottom two rows of V^T. This gives us:

```
-->Vt = V'
-->Vt(3:4,1:4)
   0.6324555  0.6324555 -0.3162278  0.3162278
   0.         0.         0.7071068  0.7071068
```

SVD returns an orthonormal basis. To generate a rational basis apply row reduction to these two rows to yield:

```
-->rref (kk)
 ans  =
    1.    1.    0.    1.
    0.    0.    1.    1.
```

The transpose of these two vectors is the null space of $\mathbf{N}^T$. This can be confirmed by computing the product $\mathbf{N}^T \, \mathcal{N}(\mathbf{N}^T)$ and showing that the product equals zero:

$$\begin{bmatrix} 1 & -1 & 0 & 0 \\ 0 & 1 & 1 & -1 \\ -1 & 0 & -1 & 1 \end{bmatrix} \begin{bmatrix} 1 & 0 \\ 1 & 0 \\ 0 & 1 \\ 1 & 1 \end{bmatrix} = \begin{bmatrix} 0 & 0 \\ 0 & 0 \\ 0 & 0 \end{bmatrix}$$

where $\mathcal{N}(\mathbf{N}^T)$ is the null space of $\mathbf{N}^T$. Because there are no row or column exchanges during SVD, the rows in the null space vectors correspond to the same rows in the original matrix, **N**. This makes it easy to identify the individual conservation entries in the conservation law vectors.

We can formalize the SVD algorithm using the following Scilab/Matlab code.

```
// Use SVD to estimate conservation laws
// Operate on the transpose of n

[u, s, v] = svd (n');
vt = v';
nRows = size(vt, 1);
nCols = size(vt, 2);
// Extract bottom nCols(n')-rank orthonormal rows
orthogns = Vt(r+1:nRows,1:nCols);
// Row reduce the transpose to get rational basis
ratns = rref (orthogns)';
// Display Result
ratns'
// Confirm it is the null space, ns should equal 0
ns = n'*ratns
```

Since there are no column or row exchanges during SVD, the order of the rows in the stoichiometry matrix can be used to influence the form of the final conservation laws. Just like the row reduction technique, the order of rows in the stoichiometry matrix should be such that any shared species (i.e species containing more than one moiety) be located as close to the bottom of the matrix as possible. This will ensure that negative terms will tend to not appear in the final conservation equations.

QR Factorization

The SVD method given in the last section is an excellent choice for determining the conservation laws. However, it has two downsides, the first is that it is far more computationally intensive that the simple row reduction technique described in 11.4. The second problem with the SVD approach is the need to carry out a final Gauss-Jordan elimination to obtain a rational basis for the conservation laws. Depending on the size of the stoichiometry matrix, Gauss-Jordan elimination can potentially be numerically unstable.

Methods that have both excellent stability properties and are less computationally intense than SVD are methods based on QR factorization.

The first QR method is based on computing L_0. Any $m \times n$ matrix can be factored into a product of two matrices Q and R and a permutation matrix, P:

$$AP = QR$$

Q is an $m \times m$ orthogonal matrix, that is $Q^T Q = I$, hence $Q^T = Q^{-1}$. R is a $m \times n$ upper trapezoidal matrix and P a permutation matrix. If A is the transpose of the stoichiometry matrix N^T, the permutation matrix will also reorder the columns of N^T such that the independent columns are on the left and the dependent rows on the right. This is equivalent to reordering the rows in N. This partitioning can be written as follows where R has been partitioned to match the left side (Recall that $Q^T = Q^{-1}$):

$$Q^T \begin{bmatrix} N_R^T & N_0^T \end{bmatrix} = \begin{bmatrix} R_{11} & R_{12} \\ 0 & 0 \end{bmatrix}$$

Note that the partitioned matrix has been absorbed into the reordered N^T matrix during the reordering. If we multiply out the terms we obtain:

$$\begin{bmatrix} R_{11} \\ 0 \end{bmatrix} = Q^T N_R{}^T$$

$$\begin{bmatrix} R_{12} \\ 0 \end{bmatrix} = Q^T N_0{}^T$$

Given that $N_0 = L_0 N_R$, R_{12} can be rewritten as:

$$\begin{bmatrix} R_{12} \\ 0 \end{bmatrix} = Q^T N_R{}^T L_0{}^T$$

so that:

$$\begin{bmatrix} R_{12} \\ 0 \end{bmatrix} = \begin{bmatrix} R_{11} \\ 0 \end{bmatrix} L_0{}^T$$

That is:

$$R_{12} = R_{11} L_0{}^T \tag{11.17}$$

Since the permutation matrix post-multiplies N^T, it means that the columns are reordered, this is reflected in column reordering in the R matrix such that all independent columns are moved to the left and dependent columns to the right. Row reduction of the R matrix to a reduced echelon form will therefore result in the left partition being transformed into the identity matrix, that is $R_{11} = I$. From this it follows (11.17) that the reduced left partition, $R_{12} = L_0{}^T$, which is the result we seek:

$$L_0 = R_{12}^T$$

By augmenting the L_0 matrix with an appropriately sized identity matrix, we can use this method to generate conservation laws in the standard form, that is in the form $[-L_0 \ I]$. This also means that the rows of the stoichiometry matrix will also have been reordered in the process as determined by the permutation matrix obtained from the QR factorization. Therefore, unlike the row reduction technique or SVD, it is not possible to greatly influence the kind of conservation laws generated by presetting the row order of the stoichiometry matrix, although some flexibility still exists. It is still advantageous to make sure that all the shared species are in the bottom rows. The one potential problem with the method is the final Gauss-Jordan elimination, however the reordering of the columns will make this less of an issue.

The Scilab/Matlab code below illustrates an implementation of this method. It is very important to note that the species labels attached to the columns of the conservation matrix is determined by the permutation matrix. This part of the calculation is not shown in the following code.

```
// Use QR to estimate conservation laws via Lo
// Operate on the transpose of n
[qm, rm, p] = qr (n');
nRows = size(n, 1);
nCols = size(n, 2);
mo = rank (n);
m = size(n, 1);
mmo = m - mo;
// Extract bottom nCols-rank orthonormal rows
rt = rm(1:r,1:nRows);
// Row reduce the transpose to get a rational basis
rrt = rref (rt);
Lo = rrt(1:mo,mo+1:nRows)';
// Display Lo
Lo
// Construct the conservation vectors and display
cm = [-Lo eye(mo,mo)];
cm
```

Example 11.6 ───

Compute the L_0 matrix of the following stoichiometry matrix using QR factorization.

$$N = \begin{bmatrix} 1 & 0 & -1 \\ -1 & 1 & 0 \\ 0 & 1 & -1 \\ 0 & -1 & 1 \end{bmatrix}$$

Many software tools offer standard QR factorization. In this example we use Scilab. Applying QR factorization yields the following R matrix:

```
R  =
   1.414217 -0.707107 -1.414217 -0.707107
   0.        1.224745  0.        -1.224745
   0.        0.        0.         0.
```

Since the rank of the stoichiometry matrix is 2, we extract the top two rows from R and carry out a row reduction (for example by using the `rref()` function) to yield:

```
ans  =
    1.    0.   - 1.   - 1.
    0.    1.     0.   - 1.
```

The transpose of the L_0 matrix can be found in the top right corner starting at column $m_o + 1$, where m_o equals the number of independent rows in the original stoichiometry matrix. In this case m_o equals 2, therefore the L_0 matrix (after transposition) is given by:

```
  -1    0
  -1   -1
```

We now combine the negative of this with the identity matrix to obtain the conservation vectors:

```
  1   0   1   0
  1   1   0   1
```

The only thing that remains is the species labeling for the conservation columns. These can be obtained from the original stoichiometry matrix and the permutation matrix, P. As returned by the QR factorization, P is given by:

```
P  =
    1.    0.    0.    0.
    0.    0.    1.    0.
    0.    1.    0.    0.
    0.    0.    0.    1.
```

and the original species order was ES, E,S_1, S_2. The permutation matrix shows that the new species order should be: ES, S_1, E, S_2.

The final QR method to consider is one based on rank revealing methods, sometimes called RRQR [15]. The specific algebra is described in more detail in the companion book [88], but the method uses the following formula to estimate the null space:

$$A P \begin{bmatrix} -R_{11}^{-1} R_{12} \\ I \end{bmatrix} = 0 \tag{11.18}$$

This approach is of particular interest because it generates a rational basis for the null space because of the identity matrix in the lower partition. The downside is that it requires an inversion of R_{11}, but since R_{11} is triangular, it is possible to exploit widely available and efficient routines for inverting such matrices. For any new software implementation I would probably recommend this approach.

Example 11.7

Use the RRQR based method to compute the null space for the transpose of the stoichiometry matrix:

$$\mathbf{N} = \begin{bmatrix} 1 & 0 & -1 \\ -1 & 1 & 0 \\ 0 & 1 & -1 \\ 0 & -1 & 1 \end{bmatrix}$$

From the last example we saw that QR factorization yielded the following R matrix:

```
R  =
  1.414217 -0.707107 -1.414217 -0.707107
  0.        1.224745  0.        -1.224745
  0.        0.        0.         0.
```

Since the rank of the stoichiometry matrix is 2, we can partition R into the following submatrices:

$$R_{11} = \begin{bmatrix} 1.414217 & -0.707107 \\ 0. & 1.224745 \end{bmatrix} \qquad R_{12} = \begin{bmatrix} -1.414217 & -0.707107 \\ 0. & -1.224745 \end{bmatrix}$$

We now compute $-R_{11}^{-1} R_{12}$ to obtain:

$$\begin{bmatrix} 1 & 1 \\ 0 & 1 \end{bmatrix}$$

Combining this with an appropriately sized identity matrix gives the null space:

$$\begin{bmatrix} 1 & 1 \\ 0 & 1 \\ 1 & 0 \\ 0 & 1 \end{bmatrix}$$

Like the previous method we need to be aware of the permutation matrix as this will determine the labels that are associated with the rows of the null space.

There are also ways to obtain the conservation vectors via the Q matrix and these are discussed in [112].

For a completely different approach to computing the conservation laws, the reader is referred to the work by Schuster and colleagues. In this work, convex analysis [97] is used to determine the conservation laws. It is used primarily to generate conservation laws that only contain (where possible) positive entries.

Most modern simulation applications either use the simpler row reduction technique or more common in recent years, they use the QR factorization technique based on estimating the L_0 matrix [112].

Method	Advantages	Disadvantages
Row Reduction	a) Simple b) Fast c) Row Order	Potential numerical instabilities
SVD	a) Robust b) Expensive on large systems	Requires one final Gauss-Jordan step
QR by L_0	a) Robust b) Faster than SVD	Requires one final Gauss-Jordan step
QR by RRQR	a) Robust b) Row order	No Gauss-Jordan step required

Table 11.1 Comparison of different approaches to computing conservation laws.

11.7 Design of Simulation Software

One practical implication of moiety conservation concerns the design of software for simulation and analysis. Two issues arise, one concerns increasing simulation efficiency by reducing the number of differential equations, and the second concerns numerical stability by removing the dependent species from a model.

Reduced Systems

The first concern is straightforward. Instead of solving the full set of systems equations, many simulators instead solve the following reduced set:

$$s_d = L_0 s_i + T$$
$$\frac{ds_i}{dt} = N_R v(s_i, s_d) \tag{11.19}$$

In these equations, s_i is the vector of independent species, s_d, the vector of dependent species, L_0 the link matrix, T the total mass vector, N_R the reduced stoichiometry matrix, and v the rate vector. This modified equation (11.19) constitutes the most general expression for a temporal model that uses differential equations [46, 44]. Equations (11.5) shows a typical reduced system. Note that in these equations the dependent species are first computed from the dependent species. This is followed by the evaluation of the reduced set of differential equations. The order of the calculation is crucial. The total amounts, T, can be computed at the start of a simulation by using equation (11.15) and the initial conditions. In multi-compartmental systems where the size of compartments may differ, it is important to sum the amounts, not concentrations.

One obvious advantage of reducing the model is that it lessens the computational burden to solving the full set of differential equations. Many biochemical simulation packages will automatically check for moiety conservations and perform this simplification before performing any analysis of the system equations. This is especially important for large models. For example, in an *E. coli* model downloaded from the BiGG repository http://bigg.ucsd.edu/, approximately five percent of the differential equations are redundant, that is they can be safely eliminated from the model by using moiety conservation constraints.

Numerical Stability

Although simplifying a model by eliminating the dependent species can offer speed improvements to simulations, the most important reason for model reduction is the gain in numerical stability. One of the most important metrics that arises often in the analysis of pathways (or any dynamical system for that matter) is the Jacobian matrix.

The Jacobian matrix, as we have seen before, is an $m \times m$ matrix of partial derivatives of the rates of change with respect to the species, that is:

$$J = \frac{\partial}{\partial \mathbf{s}} \left(\frac{d\mathbf{s}}{dt} \right)$$

For example, for a simple linear chain shown in Figure 11.15 the differential equations are given by:

$$\frac{ds_1}{dt} = v_1 - v_2$$

$$\frac{ds_2}{dt} = v_2 - v_3$$

The Jacobian matrix is then given by:

$$J = \begin{bmatrix} \dfrac{\partial(v_1 - v_2)}{\partial s_1} & \dfrac{\partial(v_1 - v_2)}{\partial s_2} \\ \dfrac{\partial(v_2 - v_3)}{\partial s_1} & \dfrac{\partial(v_2 - v_3)}{\partial s_2} \end{bmatrix} = \begin{bmatrix} -k_2 & 0 \\ k_2 & -k_3 \end{bmatrix}$$

The Jacobian is used in many ancillary calculations, for example, solving differential equations (particularly stiff equations), solving for the steady state, calculating sensitivities, frequency analysis, certain optimization algorithms and others. In many of these cases the calculation involves the inversion of the Jacobian. In the case of the linear pathway, there will always be an inverse so long as the rate constants are non-zero. However if we consider a simple cycle such as the one shown in Figure 11.16, then the Jacobian matrix is given by:

$$J = \begin{bmatrix} \dfrac{\partial(v_1 - v_2)}{\partial s_1} & \dfrac{\partial(v_1 - v_2)}{\partial s_2} \\ \dfrac{\partial(v_2 - v_1)}{\partial s_1} & \dfrac{\partial(v_2 - v_1)}{\partial s_2} \end{bmatrix} = \begin{bmatrix} -k_1 & k_2 \\ k_1 & -k_2 \end{bmatrix}$$

This shows that the row dependencies in the stoichiometry matrix reappear as dependencies in the Jacobian. This means that the Jacobian cannot be inverted and any calculations that require the inversion of the Jacobian will fail. The solution is to work with the reduced model. This eliminates the dependent species from the stoichiometry matrix which in turn makes sure that the Jacobian is once again invertible.

Multicompartment Systems

Up to now we have not mentioned the fact that models may include multiple compartments, that is separate volume spaces where the movement of mass between volumes is via specific transporter proteins. Surprisingly the literature is not very clear or extensive in discussing the modeling of multicompartment systems within cellular system, however one crucial point to bear in mind when conservation laws cross compartment boundaries is that the sum must be with respect to the total mass. For convenience, models will often assume a unit volume for a compartment such that any conserved cycles within the compartment are expressed as the sum of concentrations. In such situations it is easy to forget that what is actually conserved is in fact mass, not concentration. In general a conservation law is therefore expressed in the form:

$$\sum V_i s_i = T$$

where V_i is the volume that the concentration of species S_i resides.

Further Reading

1. Reich, JG, and Selkov EE (1981) Energy metabolism of the cell. Academic Press, London.

2. Hofmeyr JH, Kacser H, van der Merwe KJ. (1986) Metabolic control analysis of moiety-conserved cycles. Eur J Biochem. 155(3):631-41.

3. Sauro HM and Ingalls B (2004) Conservation analysis in biochemical networks: computational issues for software writers, Biophys Chem, 109, 1–15.

4. Sauro HM (2015) Introduction to Linear Algebra for Systems Biology. Ambrosius Publishing.

5. Vallabhajosyula RR, Chickarmane V and Sauro HM (2006) Conservation analysis of large biochemical networks, Bioinformatics, 22(3), 346-353.

6. Cornish-Bowden and Hofmeyr J-HS (2002) The Role of Stoichiometric Analysis in Studies of Metabolism: An Example J. theor. Biol 216, 179-191.

7. Ehlde M, and Zacchi G (1997) Chemical Engineering Science 52:2599–2606(8).

Exercises

1. Define the terms *moiety* and *conserved moiety*.

2. Write the differential equations for the following model and show that some of the rates of change are dependent on others:

   ```
   S1 -> S2
   S2 -> S3
   S3 -> S2
   S2 -> S1
   ```

3. The following paper: Eisenthal R, and Cornish-Bowden A. 1998. Journal of Biological Chemistry 273 (10):5500-5505, describes a model of Trypanosome energy metabolism. Read the paper and identify the conserved moiety cycles described in the model.

4. Analyze the following model using matrix reduction techniques to determine the conserved cycle. Assume Hexose is a fixed species.

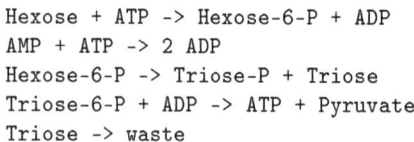

   ```
   Hexose + ATP -> Hexose-6-P + ADP
   AMP + ATP -> 2 ADP
   Hexose-6-P -> Triose-P + Triose
   Triose-6-P + ADP -> ATP + Pyruvate
   Triose -> waste
   ```

5. Use Matlab, Scilab or Octave to determine the conserved cycles in the model by Eisenthal R, and Cornish-Bowden A. 1998. Journal of Biological Chemistry 273 (10):5500-5505.

Proofs

A general non-symmetric, $m \times n$ matrix A can be as a product of two matrices Q and R as:

$$A P = Q R$$

where Q is an $m \times m$ orthogonal matrix such that $Q^T Q = I$, P is an $m \times n$ permutation matrix that indicates column changes in A and R is an $m \times n$ upper trapezoidal matrix.

Box 11.2 Elementary Matrices - Recap

Elementary matrix operations such as row exchange, row scaling or row replacement can be represented by simple matrices called elementary matrices. They are called Type I, II and III, respectively. Elementary matrices can be constructed from the identity matrix. For example, a scaling operation can be represented by replacing one of the elements of the main diagonal of an identity matrix by the scaling factor. The following matrix represents a Type II matrix which will scale the second row of a given matrix by the factor k:

$$\begin{bmatrix} 1 & 0 & 0 \\ 0 & k & 0 \\ 0 & 0 & 1 \end{bmatrix}$$

Type I elementary matrices will exchange two given rows in a given matrix and are constructed from an identity matrix where rows in an identity matrix are exchanged that correspond to the rows exchanged in the target matrix. The following Type I matrix will exchange rows 2 and 3 in a target matrix:

$$\begin{bmatrix} 1 & 0 & 0 \\ 0 & 0 & 1 \\ 0 & 1 & 0 \end{bmatrix}$$

Type III elementary matrices will add/subtract a given row in a target matrix to another row in the same matrix. Type III matrices are constructed from an identity matrix where a single off diagonal element is set to the multiplication factor, and the specific location represents the two rows to combine. If an elementary matrix adds a row i to a row j multiplied by a factor α, then the identity matrix with entry i, j is set to α. In the following example, the Type III elementary matrix will subtract five times the 1st row from the 2nd row.

$$\begin{bmatrix} 1 & 0 & 0 \\ 0 & 1 & 0 \\ -5 & 1 & 0 \end{bmatrix}$$

A particularly important property of elementary matrices is that they can all be inverted. In addition, pre-multiplying by an elementary matrix will modify the rows of a target matrix while post-multiplying will operate on the columns.

Box 11.3 Echelon Forms - Recap

There are two kinds of matrices that one frequently encounters in the study of linear equations. These are the **row echelon** and **reduced echelon forms**. Both matrices are generated when solving sets of linear equations. The row echelon form is derived using forward elimination and the reduced echelon form by Gauss-Jordan Elimination.

A **row echelon matrix** is defined as follows:

1. All rows that consist entirely of zeros are at the bottom of the matrix.
2. In each non-zero row, the first non-zero entry is a 1, called the leading one.
3. The leading 1 in each row is to the right of all leading 1's above it. This means there will be zeros below each leading 1.

The following three matrices are examples of row echelon forms:

$$\begin{bmatrix} 1 & 4 & 3 & 0 \\ 0 & 0 & 1 & 7 \\ 0 & 0 & 0 & 0 \end{bmatrix} \quad \begin{bmatrix} 1 & 1 & 0 \\ 0 & 1 & 0 \end{bmatrix} \quad \begin{bmatrix} 1 & 5 & 3 & 0 \\ 0 & 1 & 7 & 2 \\ 0 & 0 & 0 & 1 \end{bmatrix}$$

The **reduced echelon form** has one additional characteristic:

4. Each column that contains a leading one has zeros above and below it. The following three matrices are examples of reduced echelon forms:

$$\begin{bmatrix} 1 & 0 & 4 & 0 \\ 0 & 1 & 1 & 7 \\ 0 & 0 & 0 & 0 \end{bmatrix} \quad \begin{bmatrix} 1 & 0 & 0 \\ 0 & 1 & 0 \end{bmatrix} \quad \begin{bmatrix} 1 & 0 & 0 \\ 0 & 1 & 0 \\ 0 & 0 & 1 \end{bmatrix}$$

Sometimes the columns of a reduced echelon can be ordered such that each leading one is immediately to the right of the leading one above it. This will ensure that the leading 1's form an identity matrix at the front of the matrix. The reduced echelon form will therefore have the following general block structure:

$$\begin{bmatrix} I & A \\ 0 & 0 \end{bmatrix}$$

It is always possible to reduce any matrix to its echelon or reduced echelon form by an appropriate choice of elementary operations. The function `rref()` implemented in many math software applications and modern hand calculators, will generate a reduced row echelon.

```
// Compute Conservation Laws
// ------------------------

// Enter the stoichiometry matrix first

n = [1 0 -1; 0 -1 1; -1 1 0; 0 1 -1];
nRows = size(n, 1);
// Create the augmented matrix
ni = [n, eye(nRows,nRows)];
// Carry out row reduction
rni = rref (ni);
r = rank (n);
// Extract the conservation rows
c = rni(r+1:nRows,size(n,2)+1:size(ni,2));
// Display result
c
```

Figure 11.10 General purpose Scilab/Matlab code to determine conservation laws using row reduction.

```
# Compute Conservation Laws

import numpy as np
import sympy

def getNumColumns (m): return m.shape[1]
def getNumRows (m):    return m.shape[0]

n = sympy.Matrix ([[1, 0, -1], [0, -1, 1], [-1, 1, 0], [0, 1, -1]]);
nRows = getNumRows (n)
# Create the augmented matrix
ni = sympy.Matrix.hstack (n, sympy.Matrix.eye(nRows));
# Carry out the row reduction
rni = ni.rref ();
# Compute the rank
r = sympy.Matrix (n).rank();
# Extract the conservation rows
c = rni[0][r:nRows, getNumColumns (n):getNumColumns (ni)];
print c
```

Figure 11.11 General purpose Python code to determine conservation laws using row reduction.

```
import tellurium as te
import roadrunner

r = te.loada("""
    S1 -> S2; k1*S1;
    S2 -> S1; k2*S2;

    k1 = 0.5; k2 = 0.5; S1 = 10;
""")

r.conservedMoietyAnalysis = True;
print r.getConservationMatrix()
```

Figure 11.12 Computing the conservation laws using Tellurium.

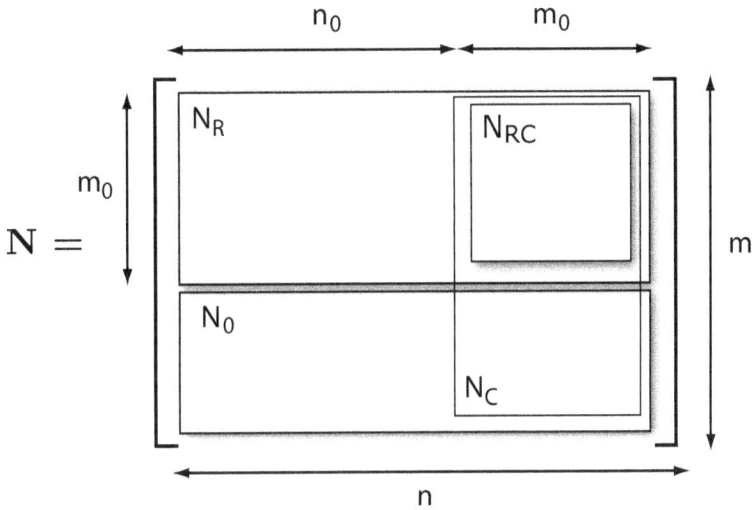

Figure 11.13 Partitioning of the Stoichiometry Matrix into Four Fundamental Partitions.

```
import tellurium as te
import roadrunner

r = te.loada("""
    S1 -> S2; v;
    S2 + E -> ES; v;
    ES -> E + S1; v

    v = 0;
""")
print r.getLinkMatrix()
```

Figure 11.14 Computing the link matrix, L, using Tellurium.

$$v_1 = k_1 x_o \qquad v_2 = k_2 s_1 \qquad v_3 = k_3 s_2$$

$$X_o \longrightarrow S_1 \longrightarrow S_2 \longrightarrow X_1$$

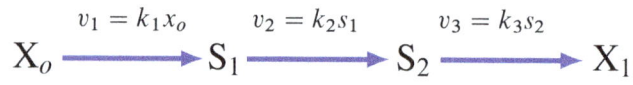

Figure 11.15 Linear Chain of Three Reactions.

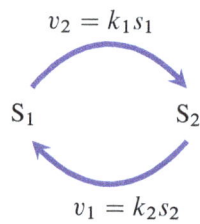

$$v_2 = k_1 s_1$$

$$S_1 \qquad\qquad S_2$$

$$v_1 = k_2 s_2$$

Figure 11.16 Conserved Cycle

bfseries

12

Moiety Conserved Cycles

12.1 Moiety Conserved Cycles

In this chapter the topic of moiety conserved cycles and their impact on behavior will be examined.

Chapter 11 introduced the idea of moiety conserved cycles. An example of a cycle is where there are two species, one might be a protein and the other, the phosphorylated protein. If we assume that the synthesis and degradation rate for the proteins are small in comparison to the timescale of phosphorylation and dephosphorylation, then we can assume that the total amount of protein, phosphorylated plus unphosphorylated, is constant over the period of study. Figure 12.3 from the previous chapter shows a simple conserved cycle.

Constraining Species Levels

One of the simplest effects of a moiety-conserved cycle is that it puts upper limits on the concentrations of the participants. In the simple cycle shown in Figure 11.2 where the total number of moles in the cycle is fixed at $s + p$, the upper limit that either S or P can reach is $s + p$. This effect was made very clear in a study of glycolysis in *Trypanosoma brucei* (Figure 12.1). What is unusual about the pathway is that much of the glycolytic pathway resides in a single membrane organelle called the glycosome. Many of the metabolites in the glycosome are phosphorylated, for example, glucose-6-phosphate, glyceraldehyde-3-phosphate etc. which creates a constraint on the level of phosphate. In addition to the glycosome, the mitochondrion of *Trypanosoma brucei* appears to do very little other than oxidize glycerol 3-phosphate via oxygen utilization.

Interestingly, an analysis of the network using the techniques described in Chapter 11 indicates the presence of

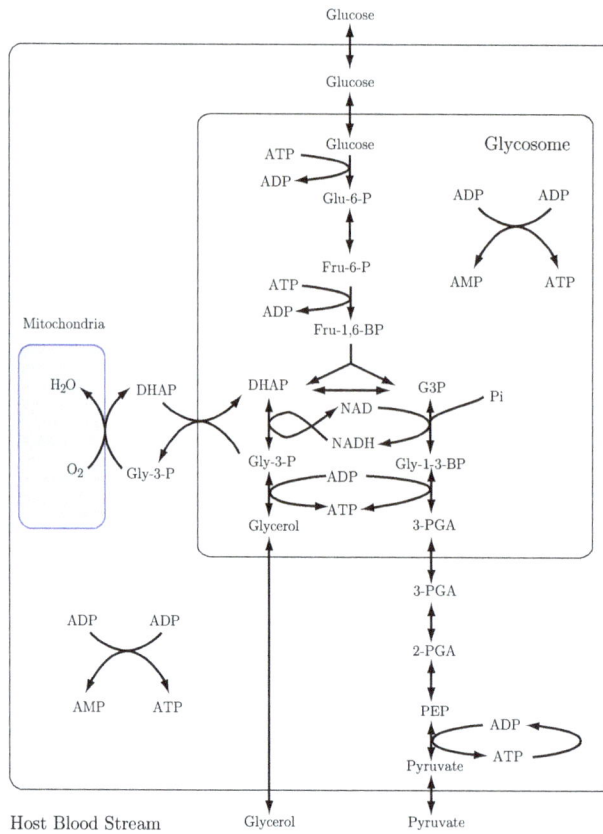

Figure 12.1 Energy mfetabolism of *Trypanasoma brucei.*

four conservation laws, these include:

1: $\text{ATP}_c + \text{ADP}_c + \text{AMP}_c$

2: $\text{ATP}_g + \text{ADP}_g + \text{AMP}_g$

3: $\text{NAD}_g + \text{NADH}_g$

4: glycerol 3-phosphate$_c$ + dihydroxyacetone phosphate$_c$ +

     glycerol 3-phosphate$_g$ + dihydroxyacetone phosphate$_g$ +

     glucose 6-phosphate$_g$ + fructose 6-phosphate$_g$ +

     fructose 1,6-bisphosphate$_g$ + glyceraldehyde 3-phosphate +

     1,3-bisphosphoglycerate + ATP_g + ADP_g

where the subscript $_c$ means cytoplasm and $_g$ means glycosome. Figure 12.2 shows the same metabolic map as Figure 12.1 but with the conserved moieties highlighted with ellipses. The fact that phosphate is a conserved moiety means that any species that includes the moiety will be constrained by the total amount of phosphate. As pointed out by Eisenthal and Cornish-Bowden [26] in relation to the work of Bakker et al [3], there are two possible ways to disrupt an organism metabolically. One can either reduce a flux to a very low level, or increase one or more metabolite levels to such high levels that they become toxic. Bakker's analysis [3] of Trypanosoma metabolism showed that much of the flux control was on glucose transport. This limits the number of potential sites for flux disruption. As for disrupting concentrations, only one step had a significant concentration control coefficient, pyruvate transport [26], again limiting the choice for drug targets. The reason why pyruvate transport is a susceptible target is because it is one of the few steps where the reactants are not involved in the conservation laws.

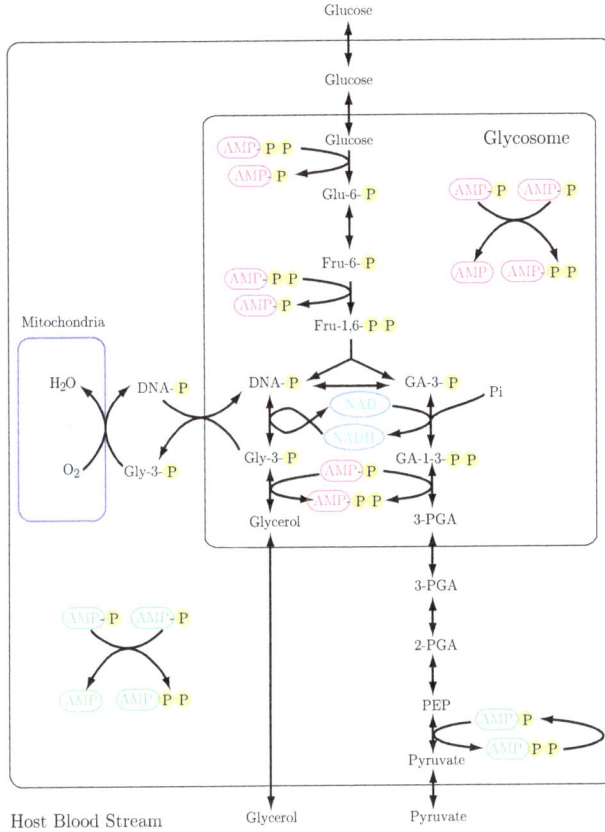

Figure 12.2 Energy metabolism of *Trypanasoma brucei* showing conserved species as ellipses.

12.2 MCA of Conserved Cycles

The MAPK (mitogen-activated protein kinase) pathways, are highly conserved and common components in signal transduction pathways (Chang & Karin, 2001). Virtually all eukaryotic cells that have been examined (ranging from yeast to man) possess multiple MAPK pathways, each of which responds to multiple inputs. In mammalian systems MAPK pathways are activated by a wide range of input signals including a variety of growth factors and environmental stresses such as osmotic shock and ischemic injury (Kyriakis & Avruch, 2002; Gomperts et al., 2002). Once the MAPK pathways have integrated these signals, they coordinately activate gene transcription with resulting changes in protein expression leading to cell division, cell death and cell differentiation.

Consider the simple conserved cycle shown in Figure 12.3. As discussed in Chapter 11, the two species, S and P, are conserved because the total $s + p$ remains constant over time (at least over a time scale shorter than protein synthesis and degradation). Let us assume that the kinetics governing each cycle arm is simple first order mass-action kinetics.

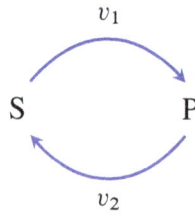

Figure 12.3 Simple Conserved cycle where $s + p = $ constant $= T$ where $v_1 = k_1 s$ and $v_2 = k_2 p$.

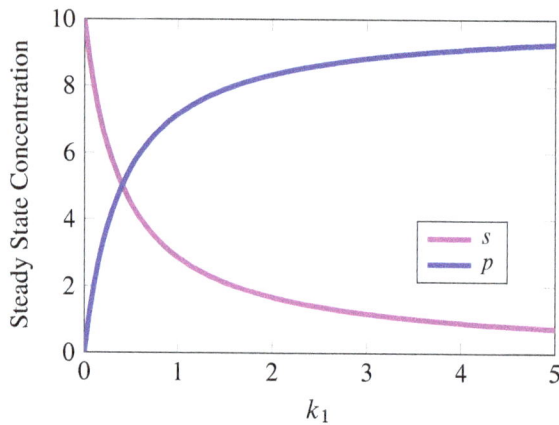

Figure 12.4 Simulation of the simple cycle with linear kinetics. Plot shows the steady state concentration of each species as a function of k_1. Model: `S -> P; k1*S; P -> S; k2*P; S=10; k1=0.1; k2=0.4`

If we plot the steady state concentration of S and P versus the first-order kinetic constant k_1, we get the response curves shown in Figure 12.4. The response curves appear hyperbolic. For example, P rises linearly then levels off to 10 concentration units at the limit. As k_1 increases, more and more S is converted to P leading to a rise in P and a fall in S. The limit is reached because there is only a fixed amount of mass in the cycle.

Simple Cycle with Non-Linear Kinetics

If we now modify the simple cycle model and instead of linear kinetics we use saturable kinetics, such as Michaelis-Menten kinetics on the forward and reverse arms, then additional changes in behavior will be observed.

The response is now sigmoidal rather than hyperbolic (Figure 12.5). The reason for this is explained in Figure 12.6. The plot shows v_1 and v_2 plotted against s and $T - s$ (which is p). The intersection points of the two curves marked by a red marker represents the corresponding steady state point when $v_1 = v_2$. A perpendicular dropped from this point indicates the corresponding steady state concentration of S. If the activity of v_1 is increased by increasing k_1 by 20%, then the v_1 curve moves up (orange). The left intersection point indicates how much the steady state concentration moves as a result, shown by Δs. The closer the steady state point is to the saturated point of the curve, the more the steady state will move. This shows that the response

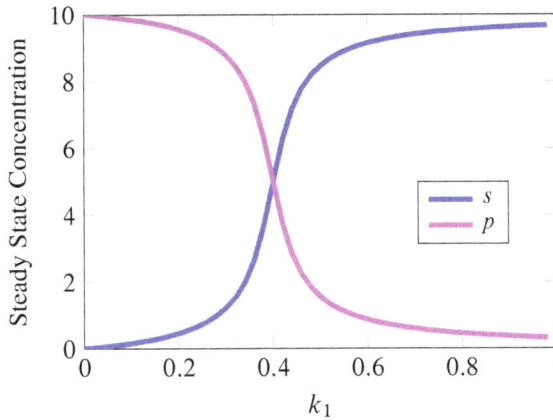

Figure 12.5 Simulation of the simple cycle with saturable kinetics illustrating sigmoid or ultrasensitive behavior. Model: S -> P; k1*S/(Km1+S); P -> S; k2*P/(Km2+P); S=10; k1=0.1; Km1=0.5; k2=0.4; Km2=0.5

in S can be very sensitive in changes in k_1. Because k_1 is a linear term in the rate law, we could replace it with the concentration of the enzyme in the Michaelis-Menten law. In practice such a cycle could represent a phosphorylation/dephophsorylation cycle where the implied enzyme is now a kinase. The kinase in turn could be controlled by other processes so that changes in the kinase activity result in sigmoid (or switch like) behavior in the cycle dynamics. In the literature such behavior was studied by [36, 37] and has been observed experimentally [50].

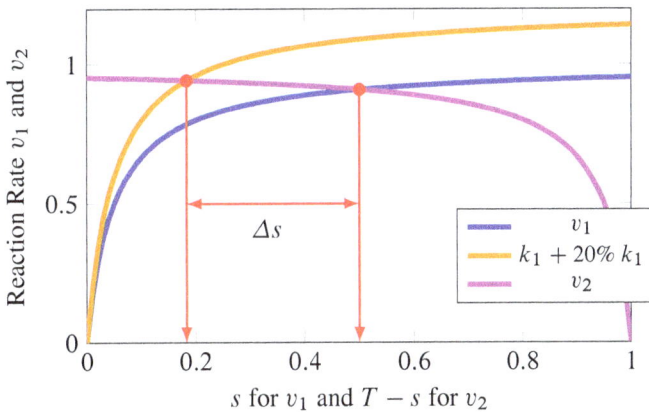

Figure 12.6 Plots the two cycle rates, v_1 and v_2 for the simple cycle with saturable kinetics. Model: S -> P; k1*S/(Km1+S); P -> S; k2*P/(Km2+P); S=1; k1=1; Km1=0.05; k2=1; Km2=0.05. The intersection points marked by a red marker represents the steady state point ($v_1 = v_2$). See main text for explanation.

The sigmoid behavior observed in a moiety conserved cycle has been termed **ultrasensitivity** and is defined in terms of the gain between an input and output. Traditionally the gain of a biological system, especially in the context of a cooperative system, has been the fold change in ligand required to change the response from 10%

to 90% of maximum, termed the response coefficient [36][1]. It is defined as:

$$R = \frac{S_{0.9}}{S_{0.1}}$$

where $S_{0.1}$ is the ligand concentration required to reach 10% of the maximum response. Using this definition, it is possible [86] to relate R to the standard Hill equation such that if:

$$v = V_m \frac{s^n}{K_H^n + s^n} \qquad \text{then} \qquad R = 81^{1/n}$$

If $n = 1$ then the system is hyperbolic and is considered not ultrasensitive and $R = 81$. Qualitatively it means that the ligand concentration must change 81 fold in order to go from 10% to 90% of the response. A system with a Hill coefficient of 4 will have a response, R, equal to 3. That is, a ligand must change 3 fold in order to go from 10% to 90% of the response. A response, R, **less** than 81 indicates that the system is ultrasensitive and tends to one as the sigmoidicity reaches its asymptotic limit. Although experimentally accessible, the definition of R is a little ad-hoc and instead, it is more useful from the point of view of theory to define the gain as the ratio of the fractional change between an output, Y, and input, X. That is:

$$R_X^Y = \frac{d \ln Y}{d \ln X}$$

For example, if the input, X, is changed by 1% and this results in an output change in Y of 2%, then the input is amplified two fold. In engineering this is often referred to as the **gain** of the system. Amplification of signals is a fundamental operation in both biology and engineering. Many inputs in biology are small magnitude and must be amplified before they can be acted upon. Examples include changes in hormonal levels and amplification of sensory inputs from external stimuli such as nutrient gradients or light levels in the retina.

It should be apparent that R_X^Y is related to the response coefficient introduced in section 4.3. In contrast to the ratio, $S_{0.9}/S_{0.1}$, a system is ultrasensitive if $R_X^Y > 1$. In the next section the response coefficient, R_X^Y, will be determined for a cycle in terms of the cycle elasticities.

12.3   Using MCA to Understand Ultrasensitivity

It is possible to use the machinery of metabolic control analysis derive the conditions for ultrasensitivity [98] in a covalent modification cycle without recourse to specific kinetic laws [36, 37].

Consider again the simple cycle shown in Figure 12.3 where v_1 is catalyzed by a kinase E_1. To investigate ultrasensitivity, the concentration control coefficient, $C_{e_1}^P$ will be evaluated, that is, how sensitive P is to changes in the activity of the enzyme (often a kinase) that catalyzes the conversion of S to P. There are different but related ways to approach this derivation. The first is to look at the effect of perturbations on the cycle.

To make matters simpler, assume that both v_1 and v_2 are irreversible and not product inhibited. Let us make a small change, δe_1, to e_1 that catalyzes the reaction rate v_1. We can write down the two local equations as follows:

$$\frac{\delta v_1}{v_1} = \frac{\delta e_1}{e_1} + \varepsilon_s^1 \frac{\delta s}{s} \quad \text{and} \quad \frac{\delta v_2}{v_2} = \varepsilon_p^2 \frac{\delta p}{p}$$

At steady state $\delta v_1/v_1 = \delta v_2/v_2$, in addition, the changes in s and p must be constrained by $s + p = T$. Note we haven't changed the total T. This means that it must be true that $\delta s = -\delta p$. Substituting δs with $-\delta p$ and equating the two local equations yields:

$$\frac{\delta e_1}{e_1} + \varepsilon_s^1 \frac{(-)\delta p}{s} = \varepsilon_p^2 \frac{\delta p}{p}$$

---

[1]Not to be confused with the response coefficient defined in MCA.

This equation can be rearranged to:

$$\frac{\delta e_1}{e_1} = \varepsilon_s^1 \frac{\delta p}{p} \frac{p}{s} + \varepsilon_p^2 \frac{\delta p}{p} = 0$$

Collecting $\delta p / p$:

$$\frac{\delta e_1}{e_1} = \frac{\delta p}{p} \left(\varepsilon_s^1 \frac{p}{s} + \varepsilon_p^2 \right)$$

Dividing both sides by $\delta e_1 / e_1$ and noting that $(\delta p / p)/(\delta e_1 / e_1)$ in the limit is equal to $C_{e_1}^p$, we obtain after rearrangement:

$$C_{e_1}^p = \frac{s}{p \varepsilon_s^1 + s \varepsilon_p^2} \tag{12.1}$$

Instead of working with actual concentrations, divide by the total, T to use molar fractions, $M_s = s/T$ and $M_p = p/T$:

$$C_{e_1}^p = \frac{M_s}{M_p \varepsilon_s^1 + M_s \varepsilon_p^2} \tag{12.2}$$

The equation can be modified to indicate how an arbitrary effector of v_1 influences P. Recalling the response coefficient relationship (4.15), the response of an effector X on the concentration, p is given by:

$$R_x^p = C_{e_1}^p \varepsilon_x^1 = \varepsilon_x^1 \frac{M_s}{M_p \varepsilon_s^1 + M_s \varepsilon_p^2} \tag{12.3}$$

If it is assumed that v_1 and v_2 operate far below saturation then the elasticities ε_1^1 and ε_2^2 are approximately one, that is:

$$C_{e_1}^p = \frac{M_s}{M_p + M_s}$$

This value will always be less than or equal to one. Therefore there is no possibility of ultrasensitivity when both enzymes are operating in their first-order regime. However if the enzymes are operating near saturation, then $\varepsilon_s^1 < 1$ and $\varepsilon_p^2 < 1$. For example, if $s = 9$ and $p = 1$ and both elasticities are equal to 0.5, then the control coefficient, $C_{e_1}^p = 1.8$. If we reduce the elasticities further, $\varepsilon_s^1 = 0.2$ and $\varepsilon_p^2 = 0.2$, the response of the system rises to 4.5. That is a 1% increase in E_1 will result in a 4.5% increase in P (Figure 12.7). In most cases like this, P is itself a protein, often a kinase. So a small change in one protein, E_1, can have a large effect on another protein, P.

The advantage of this approach is that it is kinetic mechanism independent. That is, in formulating (12.2) nothing was assumed about the details of the mechanism by which the cycle steps operated. The only thing that was assumed was the local sensitivity of the reaction rate to their respective substrates.

If we set the initial steady state value for M_1 to 0.9 and M_2 to 0.1 it is straightforward to show that the threshold where the system starts to display ultrasensitivity is when both elasticities equal 0.9. In general the threshold is equal to the initial steady state level of the mole fraction of S_1.

To complete the discussion we can also derive the effect e_1 has on the cycling flux, J. To do this we recall that:

$$\frac{\delta v_2}{v_2} = \varepsilon_p^2 \frac{\delta p}{p}$$

At steady state, $\delta v_2 / v_2$ is equal to the change in the steady state cycling flux, $\delta J / J$. Dividing both sides by $\delta e_1 / e_1$ we obtain:

$$C_{e_1}^J = \varepsilon_p^2 C_{e_1}^p = \varepsilon_p^2 \frac{M_s}{M_p \varepsilon_s^1 + M_s \varepsilon_p^2}$$

This relationship shows that the cycling flux is less sensitive to e_1 than p. This is because the elasticity, ε_p^2 will be small when the cycle is operating in the ultrasensitive mode. $C_{e_1}^s$, $C_{e_2}^s$ and $C_{e_2}^p$ can be derived in a similar manner. A more formal approach to deriving $C_{e_1}^p$, is write down the dynamical equations for the system and implicitly differentiate with respect to the parameter of interest.

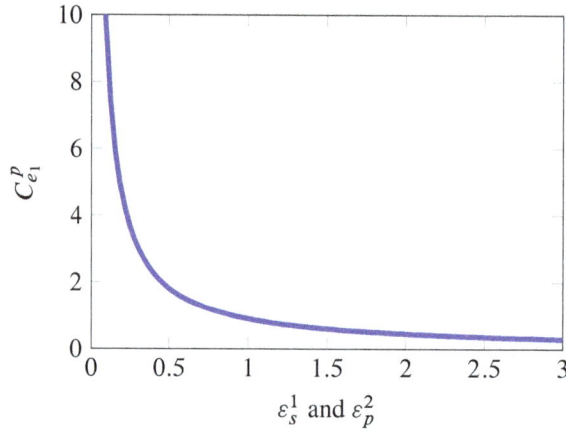

Figure 12.7 Plot of $C_{e_1}^p$ as a function of the two cycle elasticities, ε_s^1 and ε_p^2. Both elasticities have the same value on the x axis. The plot shows the rapid rise in sensitivity as the elasticities are reduced.

Derivation by Implicit Differentiation   We can also derive the $C_{e_1}^p$ term using implicit differentiation. We start by defining the differential equation for P:

$$\frac{dp}{dt} = v_1(e_1, s(e_1)) - v_2(p(e_1))$$

We assume the rate of reaction for v_1 is a function of e_1 which is the kinase catalysing v_1 and the unphosphorylated protein s. We assume that p has no effect on v_1. The key observation here is that because S is the dependent variable, we must replace it with the function $s(e_1)$, indicating that the concentration of S is also a function of E_1. Likewise for v_2, we assume it is only a function of P which in turn is also a function of E_1. Once in this form we set the equation to zero to indicate steady state and differentiate the equation with respect to our chosen parameter E_1:

$$0 = \frac{\partial v_1}{\partial e_1} + \frac{\partial v_1}{\partial s}\frac{ds}{de_1} - \frac{\partial v_2}{\partial p}\frac{dp}{de_1}$$

Noting the conservation law $s + p = T$ and differentiating this with respect to e_1, we can state that:

$$\frac{ds}{de_1} = -\frac{dp}{de_1}$$

Substituting ds/de_1 with $-dp/de_1$ we obtain:

$$0 = \frac{\partial v_1}{\partial e_1} - \frac{\partial v_1}{\partial s}\frac{dp}{de_1} - \frac{\partial v_2}{\partial p}\frac{dp}{de_1}$$

Multiplying both sides by e_1, dividing both sides by p and noting that at steady state $v_1 = v_2$, and dividing both sides by v_1 or v_2 as appropriate, we can rewrite the above as:

$$0 = \frac{\partial v_1}{\partial e_1}\frac{e_1}{v_1} - \frac{\partial v_1}{\partial s}\frac{s}{v_1}\frac{p}{s}\frac{dp}{de_1}\frac{e_1}{p} - \frac{\partial v_2}{\partial p}\frac{p}{v_2}\frac{dp}{de_1}\frac{e_1}{p}$$

Translating the terms into elasticities and $C_{e_1}^p$ we obtain:

$$0 = \varepsilon_{e_1}^1 - \varepsilon_s^1 \frac{p}{s_1} C_s^p - \varepsilon_p^2 C_{e_1}^p$$

Letting $\varepsilon_{e_1}^1 = 1$ and solving for $C_{e_1}^p$ we obtain:

$$C_{e_1}^p = \frac{1}{\varepsilon_s^1 \dfrac{p}{s} - \varepsilon_p^2}$$

Multiplying top and bottom by s we arrive at:

$$C_{e_1}^p = \frac{s}{\varepsilon_s^1 p - \varepsilon_p^2 s}$$

This is the same result as equation (12.1).

Sensitivity of the Cycle to Changes in T

Let the total amount of species in the cycle be $T = s + p$. Make a change in T, δT, by adding externally some S or P. This will cause both rates, v_1 and v_2 to change as well as changes to the steady state levels of S and P. To make matters simpler, assume as before that both v_1 and v_2 are irreversible and not product inhibited, this allows us to write the following statements:

$$\frac{\delta v_1}{v_1} = \varepsilon_s^1 \frac{\delta s}{s}, \qquad \frac{\delta v_2}{v_2} = \varepsilon_p^2 \frac{\delta p}{p}$$

Note that there have been no changes to e_1 or e_2. At steady state the changes $\delta v_1/v_1$ and $\delta v_2/v_2$ will equal each other. Therefore:

$$\varepsilon_s^1 \frac{\delta s}{s} - \varepsilon_p^2 \frac{\delta p}{p} = 0$$

Given that T has been changed by δT, it must be that case that: $\delta T = \delta s + \delta p$. That is, $\delta s = \delta T - \delta p$. Substituting this into the previous equation to eliminate δs, we obtain:

$$\varepsilon_s^1 \frac{\delta T - \delta p}{s} - \varepsilon_p^2 \frac{\delta p}{p} = 0$$

Dividing both sides by $\delta T/T$ and noting that in the limit $(\delta p/p)/(\delta T/T) = R_T^p$,[2] we obtain after rearrangement:

$$R_T^p = \frac{T \varepsilon_s^1}{\varepsilon_s^1 p + \varepsilon_p^2 s}$$

As before, the absolute concentrations in s and p can be converted to molar fractions by dividing top and bottom by T to yield:

$$R_T^p = \frac{\varepsilon_s^1}{\varepsilon_s^1 M_p + \varepsilon_p^2 M_s} \tag{12.4}$$

The sensitivity of the cycling flux with respect to T can be obtained as follows. The change in the cycling rate as a result of increasing T by δT is given by:

$$\frac{\delta v_2}{v_2} = \varepsilon_p^2 \frac{\delta p}{p}$$

However $\delta v_2/v_2$ is also the change in the cycling flux, J. Dividing both sides of the equation by $\delta T/T$ and substituting C_T^p that was derived above yields:

$$R_T^J = \varepsilon_p^2 R_T^p = \varepsilon_p^2 \frac{\varepsilon_s^1}{\varepsilon_s^1 M_p + \varepsilon_p^2 M_s} \tag{12.5}$$

---

[2]We are treating this coefficient as a response coefficient because we consider T to be an external factor.

12.4   Cycle Connectivity Theorems

Moiety conserved cycles possess modified connectivity theorems. This is because the species in the cycle cannot have arbitrary values but are constrained by the total mass in the cycle [32, 48].

Consider again the simple cycle shown in Figure 11.2. Assume that a small change, δs is made in s. In order to maintain a constant amount of $s + p$, we also make a compensating change in δp equal to $-\delta s$. These changes will cause both v_1 and v_2 to change, however we can make changes to E_1 and E_2 such that the reaction rates are unchanged. We can express this thought experiment using the following local equations:

$$\frac{\delta v_1}{v_1} = \frac{\delta e_1}{e_1} + \varepsilon_s^1 \frac{\delta s}{s} + \varepsilon_p^1 \frac{\delta p}{p} = 0$$

$$\frac{\delta v_2}{v_2} = \frac{\delta e_2}{e_2} + \varepsilon_p^2 \frac{\delta p}{p} + \varepsilon_s^2 \frac{\delta s}{s} = 0$$

To make things simpler, let us assume that each forward reaction is irreversible and product insensitive, that is $\varepsilon_1^2 = 0$ and $\varepsilon_2^1 = 0$ so that the local equations are reduced to:

$$\frac{\delta e_1}{e_1} = -\varepsilon_s^1 \frac{\delta s}{s}$$

$$\frac{\delta e_2}{e_2} = -\varepsilon_p^2 \frac{\delta p}{p}$$

We can also express the thought experiment in terms of the systems equation:

$$\frac{\delta p}{p} = C_{e_1}^p \frac{\delta e_1}{e_1} + C_{e_2}^p \frac{\delta e_2}{e_2}$$

We can now substitute the local equations into the systems equation:

$$-\frac{\delta p}{p} = C_{e_1}^p \varepsilon_s^1 \frac{\delta s}{s} + C_{e_2}^p \varepsilon_p^2 \frac{\delta p}{p} \tag{12.6}$$

The thought experiment included the constraint that $\delta s = -\delta p$, so that:

$$-\frac{\delta p}{p} = -C_{e_1}^p \varepsilon_s^1 \frac{\delta p}{p} \frac{p}{s} + C_{e_2}^p \varepsilon_p^2 \frac{\delta p}{p}$$

Canceling the $\delta p/p$ terms yields the covalent modification connectivity theorem:

$$-1 = -C_{e_1}^p \varepsilon_s^1 \frac{p}{s} + C_{e_2}^p \varepsilon_p^2$$

One way to reexpress this is to divide both sides by p and rearrange so that:

$$\frac{1}{p} = C_{e_1}^p \varepsilon_s^1 \frac{1}{s} - C_{e_2}^p \varepsilon_p^2 \frac{1}{p} \tag{12.7}$$

The above equation is the modified connectivity theorem for the moiety conserved cycle. It is possible to combine this equation with the summation theorem (which is unmodified):

$$C_{e_1}^p + C_{e_2}^p = 0$$

and solving for $C_{e_1}^p$. This is an alternative way to determine $C_{e_1}^p$.

Other Relationships

There are a number of other relationships related to moiety conserved cycles that are worth mentioning. Consider a conserved moiety cycle at steady state of m species, $S_1, S_2, \ldots, S_m$ constrained by the conservation law, $s_1 + s_2 + \ldots + s_m = T$ (Figure 12.8).

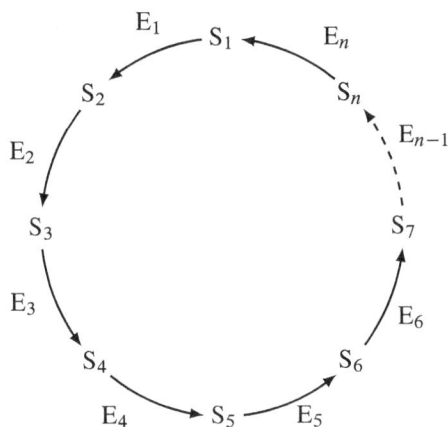

Figure 12.8 Cyclic pathway

Let us make a small change to an enzyme that catalyzes one of the reaction steps in the cycle. For example, change e_i by an amount δe_i. This will result in the system changing to a new steady state where changes in S_1, S_2, etc. are observed. Since the total number of moles in the cycle has not changed, it must be true that:

$$\delta s_1 + \delta s_2 + \ldots + \delta s_m = 0$$

Scaling by each species yields:

$$s_1 \left(\frac{\delta s}{s} \right) + s_2 \frac{\delta s_2}{s_2} + \ldots + s_m \frac{\delta s_m}{s_m} = 0$$

Finally dividing each term by the relative change in e_i:

$$s_1 \left(\frac{\delta s_1/s_1}{\delta e_i/e_i} \right) + s_2 \left(\frac{\delta s_2/s_2}{\delta e_i/e_i} \right) + \ldots + s_m \left(\frac{\delta s_m/s_m}{\delta e_i/e_i} \right) = 0$$

which can be rewritten as:

$$s_1 C_{e_i}^{s_1} + s_2 C_{e_i}^{s_2} + \ldots + s_m C_{e_i}^{s_m} = 0$$

Note the difference with the usual concentration summation theorem where the summation is over a single species and all enzymes. Here the summation is over all cycle species and a single enzyme. The relationship can be easily modified to use fractional molar amounts instead:

$$M_1 C_{e_i}^{s_1} + M_2 C_{e_i}^{s_2} + \ldots + M_m C_{e_i}^{s_m} = 0$$

where $M_i = s_i/T$.

There is also a summation with respect to $R_T^{s_i}$. Consider making a change to the total number of moles in a cycle by an amount δT. This will result in a change to the steady state such that:

$$\delta s_1 + \delta s_2 + \ldots + \delta s_m = \delta T$$

As before, each term can be scaled by the corresponding s_i concentration and each term divided by the relative change in T:

$$\frac{s_1}{T}\left(\frac{\delta s_1/s_1}{\delta T/T}\right) + \frac{p}{T}\left(\frac{\delta s_2/s_2}{\delta T/T}\right) + \ldots + \frac{s_m}{T}\left(\frac{\delta s_m/s_m}{\delta T/T}\right) = 1$$

which can be simplified to:

$$M_1 R_T^{s_1} + M_2 R_T^{s_2} + \ldots + M_m R_T^{s_m} = 1$$

The two results are summarised below:

$$M_1 C_{e_i}^{s_1} + M_2 C_{e_i}^{s_2} + \ldots + M_m C_{e_i}^{s_m} = 0 \qquad (12.8)$$

$$M_1 R_T^{s_1} + M_2 R_T^{s_2} + \ldots + M_m R_T^{s_m} = 1 \qquad (12.9)$$

Dual Cycles - First-Order Ultrasensitivity

In many signaling pathways, for example the MAPK pathway, one will often find dual phosphorylation cycles (Figure 12.9). That is, cycles where a protein is doubly phosphorylated, often in a strict order.

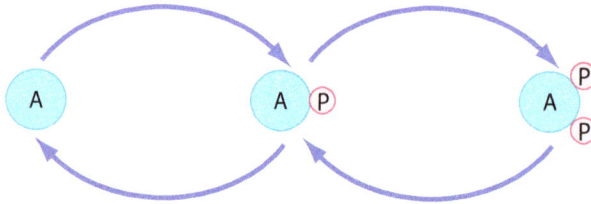

Figure 12.9 A Dual phosphorylation and dephosphorylation cycle.

Consider a dual cycle shown in Figure 12.10 where S is a signal (e.g. a kinase) that catalyzes v_1 and v_3.

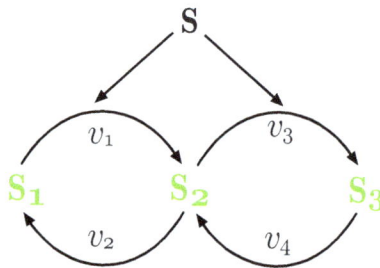

Figure 12.10 Two cycles connected by a common intermediate, S_2. The rate laws for each step is given by $v_1 = k_1 s_1$, $v_2 = k_2 s_2$, $v_3 = k_3 s_2$, and $v_4 = k_4 s_3$. S is the stimulus signal which acts by increasing k_1 and k_3 by the same factor.

The stoichiometry matrix for the dual cycle is given by:

$$\mathbf{N} = \begin{bmatrix} -1 & 1 & 0 & 0 \\ 1 & -1 & -1 & 1 \\ 0 & 0 & 1 & -1 \end{bmatrix}$$

For a dual cycle it is possible to show that there is one conservation law given by the relation:

$$s_1 + s_2 + s_3 = T$$

If we assume simple linear mass-action kinetics for each of the reactions, simulation will reveal that the concentration of S_3 shows sigmoid behavior with respect to the stimulus signal S. We can assume that the stimulus signal, S, operates on the rate constants, k_1 and k_3 by the same factor, that is, an increase in S by $x\%$ results in a change in k_1 and k_3 by $x\%$. What is interesting about this case is that a sigmoid response can be generated from first-order reactions. This can be called first-order ultrasensitivity to distinguish it from zero-order ultrasensitivity that is generated from a single cycle using saturable kinetics.

We can show that the dual cycle can display ultrasensitivity by deriving $R_s^{s_3}$. To keep things simple it will be assumed that the reactions are irreversible and not product inhibited. The strategy is to first derive the terms $C_{e_1}^{s_3}$ and $C_{e_3}^{s_3}$. The response to signal S is the sum of $C_{e_1}^{s_3}$ and $C_{e_3}^{s_3}$. It is assumed that the elasticity of S towards each reaction is one, which is reasonable if the signal is a protein catalyst such as a kinase.

The derivation will be illustrated for $C_{e_1}^{s_3}$. $C_{e_3}^{s_3}$ can be derived in a similar manner by perturbing e_3. To derive $C_{e_1}^{s_3}$, perturb e_1 by an amount δe_1. This will change the steady state from which the following local equations can be obtained:

$$\frac{\delta v_1}{v_1} = \varepsilon_1^1 \frac{\delta s_1}{s_1} + \frac{\delta e_1}{e_1}, \quad \frac{\delta v_2}{v_2} = \varepsilon_2^2 \frac{\delta s_2}{s_2}$$

$$\frac{\delta v_3}{v_3} = \varepsilon_2^3 \frac{\delta s_2}{s_2}, \quad \frac{\delta v_4}{v_4} = \varepsilon_3^4 \frac{\delta s_3}{s_3}$$

At steady state $v_1 = v_2$ and $v_3 = v_4$, though it is **not** necessarily the case that $v_1 = v_3$. This means that when the steady state changes $\delta v_1 = \delta v_2$ and $\delta v_3 = \delta v_4$. In relative terms we state that: $\delta v_1/v_1 = \delta v_2/v_2$ and $\delta v_3/v_3 = \delta v_4/v_4$. By equating the local equations $\delta v_1/v_1$ and $\delta v_2/v_2$ we obtain:

$$\varepsilon_2^1 \frac{\delta s}{s} + \frac{\delta e_1}{e_1} = \varepsilon_2^2 \frac{\delta s_2}{s_2}$$

Both sides of the equation can be divided by $\delta e_1/e_1$ to give:

$$\varepsilon_2^1 C_{e_1}^{s_1} + 1 = \varepsilon_2^2 C_{e_1}^{s_2}$$

A similar equation can be derived for $C_{e_1}^{s_2}$ and $C_{e_1}^{s_1}$ using the v_3, v_4 pair of local equations. In this case the result is simpler:

$$C_{e_1}^{s_2} = C_{e_1}^{s_3}$$

As a result we have two equations and three unknowns $C_{e_1}^{s_1}, C_{e_1}^{s_2}$, and $C_{e_1}^{s_3}$. To solve for the three unknowns, a third equation is necessary. The dual cycle has a single conservation equation, $s_1 + s_2 + s_3 = T$. Perturbing e_1 by δe_1 does not disturb the total T but will change the distribution of species such that the change in species must be constrained by $\delta s_1 + \delta s_2 + \delta s_3 = 0$. Scaling each term:

$$s_1 \frac{\delta s}{s} + s_2 \frac{\delta s_2}{s_2} + s_3 \frac{\delta s_3}{s_3} = 0$$

and dividing throughout by $\delta e_1/e_1$ yields:

$$s_1 C_{e_1}^{s_1} + s_2 C_{e_1}^{s_2} + s_3 C_{e_1}^{s_3} = 0$$

Note that this is the same equation as (12.9). We now have three equations in three unknowns which can be solved. For example, solving for $C_{e_1}^{s_3}$ gives:

$$C_{e_1}^{s_3} = \frac{s_1 \varepsilon_2^3}{s_1 \varepsilon_2^2 \varepsilon_3^4 + s_2 \varepsilon_1^1 \varepsilon_3^4 + s_3 \varepsilon_1^1 \varepsilon_2^3}$$

Using the same technique, a solution to C_{e3}^{s3} can also be found as follows:

$$C_{e1}^{s3} = \frac{s_1 \varepsilon_2^2 + s_2 \varepsilon_1^1}{s_1 \varepsilon_2^2 \varepsilon_3^4 + s_2 \varepsilon_1^1 \varepsilon_3^4 + s_3 \varepsilon_1^1 \varepsilon_2^3}$$

The influence an external signal, s, is the sum of its interactions (See 4.15) therefore $R_s^{s3} = C_{e1}^{s3} + C_{e3}^{s3}$ where we assume that the elasticity of the signal on v_1 and v_3 is one. This gives us the total response of s_3 due to changes in the signal s. The sum is given by the equation (12.10):

$$R_s^{s3} = \frac{s_1(\varepsilon_2^3 + \varepsilon_2^2) + s_2 \varepsilon_1^1}{s_1 \varepsilon_2^2 \varepsilon_3^4 + s_2 \varepsilon_1^1 \varepsilon_3^4 + s_3 \varepsilon_1^1 \varepsilon_2^3} \tag{12.10}$$

Equation (12.10) looks a little complicated but can be simplified by assuming all reactions are first-order. Under these conditions all the elasticities equal one so that the equation reduces to something much more manageable:

$$R_s^{s3} = \frac{2s_1 + s_2}{s_1 + s_2 + s_3}$$

This indicates that given the right ratios for s_1, s_2 and s_3, it is possible for $R_s^{s3} > 1$. The maximum value the equation can reach is when s_2 and s_3 are zero, at this point $R_{e1}^{s3} = 2$. This shows ultrasensitivity.

Unlike the case of a single cycle where near saturation is required to achieve ultrasensitivity, multiple cycles can achieve ultrasensitivity with simple linear kinetics.

> Doubly phosphorylated cycles can generate ultrasensitivity using linear kinetic laws. This is called first-order ultrasensitivity.

Figure 12.11 shows a simulation that plots the steady state output, s_3 versus the activity of steps v_1 and v_2.

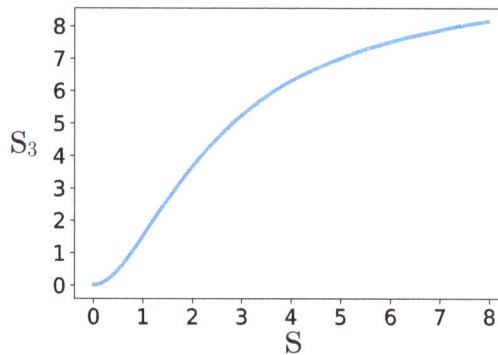

Figure 12.11 Ultrasensitivity seen in a dual phosphorylation cycle with first-order kinetics on all reactions. The steady state output, s_3 is plotted as a function of the two forward rate constants on v_1 and v_3. The Tellurium script that generated this plot can be found in listing 12.3.

The previous results can be easily generalized to systems with any number of cycles (Figure 12.12. This is more easily accomplished by setting all elasticities to one at the start and invoking theorem (12.8) directly. The response of the output species, S_m to the stimulate S is the sum of the individual responses:

$$R_s^{Sm} = R_{e1}^{Sm} + R_{e2}^{Sm} + \ldots + R_{em+1}^{Sm}$$

By assuming that the elasticity of the stimulus on each forward arm is equal to one, R_s^{Sm} can be simplified to:

$$R_s^{Sm} = C_{e1}^{Sm} + C_{e2}^{Sm} + \ldots + C_{em+1}^{Sm}$$

Assuming as before that $v_1 = v_2, v_3 = v_4$ etc, a perturbation to E_1 yields the following relationships:

$$C_{e_1}^{s_1} = C_{e_1}^{s_2} - 1, \quad C_{e_1}^{s_2} = C_{e_1}^{s_3}, \quad \ldots, \quad C_{e_1}^{s_{m-1}} = C_{e_1}^{s_m}$$

Using theorem (12.8):

$$M_1 C_{e_1}^{s_1} + M_2 C_{e_1}^{s_2} + \ldots + M_m C_{e_1}^{s_m} = 0$$

we can eliminate all the control coefficients but one by rewriting them in terms of $C_{e_1}^{s_m}$ to give:

$$M_1(C_{e_1}^{s_m} - 1) + M_2 C_{e_1}^{s_m}) + \ldots + M_m C_{e_1}^{s_m} = 0$$

Solving for $C_{e_1}^{s_m}$ yields:

$$C_{e_1}^{s_m} = \frac{M_1}{M_1 + M_2 + \ldots + M_m}$$

For a perturbation in the second forward arm, E_3, we obtain:

$$C_{e_3}^{s_1} = C_{e_3}^{s_2} - 1, \quad C_{e_3}^{s_2} = C_{e_3}^{s_3} - 1, \quad \ldots, \quad C_{e_3}^{s_{m-1}} = C_{e_3}^{s_m}$$

Again we invoke theorem (12.8) and eliminate all the control coefficients except $C_{e_3}^{s_m}$:

$$M_1(C_{e_3}^{s_m} - 1) + M_2(C_{e_3}^{s_m} - 1) + \ldots + M_m C_{e_3}^{s_m} = 0$$

Solving for $C_{e_3}^{s_m}$:

$$C_{e_3}^{s_m} = \frac{M_1 + M_2}{M_1 + M_2 + \ldots + M_m}$$

Looking closely at these results we can see a pattern. For the nth forward cycle arm, the control coefficient $C_{e_n}^{s_m}$ is given by:

$$C_{e_n}^{s_m} = \frac{M_1 + M_2 + \ldots + M_{m-1}}{M_1 + M_2 + \ldots + M_m}$$

The response is the sum of all forward cycle control coefficients. To illustrate this, consider a system with three cycles where the output species is s_4. The individual control coefficients are given by:

$$C_{e_1}^{s_4} = \frac{M_1}{M_1 + M_2 + M_3 + M_4}$$

$$C_{e_3}^{s_4} = \frac{M_1 + M_2}{M_1 + M_2 + M_3 + M_4}$$

$$C_{e_5}^{s_4} = \frac{M_1 + M_2 + M_3}{M_1 + M_2 + M_3 + M_4}$$

The sum of the three control coefficients is:

$$R_s^{s_4} = \frac{3M_1 + 2M_2 + M_3}{M_1 + M_2 + M_3 + M_4}$$

The maximum value that $R_s^{s_4}$ can reach in this case is three. This can be generalized so that a system with $n - 1$ cycles will display a maximum response of $n - 1$. For example a system that has six cycles will display a maximum response of six.

12.5 Sequestration

The cyclic models considered so far assume negligible sequestration of the cycle species by the catalyzing kinase and phosphatase. In reality this is not likely to be the case because experimental evidence indicates that the concentrations of the catalyzing enzymes and cycle species are comparable (See [6] for a range of illustrative data). In such situations additional effects are manifest [33, 85]. Of particular interest is the emergence of new regulatory feedback loops which can alter the behavior quite markedly (See [65] and [73]).

Interlocked cycles that also incorporate sequestration effects are likely to be able to display an extremely wide range of behaviors. This is an area of theoretical analysis has not received much attention in the literature [114].

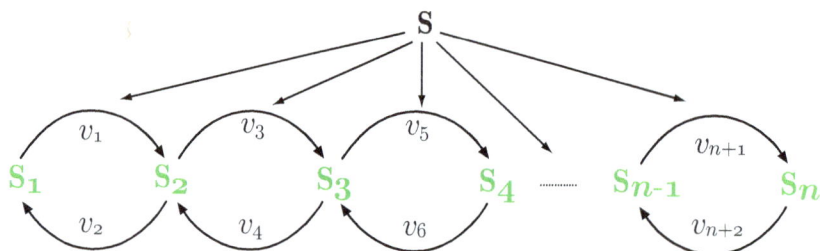

Figure 12.12 Multiple cycles with S as the stimulus signal.

Ultrasensitivity via Sequestration

The simplest sequestration mechanism that can generate an ultrasensitive response involves the dimerization of two dissimilar species. The effect depends on a conservation law between the active participant, A and a sequester molecule B to form complex AB. The reaction is simply:

$$A + B \rightleftharpoons AB$$

The response begins with a low level of active molecule A. As A is increased, most of the added A is sequestered by B. However eventually enough A is added so that there remains very little of B left that can sequester A. At this point the concentration of A rises rapidly. This is the point where ultrasensitivity is seen. Figure 12.13 illustrates a plot that shows ultrasensitivity near 80 and 120 on the x axis [11, 10].

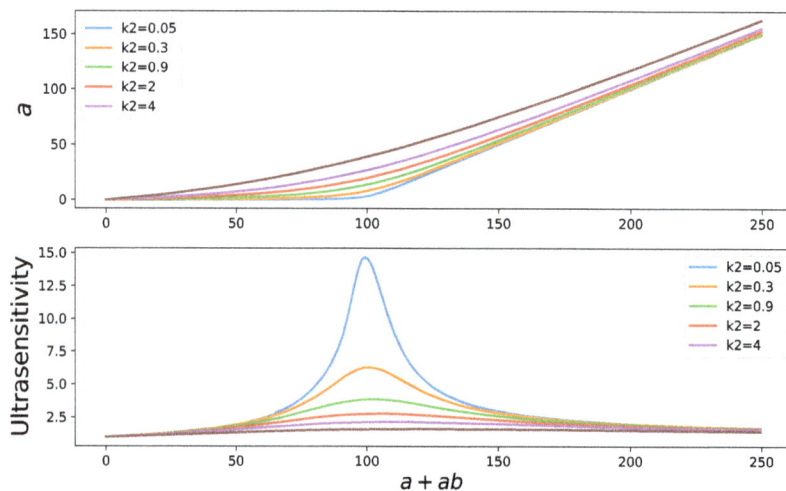

Figure 12.13   Ultrasensitivity seen in a simple sequestration model for different values of k_2. A + B <-> AB; k1*A*b - k2*AB; a = 0.001; B = 100; AB = 0; k1 = 0.4; k2 = 1. Ultrasensitivity is measured by computing R_T^a. The Tellurium script that generated this plot can be found in listing 12.1.

To illustrate another example where conservation laws contribute to new behavior, we will look at a very simple linear pathway where there is a dead-end leak caused by complex formation (see Figure 12.14). The observed ultrasensitivity is in response to a change in the stimulus signal and originates from a combination of kinetic and conservation factors. Sigmoid behavior can be observed in both the free species, X and the complex XI forms. Saturation in the level of XI is due to a conservation law involving the I moiety. To achieve a saturating effect in X, the second step, v_2 should be modeled using a Michaelis-Menten rate law (itself based on a conservation law between free enzyme and enzyme substrate complex) and the first step, v_1 should be reversible to ensure that a steady state exists at high stimulus levels (X would go to infinity otherwise). Figure 12.15 shows an example simulation that illustrates ultrasensitivity in this more complex sequestration model.

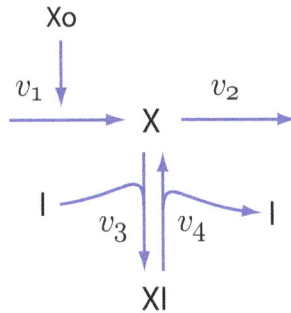

Figure 12.14 Sequestration steps attached to a pathway with a signal input X_o to v_1.

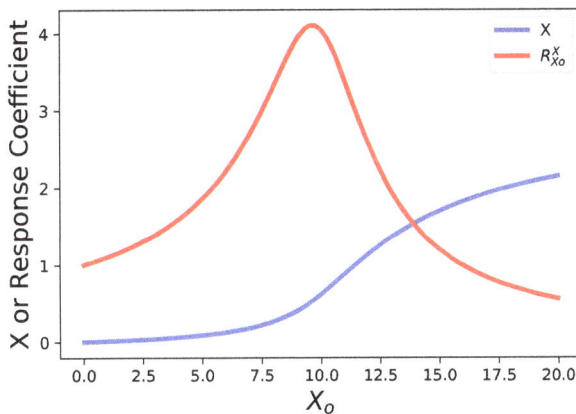

Figure 12.15 Simulation of model shown in Figure 12.14. The degree of ultrasensitivity is measured by the response coefficient which achieved a maximum of four (upper curve). The Tellurium script that generated this plot can be found in listing 12.2.

The Markevich Switch

The next example will illustrate a fairly complex set of interlinked conservation laws that lead to quite elaborate behavior. This system, first discovered by Kholodenko and co-workers *et al.*, will be referred to as the **Markevich Switch** after the first author on the original paper [65].

The system involves a double cycle but with secondary sequestration effects occurring on the limbs. Figure 12.16 illustrates the full pathway. The model describes the catalysis of the conversion of S_1 through two

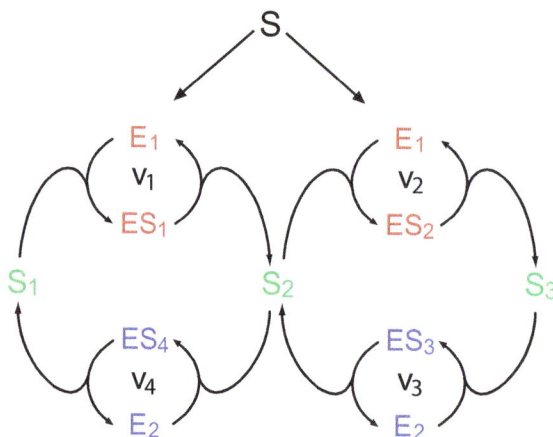

Figure 12.16   A complex interlinked set of conserved cycles that describes the Markevich switch [65]. S controls the activity of the pathway by controlling the amount of total E_1.

enzyme catalyzed reactions, v_1 and v_2. The individual catalytic cycles are made explicit in this model, that is, the binding of S_1 to enzyme E_1 to form complex, and dissociation to form product, S_2, is explicitly modeled. In addition there is the reverse conversion of S_3 back to S_1, again by a sequence of two enzyme catalyzed reactions, v_3 and v_4, also in explicit form. The stimulus, S, acts by adding more total E_1 to the upper limbs.

This pathway has multiple conservation laws stemming from the two different enzymes and a separate substrate cycle. These conservation laws include:

$$S_1 + S_2 + S_3 + ES_1 + ES_2 + ES_3 + ES_4 = T_1 \tag{12.11}$$

$$E_1 + ES_1 + ES_2 = T_2 \tag{12.12}$$

$$E_2 + ES_3 + ES_4 = T_3 \tag{12.13}$$

Figure 12.17 illustrates graphically the three conservation laws. The presence of the conservation laws leads to the emergence of bistable behavior. That is, given a particular set of parameters, there exists three possible steady states, two stable and one unstable (sometimes called metastable). We can see this depicted in the steady state plot (Figure 12.18) that shows the concentration of S_3 versus total E_1 ($E_1 + ES_1$). At a certain range of total E_1, the curve shows three possible steady states. A high stable state, a low stable state and an intermediate unstable state (thin line in the graph). In principle, the unstable state could be achieved and maintained indefinitely, but random fluctuations at the molecular level would move the network to one of the two stable steady state. The question is, how does this come about given there is no obvious positive feedback in the network which is often the cause of bistability?

A major part of the answer lies in the constraints imposed by the conservation laws. Consider the following scenario. If the activity of the two forward limbs, v_1 and v_2 is increased, this will cause more S_2 and S_3 to be made. These changes have a number of consequences. To begin with, the additional S_3 will bind to more E_2 to form complex ES_3. However because ES_3 is linked by way of a conservation law (12.13) to the levels of ES_4 and E_2, these concentrations will therefore decline. This effectively makes S_3 compete with S_2 for E_2. The result is that there is less E_2 to catalyze v_4 resulting in an effective inhibition of v_4 by S_3. This kind of inhibition can be called apparent regulation because there is no direct molecular mechanism involved, such as allosteric regulation, it is simply an effect brought about by competitive sequestration. There are other factors

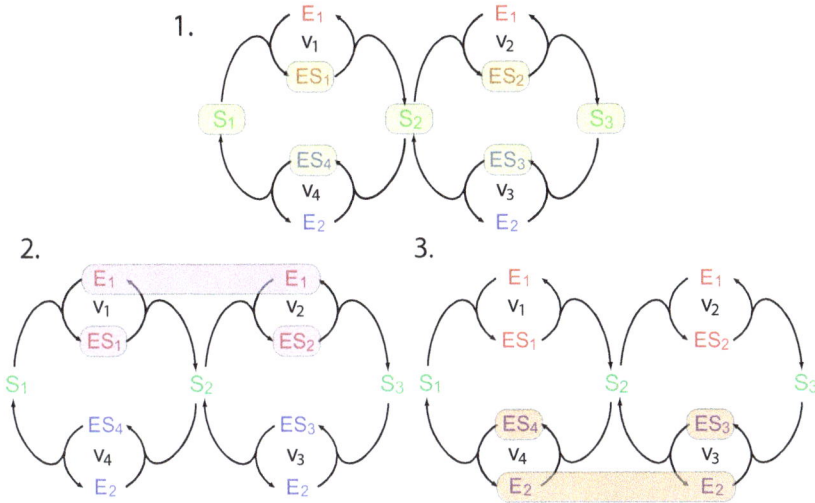

Figure 12.17 Three conservation laws in the extended dual cycle network with the conserved species highlighted in rounded squares.

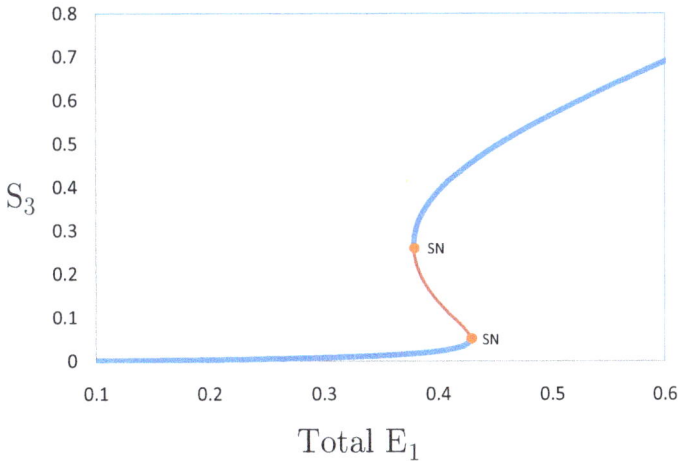

Total E_1

Figure 12.18 Bifurcation plot illustrating bistability in the concentration of S_3 as a function of E_1. The symbol SN indicates a turning point, i.e. a change in stability. Thick lines represent stable branches and the thinner central line an unstable branch. Simulations were carried out by the Oscill8 Tool (oscill8.sf.net), the model was obtained from [78] as a SBML file via the BioModels Database (http://www.ebi.ac.uk/biomodels-main/).

at play as well, for example the degree of saturation (see [65] for details), however the constraints imposed by the conservation laws are critical to the observed bistability.

To continue, given that S_2 and S_3 have both increased, then S_1 is likely to have decreased (12.12). If this is the case then there is less binding of S_1 to E_1. This results in a greater availability of E_1 which can be used to increase v_2. If we invert the logic here then we see that **increases** in S_1 will lead to **decreases** in v_2. This is another example of apparent regulation due to conservation law constraints, in this case due to equation (12.12). We can therefore redraw the pathway in a more simplified way as depicted in Figure 12.19.

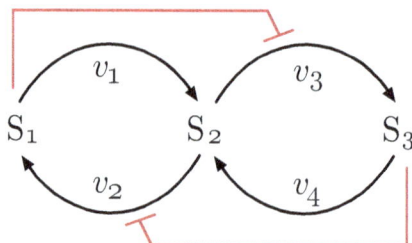

Figure 12.19 Two apparent regulatory loops in the Markevich pathway.

We can simplify this diagram even further by removing the central link, S_2 to give the diagram shown in Figure 12.20. This shows more clearly the opposing repression loops that surround the pathway.

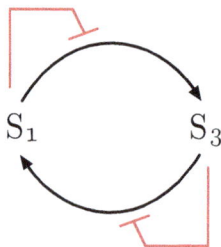

Figure 12.20 A highly simplified version of the Markevich pathway showing the opposing repression loops that surround the pathway.

In essence, what we have here is a toggle switch. Consider the possible states that can exist in the pathway shown in Figure 12.20. If the concentration of S_1 is low, then this relieves the inhibition on the forward limb converting S_1 into S_2 and thus maintaining S_1 in the low state. S_2 is now at a higher concentration and its effect is to repress the low limb. This state of affairs is therefore stable. If, on the other hand, we start S_1 at a high concentration, the reverse logic applies. The forward limb is now repressed this stabilizing S_1 at its high state. In contract S_2 must now be at a low concentration where the repression it applies to the lower limb is now released, thus stabilizing its low level.

12.6 Cascades

It is common in eukaryotic organisms to find cascades of covalent modification cycles. A single cycle will comprise of two or more interconvertible forms of a signaling protein, often tagged with phosphate groups to distinguish the forms. Phosphatases and kinases are responsible for dephosphorylating or phosphorylating the different forms. One or more of the forms within a cycle will be active, meaning that they are themselves capable of phosphorylating proteins in other cycles (Figure 12.21). A sequence of such cycles is referred to as a cascade. Although Figure 12.21 shows only a single phosphorylation event per cycle, in many cases there will be multiple phosphorylation events.

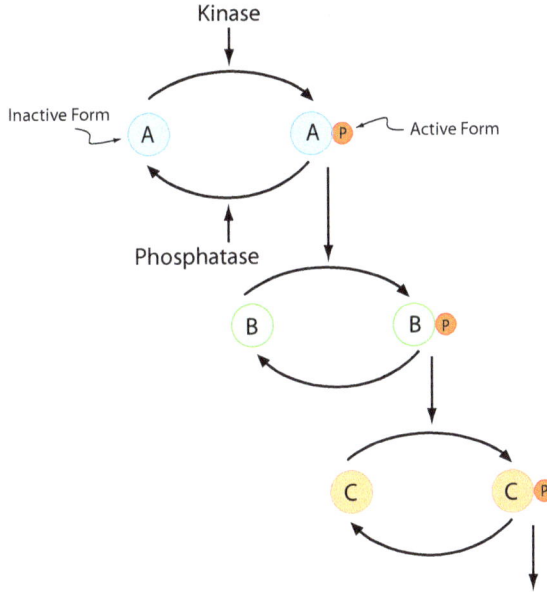

Figure 12.21 A cascade of phosphorylation/dephosphorylation cycles.

For the purpose of this study we will focus only on cycles that involve a single phosphorylation event. In order to keep track of the various entities and terms in the subsequent analysis, a specific notation will be used. A layer will be defined as a single cycle within a cascade. For example Figure 12.22 illustrates a cascade with two layers. Layers will be numbered from top to bottom. The top layer will be layer one, the second layer, layer two and so on.

Within a layer there is a cycle comprising two protein forms termed S and P. The P form refers to the phosphorylated and active form (active in the sense that it can phosphorylate other proteins). The two forms will have subscripts to denote which layer they belong to. For example, S_2 and P_2 refers to the protein forms in layer two.

The phosphorylation and dephosphorylation steps are also indicated using a specific notation. The rate of the phosphorylation step, or forward step, will be designated v_{if} where i will refer to the particular layer it belongs to. The rate of the dephosphorylation step, or reverse step, will be designated v_{ir} where i refers to the layer it belongs to.

For example, v_{2f} and v_{2r} refer to the forward and reverse rates for the steps in layer two. Figure 12.22 shows a cascade made from two cycles that use this notation.

Finally, the relative molar amounts of protein in a given layer will be designated M_{si} or M_{pi} where i is the corresponding layer. $M_{si} = s_i / T_i$ and $M_{pi} = p_i / T_i$ where T_i is the total amount of protein in the ith layer.

Effect of Modulating the Input Signal

The signal S alters the activity of v_{1f}, perhaps by phosphorylating S_1. The actual mechanism of action is unimportant at this stage. The effectiveness of the signal on v_{1f} can be summarized by the elasticity $\varepsilon_s^{v_{1f}}$.

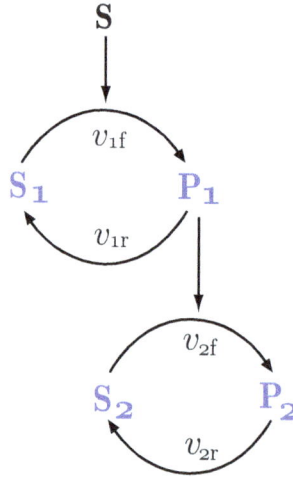

Figure 12.22 Two cycles in a cascade.

Sequestration effect of signal binding are ignored in this model.

One question to ask is what is the sensitivity of P_2 to changes in signal S? To answer this question, $R_s^{p_1}$ and $R_s^{p_2}$ will be evaluated in turn. We already know from a previous section (12.2), that:

$$C_{e_1}^{p_1} = \frac{M_{s_1}}{M_{p_1} \, \varepsilon_{s_1}^1 + M_{s_1} \, \varepsilon_{p_1}^2} \tag{12.14}$$

So that:

$$R_s^{p_1} = \varepsilon_s^{v1f} \, C_{e_1}^{p_1}$$

Consider now evaluating $R_s^{p_2}$. v_{2f} is a function of p_1 and s_2 and v_{2r} a function of p_2. We assume irreversibility in all reaction steps to make matters simpler, and also assume negligible sequestration of P_1 by binding with S_2. With these in mind we state:

$$v_{2f} = f(s_2, p_1) \quad \text{and} \quad v_{2r} = g(p_2)$$

Making a perturbation to the signal, S by δs, the relative change in v_{2f} and v_{2r} can be determined by implicit differentiation and scaling of the previous equations.

$$\frac{\delta v_{2f}}{v_{2f}} = \frac{\partial v_{2f}}{\partial s_2} \frac{p_1}{v_{2f}} \frac{\delta s_2}{p_1} + \frac{\partial v_{2f}}{\partial p_1} \frac{p_1}{v_{2f}} \frac{\delta p_1}{p_1}$$

$$\frac{\delta v_{2r}}{v_{2r}} = \frac{\partial v_{2r}}{\partial p_2} \frac{p_2}{v_{2r}} \frac{\delta p_2}{p_2}$$

At steady state the change in rates will be equal, $\delta v_{2f}/v_{2f} = \delta v_{2r}/v_{2r}$. In addition, due to conservation of S_2 and P_2, $\delta s_2 = -\delta p_2$. Therefore:

$$\frac{\partial v_{2r}}{\partial p_2} \frac{p_2}{v_{2r}} \frac{\delta p_2}{p_2} = \frac{\partial v_{2f}}{\partial p_1} \frac{p_1}{v_{2f}} \frac{\delta p_1}{p_1} + \frac{\partial v_{2f}}{\partial s_2} \frac{s_2}{v_{2f}} \frac{(-)\delta p_2}{p_2} \frac{p_2}{s_2}$$

Divide both sides by $\delta s_2/s_2$:

$$\varepsilon_{p_2}^{2r} R_s^{p_2} = \varepsilon_{p_1}^{2f} R_s^{p_1} - \varepsilon_{s_2}^{2f} R_s^{p_2} \frac{p_2}{s_2}$$

Solving for R_s^{p2} and replacing the absolute amounts by relative amounts yields:

$$R_s^{p2} = \frac{M_{s2}\, \varepsilon_{p1}^{2f}\, R_s^{p1}}{M_{s2}\varepsilon_{p2}^{2r} + M_{p2}\varepsilon_{s2}^{2f}} \tag{12.15}$$

The subexpression $(M_{s2}\varepsilon_{p1}^{2f})/(M_{s2}\varepsilon_{p2}^{2r} + M_{p2}\varepsilon_{s2}^{2f})$ will be recognized from (12.3) as the sensitivity of p_2 to changes in p_1. Following [57] we will refer to these 'local' sensitivities using the expression:

$$r_{p1}^{p2} \tag{12.16}$$

For a cascade with multiple layers, the local sensitivity of the ith layer with respect to the proceeding signal will be denoted by:

$$r_{p_{i-1}}^{p_i}$$

Equation (12.15) can be generalized for any length cascade so that the sensitivity of the ith layer to changes in the proceeding inputm p_{i-1}, is given by:

$$r_{p_{i-1}}^{p_i} = \frac{M_{si}\, \varepsilon_{p_{i-1}}^{vif}}{M_{si}\varepsilon_{p_i}^{vir} + M_{pi}\varepsilon_{s_i}^{vif}} \tag{12.17}$$

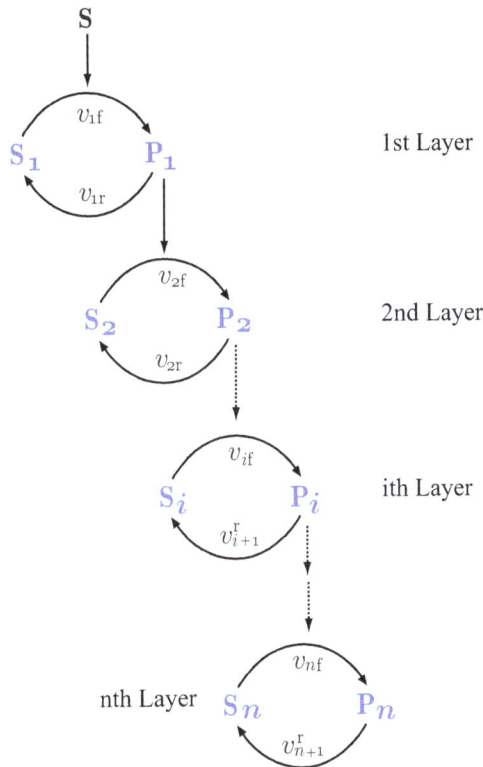

Figure 12.23 Arbitrary number of cycles or layers in a cascade.

We can generalize (12.15) so that for the ith layer we can write:

$$R_s^{p_i} = r_{p_{i-1}}^{p_i} R_s^{p_{i-1}}$$

This means that one step earlier has the response:

$$R_s^{p_i-1} = r_{p_i-2}^{p_i-1} R_s^{p_i-2}$$

and so on until we reach the top layer. To illustrate this formalism consider the two cycle layer in Figure 12.22. Let's investigate the sensitivity of P_2 to changes in the signal that is $R_s^{p_2}$. Since there are two layers, there will be two local sensitivities, $r_s^{p_1}$ and $r_{p_1}^{p_2}$. To obtain the overall sensitivity, we form the product of the local sensitivities:

$$R_s^{p_2} = r_s^{p_1} r_{p_1}^{p_2} \qquad (12.18)$$

For a cascade of n layers (Figure 12.23), the overall sensitivity will be:

$$R_s^{p_n} = r_s^{p_1} r_{p_1}^{p_2} r_{p_2}^{p_3} \dots r_{p_{n-1}}^{p_n}$$

This derivation assumes that there are no regulatory feedback or feedforward loops, and that there is no sequestration between cycles. It should be apparent that if each cycle is operating in an ultrasensitivity mode where an individual r will be high, the overall sensitivity will be significantly higher. For example, if we consider cascade of three cycles, where each individual cycle sensitivity, r, is of the order of 4.0, then the overall sensitivity will be $4 \times 4 \times 4 = 64$. As a result, layered cascades of protein cycles have the potential to generate very strong ultrasensitivity.

Effect of Modulating the Cycle Totals

Another perturbation that can be made to the cascade is to change the total mass in a given cycle and investigate how that affects the output signal. Therapeutic drugs directed a protein signaling networks effectively alter the total protein mass. A study of how the pathway responds to T is therefore an important consideration.

Previously, it was shown (12.5) that for a single cycle, the effect of changing the total on the concentration of cycle species was given by:

$$R_T^p = \frac{\varepsilon_s^1}{\varepsilon_s^1 M_p + \varepsilon_p^2 M_s}$$

Likewise, it was shown that the cycling flux as a result of changes in the total cycle mass was given by:

$$R_T^J = \varepsilon_p^2 R_T^p = \varepsilon_p^2 \frac{\varepsilon_s^1}{\varepsilon_s^1 M_p + \varepsilon_p^2 M_s}$$

Consider the two layer cascade in Figure 12.22. The response of p_1 to changes in T_1 is given by:

$$R_{T_1}^{p_1} = \frac{\varepsilon_{s_1}^1}{\varepsilon_{s_1}^1 M_{p_1} + \varepsilon_{p_1}^2 M_{s_1}}$$

The local response of p_1 on p_2 is given by equation (12.17):

$$r_{p_1}^{p_2} = \frac{M_{s_2} \varepsilon_{p_1}^{2f}}{M_{s_2} \varepsilon_{p_2}^{2r} + M_{p_2} \varepsilon_{p_2}^{2r}}$$

However equation (12.4) indicates how T_1 influences p_1, therefore $R_{T_1}^{p_2}$ can be obtained as the product:

$$R_{T_1}^{p_2} = R_{T_1}^{p_1} r_{p_1}^{p_2}$$

$$R_{T_1}^{p_2} = \frac{\varepsilon_{s_1}^{1f}}{(\varepsilon_{s_1}^{1f} M_2 + \varepsilon_{p_1}^{1r} M_1)} \cdot \frac{M_{s_2} \varepsilon_{p_1}^{2f}}{(M_{s_2} \varepsilon_{p_2}^{2r} + M_{p_2} \varepsilon_{p_2}^{2f})}$$

In general for a cascade with n layers:

$$R_{T_1}^{p_n} = R_{T_1}^{p_1} \, r_{p_1}^{p_2} \, r_{p_2}^{p_3} \, r_{p_3}^{p_4} \cdots r_{p_{n-1}}^{p_n}$$

When considering the influence of other cycle totals, for example $R_{T_2}^{p_n}$, the relationship is similar but shorter:

$$R_{T_2}^{p_2} = R_{T_2}^{p_2} \, r_{p_2}^{p_3} \, r_{p_3}^{p_4} \cdots r_{p_{n-1}}^{p_n}$$

As expected, perturbations of the cycle totals that are farther away from the output, p_n, lead to higher responses because each layer has the potential to amplify.

Cascades with Negative Feedback

Many signalling pathways include feedback loops which can have marked effects on the behavior of the pathways [56, 59]. Figure 12.24 illustrates a two layer protein cascade with a single negative feedback loop from the output signal P_2 to the reaction step where the input signal, S enters.

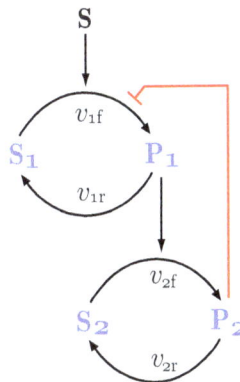

Figure 12.24 Two layer protein cascade with negative feedback loop from P_2 to v_{1f}.

The usual approach to deriving the response coefficient, $R_S^{p_2}$, is quite tedious when dealing with negative feedback. Instead, following Bruggeman [9, 8] we will write out the functional equations in the following way:

$$p_1 = p_1(p_2, s, T_1)$$
$$p_2 = p_2(p_1, T_2)$$

In these equations s_1 and s_2 are absent; this is because we invoke the conservation laws, $s_1 + p_1 = T_1$ and $s_2 + p_2 = T_2$. For example, s_1 is a function of p_1 and T_1 which allows us to omit s_1 from the function and replace it with p_1 and T_1. Likewise for the second equation. If we assume that the total amount of mass in each cycle is constant, then we can also omit T_1 and T_2, yielding:

$$p_1 = p_1(p_2, s)$$
$$p_2 = p_2(p_1)$$

Taking the total derivative of each equation yields:

$$dp_1 = \frac{\partial p_1}{\partial p_2} dp_2 + \frac{\partial p_1}{\partial s} ds$$

$$dp_2 = \frac{\partial p_2}{\partial p_1} dp_1$$

Dividing both sides by p_1 and p_2 respectively:

$$\frac{dp_1}{p_1} = \frac{\partial p_1}{\partial p_2}\frac{1}{p_1}dp_2 + \frac{\partial p_1}{\partial s}\frac{1}{p_1}ds$$

$$\frac{dp_2}{p_2} = \frac{\partial p_2}{\partial p_1}\frac{1}{p_2}dp_1$$

Multiplying top and bottom by the appropriate term p_1, p_2 and s:

$$\frac{dp_1}{p_1} = \frac{\partial p_1}{\partial p_2}\frac{p_2}{p_1}\frac{dp_2}{p_2} + \frac{\partial p_1}{\partial s}\frac{s}{p_1}\frac{ds}{s}$$

$$\frac{dp_2}{p_2} = \frac{\partial p_2}{\partial p_1}\frac{p_1}{p_2}\frac{dp_1}{p_1}$$

Finally, dividing both sides by ds/s yields:

$$R_s^{p_1} = \frac{\partial p_1}{\partial p_2}\frac{p_2}{p_1}R_s^{p_2} + \frac{\partial p_1}{\partial s}\frac{s}{p_1}$$

$$R_s^{p_2} = \frac{\partial p_2}{\partial p_1}\frac{p_1}{p_2}R_s^{p_1}$$

The partial derivative terms are the local sensitivities, r (12.16). The equations can therefore be rewritten as:

$$R_s^{p_1} = r_{p_2}^{p_1} R_s^{p_2} + r_s^{p_1}$$

$$R_s^{p_2} = r_{p_1}^{p_2} R_s^{p_1}$$

Given the two equations we can solve for $R_s^{p_2}$:

$$R_s^{p_2} = \frac{r_1^2 r_s^1}{1 - r_1^2 r_2^1} \tag{12.19}$$

The feedback term is given by r_2^1. Note the similarity to the generic negative feedback response to equation (8.1). If the feedback term is set to zero, meaning the feedback loop is absent, the equation reduces to equation (12.18):

$$R_s^{p_2} = r_1^2 r_s^1$$

It is worth noting that the term in the denominator, $r_1^2 r_2^1$, is the loop gain (8.2) within the feedback system.

The sign for the negative feedback term will be negative so that the term $-r_1^2 r_2^1$ is positive. This means that the presence of negative feedback will *reduce* the sensitivity of P_2 to signal, S. That is, negative feedback will lock P_2 into a narrow range. If the loop gain of the cascade is large, that is, $-r_1^2 r_2^1 \gg 0$, then the response equation can be simplified further to:

$$R_s^{p_2} = \frac{r_s^1}{r_2^1}$$

In this mode, the cascade acts as a **negative feedback amplifier** [90, 104] although with a gain reduced by r_1^2, but partly compensated by r_s^1. If the two gains remain relatively constant over the operating range, then the cascade acts as a **tracking device**, tracking p_2 in response to changes to s. For long cascades with say n layers:

$$R_s^{p_n} = \frac{r_s^1}{r_n^1}$$

In other words the feedback response depends only on the feedback sensitivity and the effect of the input on the first layer. This means that the performance of the cascade is independent of the middle layers. This is exactly what we saw with the analysis of generic feedback in Chapter 8. The system is therefore immune to noise or natural genetic variation in the proteins levels in the middle layers of the cascade.

Further Reading

1. Reich, J.G. and Selkov, E.E., 1981. Energy metabolism of the cell: a theoretical treatise. Academic Press.

2. Fell D.A. and Sauro, H.M. 1985. Metabolic control analysis. Additional relationships between elasticities and control coefficients Eur. J. Biochem. 148:555–561.

3. Hofmeyr, J.H.S., Kacser, H. and Merwe, K.J., 1986. Metabolic control analysis of moiety conserved cycles. The FEBS Journal, 155(3), 631-640.

4. Bluthgen, N., Bruggeman, F.J., Legewie, S., Herzel, H., Westerhoff, H.V. and Kholodenko, B.N., 2006. Effects of sequestration on signal transduction cascades. The FEBS journal, 273(5), 895-906.

Exercises

1. Create a model of single protein phosphorylation cycle and investigate using simulation the effects of changing the kinase and phosphatase K_ms on the steady state response to changes in kinase concentrations.

2. Repeat the first exercise but add additional cycles forming a multi-layered phosphorylation cascade. Show by simulation how the response coefficient of the entire cascade increases with the number of layers.

3. Create a two layered cascade and add a negative feedback loop from the last species in the cascade to the first cycle. Investigate by simulation the response of the output of the cascade to changes in the inputs as a function of the strength of the feedback.

4. Use the moiety conservation connectivity theorem and the summation theorem to derive equation 12.2.

12.A   Python/Tellurium Scripts

```python
import tellurium as te
import roadrunner
import pylab

r = te.loada('''
     var AB, A, B
     A + B -> AB; k1*A*B - k2*AB;

     A = 0.001; B = 100; AB = 0;
     k1 = 0.4; k2 = 1;
''')

r.conservedMoietyAnalysis = True;

print ("A, B, AB = ", r.A, r.B, r.AB)
print ("_CSUM0 = ", r._CSUM0)
print ("_CSUM1 = ", r._CSUM1)

# _CSUM0 = AB + A
```

```
# _CSUM1 = AB + B

pylab.figure(figsize=(10,5))
pylab.xticks(fontsize=14)
pylab.yticks(fontsize=14)

abSumX = []; aY = [];
n = 400
for i in range (n):
    r.steadyState()
    abSumX.append (r._CSUM0);
    aY.append (r.A);
    r._CSUM0 = r._CSUM0 + 0.5

pylab.plot (abSumX, aY, linewidth=3)
pylab.xlabel ("$a+ab$", fontsize=16)
pylab.ylabel ("$a$", fontsize=18)
pylab.savefig("sequestrationUltraPlotJustA.pdf")

pylab.show()
```

Listing 12.1 Ultrasensitivity by sequestration using $a + b \rightleftharpoons ab$ model.

```
import tellurium as te
import roadrunner
import pylab

r = te.loada("""
    J1: $Xo -> X; kcat*(k1*Xo - k11*X);
    J2: X -> $X1; k2*X/(X + Km);
    X + II -> XI; k3*X*II - k4*XI;

    Xo = 1; k1 = 0.1; k2 = 1;
    k3 = 3; k4 = 0.1; II = 3;
    Km = 0.001; k11 = 12; kcat = 1;
""")

r.conservedMoietyAnalysis = True;

r.Xo = 0.01;
xv = []; yv = []; yc = []
n = 400;
for i in range (n):
    r.steadyState()
    xv.append (r.Xo)
    yc.append (r.getCC ('XI', 'Xo'))
    yv.append (r.XI)
    r.Xo = r.Xo + 0.05


pylab.xlabel ("$X_o$", fontsize=16)
```

```
pylab.ylabel ("X or Response Coefficient", fontsize=16)

pylab.plot (xv, yv, color='b', label='X', linewidth=2)
pylab.plot (xv, yc, color='r', label='$R^X_{Xo}$', linewidth=2)
pylab.legend()

pylab.savefig ('UltrasensitivityMoreComplex.pdf')
```

Listing 12.2 Ultrasensitivity by sequestration showing sigmoid behavior.

```
import tellurium as te
import matplotlib.pyplot as plt

r = te.loada("""
    S1 -> S2; k11*S1;
    S2 -> S1; k2*S2;
    S2 -> S3; k11*S2;
    S3 -> S2; k4*S3;

    k2 = 3.5; k4 = 1.26;
    S1 = 10; k11 = 6.6
""")

r.conservedMoietyAnalysis = True;
print (r.getConservationMatrix())

r.k11 = 0.01;
te.saveToFile ('dual.xml', r.getSBML())
r.steadyState()
print (r.S1, r.S2, r.S3)
x = []; y = []
for i in range (200):
    r.steadyState()
    x.append (r.k11);
    y.append (r.S3)
    r.k11 = r.k11 + 0.04;

plt.plot (x, y, linewidth=3.0)
plt.grid(False)
plt.xlabel ('S', fontsize=18)
plt.ylabel ('$S_3$', fontsize=18)
plt.xticks(fontsize=16)
plt.yticks(fontsize=16)
plt.show()
```

Listing 12.3 Ultrasensitivity by sequestration showing sigmoid behavior.

Part I

Appendices

A

List of Symbols and Abbreviations

Symbols

A	Name and amount of species
a	Concentration of species A
$s_a, s_b, \ldots$	Stoichiometric amounts for species $A, B, \ldots$
c_a, c_i	Stoichiometric coefficient for species A or species i
Δ	Change
δs	Small change to species S
ε_s^v	Elasticity coefficient
h	Hill coefficient
k, k_i	Rate constant
Γ	Mass-action ratio
γ	Normalized activator concentration, A/K_A
$\Delta_r G$	Reaction free energy change
J	Flux
K_a	Association constant
K_d	Dissociation constant

K_{eq}	Equilibrium constant
M_i	Fractional molar amount of species i.
T	Total number of moles in a given conserved cycle
K_H	Half-maximal activity
K_i	Inhibition constant
K_m	Michaelis constant
K_s	Michaelis constant with respect to substrate
K_p	Michaelis constant with respect to product
L	Allosteric constant
L	Liter (US spelling)
M	Modifier
M_i	Molar amount of species i
mol	Mole
n_i	Amount of substance i
n	Number of subunits
P	Product concentration
ρ	Disequilibrium ratio
q	Equilibrium constant
R	Gas constant
σ	Modifier factor
S	Substrate concentration or entropy depending on context
T	Temperature
t	Time
v_f	Forward reaction rate
v_i	i^{th} reaction rate
v_r	Reverse reaction rate
V	Volume
V_m	Maximal velocity
Y	Fractional saturation
C_e^J	Flux Control Coefficient of flux J with respect to enzyme e
C_e^s	Concentration Control Coefficient of species s with respect to enzyme e
ε_s^v	Elasticity Coefficient of reaction rate v with respect to species s

Non-Mathematical Abbreviations

AMP, ADP, ATP	Adenine nucleotides
ATCase	Aspartate transcarbamylase
cAMP	Cyclic AMP
CTP	Cytidine triphosphate
DNA	Deoxyribonucleic acid
F6P	Fructose-6-Phosphate
KNF	Koshland, Nemethy and Filmer model
IPTG	Isopropyl β-D-1-thiogalactopyranoside
LacI	Lactose Operon Repressor
mRNA	Messenger RNA
MWC	Monod, Wyman and Changeux model
PEP	Phosphoenolpyruvate
PFK	Phosphofructokinase
RNA	Ribonucleic acid
TF	Transcription factor

B

Control Equations

B.1   Linear Pathways

Two Step pathway

$$X_o \xrightarrow{\ v_1\ } S_1 \xrightarrow{\ v_2\ } X_1$$

Figure B.1 Two Step Pathway.

$$C_{e_1}^{J} = \frac{\varepsilon_1^2}{\varepsilon_1^2 - \varepsilon_1^1}$$

$$C_{e_2}^{J} = -\frac{\varepsilon_1^1}{\varepsilon_1^2 - \varepsilon_1^1}$$

$$C_{e_1}^{s_1} = \frac{1}{\varepsilon_1^2 - \varepsilon_1^1}$$

$$C_{e_2}^{s_1} = -\frac{1}{\varepsilon_1^2 - \varepsilon_1^1}$$

Three Step pathway

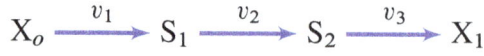

$$X_o \xrightarrow{\;v_1\;} S_1 \xrightarrow{\;v_2\;} S_2 \xrightarrow{\;v_3\;} X_1$$

Figure B.2 Three Step Pathway.

The denominator, D is given by:

$$D = \varepsilon_1^2 \varepsilon_2^3 - \varepsilon_1^1 \varepsilon_2^3 + \varepsilon_1^1 \varepsilon_2^2$$

$$C_{e_1}^J = \frac{\varepsilon_1^2 \varepsilon_2^3}{D} \qquad C_{e_2}^J = -\frac{\varepsilon_1^1 \varepsilon_2^3}{D} \qquad C_{e_3}^J = \frac{\varepsilon_1^1 \varepsilon_2^2}{D}$$

$$C_{e_1}^{s_1} = \frac{\varepsilon_2^3 - \varepsilon_2^2}{D} \qquad C_{e_1}^{s_2} = \frac{\varepsilon_1^2}{D}$$

$$C_{e_2}^{s_1} = -\frac{\varepsilon_2^3}{D} \qquad C_{e_2}^{s_2} = -\frac{\varepsilon_1^1}{D}$$

$$C_{e_3}^{s_1} = \frac{\varepsilon_2^2}{D} \qquad C_{e_3}^{s_2} = \frac{\varepsilon_1^1 - \varepsilon_1^2}{D}$$

Three Step pathway with Negative Feedback

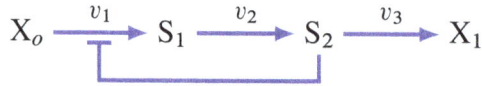

$$X_o \xrightarrow{\;v_1\;} S_1 \xrightarrow{\;v_2\;} S_2 \xrightarrow{\;v_3\;} X_1$$

Figure B.3 Three Step Pathway.

$$D = \varepsilon_1^1 \varepsilon_2^2 - \varepsilon_1^1 \varepsilon_2^3 + \varepsilon_1^2 \varepsilon_2^3 - \varepsilon_2^1 \varepsilon_1^2$$

$$C_{e_1}^J = \frac{\varepsilon_1^2 \varepsilon_2^3}{D} \qquad C_{e_2}^J = -\frac{\varepsilon_1^1 \varepsilon_2^3}{D} \qquad C_{e_3}^J = \frac{\varepsilon_1^1 \varepsilon_2^2 - \varepsilon_2^1 \varepsilon_1^2}{D}$$

$$C_{e_1}^{s_1} = \frac{\varepsilon_2^3 - \varepsilon_2^2}{D} \qquad C_{e_2}^{s_1} = -\frac{\varepsilon_2^3 - \varepsilon_2^1}{D} \qquad C_{e_3}^{s_1} = \frac{\varepsilon_2^2 - \varepsilon_2^1}{D}$$

$$C_{e_1}^{s_2} = \frac{\varepsilon_1^2}{D} \qquad C_{e_2}^{s_2} = -\frac{\varepsilon_1^1}{D} \qquad C_{e_3}^{s_2} = \frac{\varepsilon_1^1 - \varepsilon_1^2}{D}$$

Three Step Pathway with Feedforward Loop

$$D = \varepsilon_1^1 \varepsilon_2^2 + \varepsilon_1^2 \varepsilon_2^2 - \varepsilon_1^1 \varepsilon_2^3 + \varepsilon_1^2 \varepsilon_2^3$$

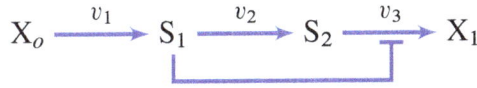

Figure B.4 Three Step Pathway.

$$C_{e_1}^{J} = \frac{\varepsilon_2^2 \varepsilon_1^3 + \varepsilon_1^2 \varepsilon_2^3}{D} \qquad C_{e_2}^{J} = -\frac{\varepsilon_1^1 \varepsilon_3^3}{D} \qquad C_{e_3}^{J} = \frac{\varepsilon_1^1 \varepsilon_2^2}{D}$$

$$C_{e_1}^{S_1} = \frac{\varepsilon_2^3 - \varepsilon_2^2}{D} \qquad C_{e_2}^{S_1} = -\frac{\varepsilon_2^3}{D} \qquad C_{e_3}^{S_1} = \frac{\varepsilon_2^2}{D}$$

$$C_{e_1}^{S_2} = \frac{\varepsilon_1^2 + \varepsilon_1^3}{D} \qquad C_{e_2}^{S_2} = \frac{\varepsilon_1^1 + \varepsilon_1^3}{D} \qquad C_{e_3}^{S_2} = \frac{\varepsilon_1^1 - \varepsilon_1^2}{D}$$

Four Step pathway

Figure B.5 Four Step Pathway.

$$D = \varepsilon_1^1 \varepsilon_2^2 \varepsilon_3^3 - \varepsilon_1^1 \varepsilon_2^2 \varepsilon_3^4 + \varepsilon_1^1 \varepsilon_2^3 \varepsilon_3^4 - \varepsilon_1^2 \varepsilon_2^3 \varepsilon_3^4$$

The denominator term D has been omitted in the following equation to ease readability.

$$C_{e_1}^{J} = -\varepsilon_1^2 \varepsilon_2^3 \varepsilon_3^4 \qquad ,C_{e_2}^{J} = \varepsilon_1^1 \varepsilon_2^3 \varepsilon_3^4 \qquad ,C_{e_3}^{J} = -\varepsilon_1^1 \varepsilon_2^2 \varepsilon_3^4 \qquad ,C_{e_4}^{J} = \varepsilon_1^1 \varepsilon_2^2 \varepsilon_3^3$$

$$C_{e_1}^{S_1} = -\varepsilon_2^2 \varepsilon_3^3 + \varepsilon_2^2 \varepsilon_3^4 - \varepsilon_2^3 \varepsilon_3^4, C_{e_2}^{S_1} = -\varepsilon_2^3 \varepsilon_3^4 \qquad ,C_{e_3}^{S_1} = -\varepsilon_2^2 \varepsilon_3^4 \qquad ,C_{e_4}^{S_1} = \varepsilon_2^2 \varepsilon_3^3$$

$$C_{e_1}^{S_2} = \varepsilon_1^2 \varepsilon_3^3 - \varepsilon_1^2 \varepsilon_3^4 \qquad ,C_{e_2}^{S_2} = -\varepsilon_1^1 \varepsilon_3^3 + \varepsilon_1^1 \varepsilon_3^4, C_{e_3}^{S_2} = -\varepsilon_1^1 \varepsilon_3^4 + \varepsilon_1^2 \varepsilon_3^4, C_{e_4}^{S_2} = \varepsilon_1^1 \varepsilon_3^3 - \varepsilon_1^2 \varepsilon_3^3$$

$$C_{e_1}^{S_3} = -\varepsilon_1^2 \varepsilon_2^3 \qquad ,C_{e_2}^{S_3} = \varepsilon_1^1 \varepsilon_2^3 \qquad ,C_{e_3}^{S_2} = -\varepsilon_1^1 \varepsilon_2^2 \qquad ,C_{e_4}^{S_2} = \varepsilon_1^1 \varepsilon_2^2 - \varepsilon_1^1 \varepsilon_2^3 + \varepsilon_1^2 \varepsilon_2^3$$

B.2   Cycles

$$M_1 = s_1/T \quad M_2 = s_2/T \quad \text{where} \quad T = s_1 + s_2$$

$$C_{e_1}^{S_2} = \frac{M_1}{M_2 \varepsilon_1^1 + M_1 \varepsilon_2^2}$$

$$C_{e_2}^{S_2} = -\frac{M_2}{M_2 \varepsilon_1^1 + M_1 \varepsilon_2^2}$$

$$C_{e_1}^J = \frac{M_1 \varepsilon_2^2}{M_2 \varepsilon_1^1 + M_1 \varepsilon_2^2}$$

$$C_{e_2}^J = -\frac{M_2 \varepsilon_1^1}{M_2 \varepsilon_1^1 + M_1 \varepsilon_2^2}$$

$$C_T^{s_2} = \frac{\varepsilon_1^1}{M_2 \varepsilon_1^1 + M_1 \varepsilon_2^2}$$

$$C_T^J = -\frac{\varepsilon_2^2 \varepsilon_1^1}{M_2 \varepsilon_1^1 + M_1 \varepsilon_2^2}$$

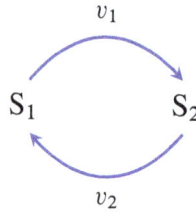

Figure B.6 Simple Conserved cycle where $s_1 + s_2 = \text{constant} = T$.

B.3   Branches

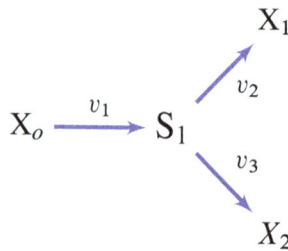

Figure B.7 Three Step Pathway.

Let $\alpha = \dfrac{v_2}{v_1}$ and $\varepsilon_1 \equiv \varepsilon_1^1 \quad \varepsilon_2 \equiv \varepsilon_1^2 \quad \varepsilon_3 \equiv \varepsilon_1^3$

$$d = \varepsilon_2 \alpha + \varepsilon_3 (1 - \alpha) - \varepsilon_1$$

$$C_{e_1}^{J_1} = \frac{\varepsilon_3(1-\alpha) + \varepsilon_2\alpha}{d} \qquad C_{e_2}^{J_2} = \frac{\varepsilon_3(1-\alpha) - \varepsilon_1}{d} > 0$$

$$C_{e_1}^{J_1} = \frac{-\varepsilon_1\alpha}{d} \qquad C_{e_1}^{J_2} = \frac{\varepsilon_2}{d} > 0$$

$$C_{e_1}^{J_1} = \frac{-\varepsilon_1(1-\alpha) + \varepsilon_2\alpha}{d} \qquad C_{e_3}^{J_2} = \frac{-\varepsilon_2(1-\alpha)}{d} < 0$$

For the concentration control coefficients:

$$C_{e_1}^s = \frac{1}{\varepsilon_2\alpha + \varepsilon_3(1-\alpha) - \varepsilon_1}$$

$$C_{e_2}^s = \frac{-\alpha}{\varepsilon_2\alpha + \varepsilon_3(1-\alpha) - \varepsilon_1}$$

$$C_{e_3}^s = \frac{-(1-\alpha)}{\varepsilon_2\alpha + \varepsilon_3(1-\alpha) - \varepsilon_1}$$

References

[1] Acerenza, L., and A. Cornish-Bowden. 1997. Biochem J **327** (**Pt 1**):217–224.

[2] Aström, K. J., and Richard M Murray. 2010. *Feedback systems: an introduction for scientists and engineers*. Princeton university press.

[3] Bakker, B. M., P. A. Michels, F. R. Opperdoes, and H. V. Westerhoff. 1997. J Biol Chem **272** (6):3207–3215.

[4] Black, H. 1977. IEEE spectrum **14** (12):55–61.

[5] Blackman, F. F. 1905. Ann. Botany **19**:281–295.

[6] Blüthgen, N., F J Bruggeman, S Legewie, H Herzel, H V Westerhoff, and B N Kholodenko. 2006. FEBS J **273** (5):895–906.

[7] Brown, G. C. 1991. Journal of theoretical biology **153** (2):195–203.

[8] Bruggeman, F., JL Snoep, and HV Westerhoff. 2008. IET systems biology **2** (6):397–410.

[9] Bruggeman, F. J., H. V. Westerhoff, J. B. Hoek, and B. N. Kholodenko. 2002. J. Theor Biol. **218**:507–20.

[10] Buchler, N. E., and Frederick R Cross. 2009. Molecular systems biology **5** (1):272.

[11] Buchler, N. E., and Matthieu Louis. 2008. J Mol Biol **384** (5):1106–1119.

[12] Bujara, M., Michael Schümperli, René Pellaux, Matthias Heinemann, and Sven Panke. 2011. Nature chemical biology **7** (5):271–277.

[13] Burrell, M. M., P. J. Mooney, M. Blundy, D. Carter, F. Wilson, J. Green, K. S. Blundy, and T. ap Rees. 1994. Planta **194**:95—101.

[14] Burton, A. C. 1936. Journal Cellular and Comparative Physiology **9(1)**:1–14.

[15] Chan, T., and P.C. Hansen. 1992. SIAM Journal on Scientific and Statistical Computing **13**:727.

[16] Chance, B., and G R Williams. 1955. J Biol Chem **217** (1):429–438.

[17] Chassagnole, C., Rais B, E. Quentin, D. A. Fell, and J.-P. Mazat. 2001. Biochem. J **356**:415–423.

[18] Christensen, C. D., Jan-Hendrik S Hofmeyr, and Johann M Rohwer. 2015. BMC systems biology **9** (1):89.

[19] Cornish-Bowden, A., and J. H. Hofmeyr. 2002. J. theor. Biol **216**:179–191.

[20] Cornish-Bowden, A., J. H. Hofmeyr, and M. Cardenas. 2002. Biochemical Society Transactions **30**:43–47.

[21] Curien, G., Stephane Ravanel, and Renaud Dumas. 2003. The FEBS Journal **270** (23):4615–4627.

[22] Davies, S. E., and Kevin M Brindle. 1992. Biochemistry **31** (19):4729–4735.

[23] De Camp, L. S. 1990. *The ancient engineers*. Barnes & Noble Publishing.

[24] Del Vecchio, D., Aaron J Dy, and Yili Qian. 2016. Journal of The Royal Society Interface **13** (120):20160380.

[25] Ehlde, M., and G. Zacchi. 1997. Chemical Engineering Science **52**:2599–2606(8).

[26] Eisenthal, R., and A. Cornish-Bowden. 1998. Journal of Biological Chemistry **273** (10):5500–5505.

[27] Eschrich, K., W. Scellenberger, and E. Hofmann. 1980. Arch. Biochem. Biophys. **205**:114–121.

[28] Eschrich, K., W. Schellenberger, and E. Hofmann. 1983. Arch. Biochem. Biophys. **222**:657–660.

[29] Eschrich, K., W. Schellenberger, and E. Hofmann. 1990. European Journal Of Biochemistry **188**:697–703.

[30] Fell, D. 1997. *Understanding the Control of Metabolism.* Portland Press., London.

[31] Fell, D., and K. Snell. 1988. Biochem. J. **256**:97–101.

[32] Fell, D. A., and H. M. Sauro. 1985. Eur. J. Biochem. **148**:555–561.

[33] Fell, D. A., and H. M. Sauro. 1990. Eur. J. Biochem. **192**:183–187.

[34] Flint, H. J., R. W. Tateson, I. B. Bartelmess, D. J. Porteous, W. D. Donochie, and H. Kacser. 1981. Biochem. J. **200**:231–246.

[35] Giersch, C. 1995. The FEBS Journal **227** (1-2):194–201.

[36] Goldbeter, A., and D. E. Koshland. 1981. Proc. Natl. Acad. Sci **78**:6840–6844.

[37] Goldbeter, A., and D. E. Koshland. 1984. J. Biol. Chem. **259**:14441–7.

[38] Goodwin, B. 1965. Advances in Enzyme Regulation **3**:425–438.

[39] Groen, A. K., R. J. A. Wanders, H. V. Westerhoff, R. van der Meer, and J. M. Tager. 1982. J. Biol. Chem. **257**:2754–2757.

[40] Hearon, J. Z. 1952. Physiol. Rev. **32**:499–523.

[41] Heinisch, J. 1986. Mol. Gen Genet. **202**:75–82.

[42] Heinrich, R., S. M. Rapoport, and T. A. Rapoport. 1977. Prog. Biophys. Molec. Biol. **32**:1–82.

[43] Heinrich, R., and T. A. Rapoport. 1974. Eur. J. Biochem. **42**:89–95.

[44] Heinrich, R., and S Schuster. 1996. *The Regulation of Cellular Systems.* Chapman and Hall.

[45] Higgins, J. 1963. Ann. N. Y. Acad. Sci. **108**:305–321.

[46] Hofmeyr, J.-H. 2001. p. 291–300. *In:* Proceedings of the Second International Conference on Systems Biology . Caltech,.

[47] Hofmeyr, J.-H., and A Cornish-Bowden. 2000. FEBS Lett **476** (1-2):47–51.

[48] Hofmeyr, J.-H., H. Kacser, and K. J. van der Merwe. 1986. Eur. J. Biochem. **155**:631–641.

[49] Hofmeyr, J. H., and J M Rohwer. 2011. Methods Enzymology **500**:533–554.

[50] Huang, C. F., and J. E. Ferrell. 1996. Proc. Natl. Acad. Sci **93**:10078–10083.

[51] Ingalls, B. 2013. *Mathematical Modeling in Systems Biology: An Introduction.* MIT Press.

[52] Ingalls, B. P. 2004. Journal of Physical Chemistry B **108**:1143–1152.

[53] Kacser, H. 1983. Biochem. Soc. Trans. **11**:35–40.

[54] Kacser, H., and J. A. Burns. 1973. *In:* Davies, D. D. , (ed.), Rate Control of Biological Processes, vol. 27 of *Symp. Soc. Exp. Biol.* p. 65–104. Cambridge University Press.

[55] Khammash, M. 2016. BMC biology **14** (1):1.

[56] Kholodenko, B. N. 2006. Nat Rev Mol Cell Biol **7** (3):165–176.

[57] Kholodenko, B. N., J. B. Hoek, H. V. Westerhoff, and G. C. Brown. 1997. FEBS Letters **414**:430–434.

[58] Klipp, E., and Reinhart Heinrich. 1999. Biosystems **54** (1-2):1–14.

[59] Kolch, W., Melinda Halasz, Marina Granovskaya, and Boris N Kholodenko. 2015. Nature reviews. Cancer **15** (9):515.

[60] Kornberg, A., *et al..* 1960. Science **131** (3412):1503–1508.

[61] Kouril, T., Dominik Esser, Julia Kort, Hans V Westerhoff, Bettina Siebers, and Jacky L Snoep. 2013. FEBS Journal **280** (18):4666–4680.

[62] LaPorte, D. C., K Walsh, and D E Koshland. 1984. J Biol Chem **259** (22):14068–14075.

[63] Liu, A. P., and Daniel A Fletcher. 2009. Nature reviews. Molecular cell biology **10** (9):644.

[64] Marín-Hernández, A., Sara Rodríguez-Enríquez, Paola A Vital-González, Fanny L Flores-Rodríguez, Marina Macías-Silva, Marcela Sosa-Garrocho, and Rafael Moreno-Sánchez. 2006. FEBS Journal **273** (9):1975–1988.

[65] Markevich, N. I., J B Hoek, and B. N. Kholodenko. 2004. J. Cell Biol. **164**:353–9.

[66] Mayr, O. 1975. *The Origins of Feedback Control*. The MIT Press.

[67] Mindell, D. 2000. Technology and Culture **41** (3):405–434.

[68] Mor, I., EC Cheung, and KH Vousden. 2011. p. 211–216. *In:* Cold Spring Harbor symposia on quantitative biology vol. 76. Cold Spring Harbor Laboratory Press,.

[69] Morales, M. F. 1921. Journal Cellular and Comparative Physiology **30**:303–313.

[70] Moreno-Sánchez, R., Alvaro Marín-Hernández, Emma Saavedra, Juan P Pardo, Stephen J Ralph, and Sara Rodríguez-Enríquez. 2014. The international journal of biochemistry & cell biology **50**:10–23.

[71] Müller, S., Friedrich K Zimmermann, and Eckhard Boles. 1997. Microbiology **143** (9):3055–3061.

[72] Niekerk, D. D., Gerald P Penkler, Francois Toit, and Jacky L Snoep. 2016. The FEBS journal **283** (4):634–646.

[73] Ortega, F., J L Garcés, F Mas, B N Kholodenko, and M Cascante. 2006. FEBS J **273** (17):3915–3926.

[74] Palsson, B. O. 2007. *Systems Biology: Properties of Reconstructed Networks*. Cambridge University Press.

[75] Papadopoulos, E. 2007. Distinguished figures in mechanism and machine science :217–245.

[76] Park, D. 1988. Computers & chemistry **12** (2):175–188.

[77] Penkler, G., Francois du Toit, Waldo Adams, Marina Rautenbach, Daniel C Palm, David D van Niekerk, and Jacky L Snoep. 2015. FEBS Journal **282** (8):1481–1511.

[78] Qiao, L., R.B. Nachbar, I.G. Kevrekidis, S.Y. Shvartsman, and A. Asthagiri. 2007. PLoS Comput Biol **3** (9):e184.

[79] Rao, C. V., H. M. Sauro, and A. P. Arkin. 2004. 7th International Symposium on Dynamics and Control of Process Systems, DYCOPS, 7 **7**:0–0.

[80] Rapoport, T. A., and R. Heinrich. 1975. Biosystems **7**:120–129.

[81] Reder, C. 1988. J. Theor. Biol. **135**:175–201.

[82] Reich, J. G., and E. E. Selkov. 1981. *Energy metabolism of the cell*. Academic Press, London.

[83] Rohwer, J. M., and J. H. Hofmeyr. 2008. J Theor Biol **252** (3):546–554.

[84] Ruijter, G., H Panneman, and J Visser. 1997. Biochimica et Biophysica Acta (BBA)-General Subjects **1334** (2):317–326.

[85] Sauro, H. M. 1994. BioSystems **33**:15–28.

[86] Sauro, H. M. 2012. *Enzyme Kinetics for Systems Biology*. Ambrosius Publishing. 2nd Edition.

[87] Sauro, H. M. 2014. *Systems Biology: An Introduction to Pathway Modeling*. Ambrosius Publishing, Seattle.

[88] Sauro, H. M. 2015. *Systems Biology: Linear Algebra for Pathway Modeling*. Ambrosius Publishing, Seattle.

[89] Sauro, H. M., and D. A. Fell. 1991. Mathl. Comput. Modelling **15**:15–28.

[90] Sauro, H. M., and Brian Ingalls. 2007. arXiv preprint arXiv:0710.5195 .

[91] Savageau, M. A. 1972. Curr. Topics Cell. Reg. **6**:63–130.

[92] Savageau, M. A. 1975. J Mol Evol **5** (3):199–222.

[93] Savageau, M. A. 1976. *Biochemical systems analysis: a study of function and design in molecular biology.* Addison-Wesley, Reading, Mass.

[94] Schaaff, I., J. Heinisch, and F. K. Zimmermann. 1989. Yeast **5** (4):285–290.

[95] Schellenberger, W., K. Eschrich, and E. Hofmann. 1978. Acta biol. med. germ. **37**:1425–1441.

[96] Schellenberger, W., Klaus Eschrich, and Eberhard Hofmann. 1981. Advances in enzyme regulation **19**:257–284.

[97] Schuster, S., and T. Hofer. 1991. J. Chem. Soc. Faraday Trans. **87**:2561–2566.

[98] Small, J. R., and D. A. Fell. 1990. Eur. J. Biochem. **191**:405–411.

[99] Smallbone, K., Hanan L Messiha, Kathleen M Carroll, Catherine L Winder, Naglis Malys, Warwick B Dunn, Ettore Murabito, Neil Swainston, Joseph O Dada, Farid Khan, *et al.*. 2013a. FEBS letters **587** (17):2832–2841.

[100] Smallbone, K., Hanan L Messiha, Kathleen M Carroll, Catherine L Winder, Naglis Malys, Warwick B Dunn, Ettore Murabito, Neil Swainston, Joseph O Dada, Farid Khan, *et al.*. 2013b. FEBS letters **587** (17):2832–2841.

[101] Snell, K., and D. A. Fell. 1990. Adv. Enzyme Reg. **30**:13–32.

[102] Stitt, M., W. P. Quick, U. Schurr, E.-D. Schulze, S. R. Rodermel, and L. Bogorad. 1991. Planta **183**:555–566. 10.1007/BF00194277.

[103] Storey, K. B. 2005. *Functional metabolism: regulation and adaptation.* John Wiley & Sons.

[104] Sturm, O. E., Richard J Orton, Joan Grindlay, *et al.*. 2010. Sci Signal **3(153)**.

[105] Szent-Györgyi, A. 1963. Annual review of biochemistry **32** (1):1–15.

[106] Thomas, S., Peter JF Mooney, Michael M Burrell, *et al.*. 1997. Biochemical Journal **322** (1):119–127.

[107] Thron, C. 1991. Bulletin of mathematical biology **53** (3):403–424.

[108] Torres, N., F Mateo, E Melendez-Hevia, and H Kacser. 1986. Biochem. J **234**:169–174.

[109] Tyson, J., and H. G. Othmer. 1978. Progress in Theoretical Biology (R. Rosen & F.M. Snell, Eds.) **5**:1–62.

[110] Tyson, J. J., K. C. Chen, and B. Novak. 2003. Current Opinion in Cell Biology **15**:221–231.

[111] Urbano, A. M., Helen Gillham, Yoram Groner, and Kevin M Brindle. 2000. Biochemical Journal **352** (3):921–927.

[112] Vallabhajosyula, R. R., V Chickarmane, and H M Sauro. 2006. Bioinformatics **22** (3):346–353.

[113] van Eunen, K., and Barbara M Bakker. 2014. Perspectives in Science **1** (1):126–130.

[114] Varusai, T. M., Walter Kolch, Boris N Kholodenko, and Lan K Nguyen. 2015. Molecular BioSystems **11** (10):2750–2762.

[115] Waley, S. G. 1964. Biochem. J. **91** (3):514–0.

[116] Walsh, K., and Daniel E Koshland. 1985. Proceedings of the National Academy of Sciences **82** (11):3577–3581.

[117] Webb, J. 1963. *Enzyme and Metabolic Inhibitors: General principles of inhibition. Volume 1.* Number v. 1. Academic Press.

[118] Woods, J. H., and H. M. Sauro. 1997. Comput Appl Biosci **13** (2):123–130.

[119] Wright, S. 1934. The American Naturalist **68**:24–53.

History

1. VERSION: 1.0

   Date: 2018-09-9

   Author(s): Herbert M. Sauro

   Title: Systems Biology: Introduction to Metabolic Control Analysis

   Modification(s): First Edition Release

2. VERSION: 1.01

   Date: 2019-13-4

   Author(s): Herbert M. Sauro

   Title: Systems Biology: Introduction to Metabolic Control Analysis

   Modification(s): Minor corrections to code in Chapter 12 and corrected subscript in loop gain definition on top of page 251.

Index

www.ingramcontent.com/pod-product-compliance
Lightning Source LLC
Chambersburg PA
CBHW081807200326
41597CB00023B/4174